QUANTITATIVE METHODS IN EDUCATION

QUANTITATIVE METHODS IN EDUCATION

A Practical Introduction to Statistics

YUKIKO MAEDA
JOHN GIPSON
SHEILA HURT
KATIE H. DUFAULT

PURDUE UNIVERSITY PRESS
WEST LAFAYETTE, INDIANA

Cataloging-in-Publication Data on file at the Library of Congress.

978-1-62671-180-8 (paperback)

978-1-62671-182-2 (epdf)

Cover image: Standard Deviation Diagram with Sample Size Chart:
Iamnee/iStock via Getty Images Plus.

CONTENTS

PREFACE

Welcome to *Quantitative Methods in Education: A Practical Introduction to Statistics*. This book is designed to equip current and future researchers and educators with a fundamental understanding of statistical tools necessary for analyzing and interpreting educational data. In today's data-driven world, the ability to apply quantitative methods is essential for making informed decisions that enhance learning outcomes and maximize student success.

Educational research frequently involves investigating patterns in student achievement, teacher effectiveness, and institutional performance. This book offers a structured approach to analyzing these patterns using both descriptive and inferential statistics, providing readers with the knowledge needed to conduct rigorous analyses and draw meaningful conclusions. While some statistical concepts can be challenging, this is not a traditional textbook. We use a conversational writing style to make the material accessible and engaging.

ORGANIZATION OF THE BOOK

The book is structured to build your statistical literacy progressively, covering essential topics with practical applications:

- **Chapter 1, From Data to Decisions: Quantitative Research Methods and Measurement,** introduces the fundamentals of quantitative research methods, including the distinction between different types of data, research designs, and the role of statistics in educational decision-making. Readers will gain an understanding of how to define research questions, collect quantitative data, and ensure the validity of their findings.
- **Chapter 2, Understanding the Center: Measures of Central Tendency,** explores measures of central tendency—mean, median, and mode—providing insight into how educators and researchers can summarize large datasets with a single representative value. This chapter discusses when each measure is most appropriate and how they impact data interpretation.
- **Chapter 3, Exploring Data Spread: Measures of Variability,** explores how range, variance, and standard deviation can be used to quantify how dispersed or clustered data points are in a dataset. Understanding variability is crucial in assessing student performance and comparing outcomes across different groups.
- **Chapter 4, Displaying Data Graphically,** discusses fundamental yet effective data visualization techniques, such as histograms, box plots, and scatterplots, that aid in interpreting and presenting data clearly. The chapter also highlights common pitfalls in data visualization and best practices for creating meaningful visual representations.
- **Chapter 5, The Normal Distribution,** introduces the foundational concept of normality, including properties of the normal curve, the empirical rule, and *z* scores, which are essential for many statistical analyses. The chapter explains how understanding the normal distribution can help in making inferences from data, particularly when assuming a certain distribution of scores, as seen in contexts such as student performance and standardized assessments.

- **Chapter 6, Sampling Distributions,** explains how sample statistics relate to population parameters, highlighting the central limit theorem and the importance of understanding sampling variability. The chapter also introduces confidence intervals and the role they play in making predictions about larger populations.
- **Chapter 7, Foundations of Statistical Inference: Understanding Hypothesis Testing,** presents a framework for making data-driven inferences, covering the logic behind null and alternative hypotheses, significance levels, and *p* values. Readers will learn how to test hypotheses and interpret results to make evidence-based conclusions in educational research.
- **Chapter 8, Evaluating Group Averages against a Benchmark: One-Sample *t* Test and Goodness-of-Fit Analysis,** introduces one-sample *t* tests and chi-square goodness-of-fit analyses for comparing observed values against known benchmarks to assess statistical significance. This chapter is particularly useful for evaluating program effectiveness and student performance relative to expected standards.
- **Chapter 9, Analyzing Differences between Two Groups,** covers independent- and related-samples *t* tests, explaining how to determine whether two groups differ significantly in their performance or characteristics. The chapter also introduces nonparametric alternatives when data do not meet normality assumptions.
- **Chapter 10, Analyzing Differences among More Than Two Groups,** introduces analysis of variance (ANOVA), including one-way and factorial ANOVA, that allows comparisons across three or more groups.
- **Chapter 11, Examining Relationships between Two Variables,** explores correlation and chi-square tests for independence, helping researchers analyze whether relationships exist between different educational variables, such as student engagement and academic performance. The chapter discusses how correlation does not imply causation and how to interpret results correctly.
- **Chapter 12, Predicting an Outcome,** introduces regression analysis techniques, including simple, multiple, and logistic regression, that help predict outcomes based on known predictor variables. These methods are useful in understanding factors influencing student success and institutional effectiveness.
- **Chapter 13, Analyzing Change and Adjusting for Covariates,** explores how repeated-measures ANOVA is used to track changes in student performance, school effectiveness, and educational trends over time. The chapter also covers key considerations for designing and interpreting analysis of covariance (ANCOVA) studies to control for covariates in educational research.
- **Chapter 14, Advanced Quantitative Methods,** introduces advanced techniques such as multilevel and structural equation modeling, providing readers with a comprehensive tool kit for tackling complex educational questions.

Each chapter provides clear explanations of statistical concepts with real-world examples and opportunities for practice to help reinforce readers' understanding of key concepts. Our goal is to equip readers with a solid theoretical foundation and a clear understanding of the real-world educational contexts in which these concepts are applicable, enabling them to confidently interpret and apply statistical methods in practice.

WHO SHOULD READ THIS BOOK?

This book is designed for students, educators, administrators, and researchers who are new to quantitative analysis and want to develop their skills. The book provides a beginner-friendly introduction to key concepts and methods, making it accessible to those with little or no prior experience in statistics or quantitative research. Whether you are conducting research, evaluating policy decisions, or analyzing student outcomes, this book offers a practical foundation for making data-informed decisions.

CONCLUSION

Quantitative methods play a crucial role in educational research and practice. By understanding and applying statistical techniques, educators can make evidence-based decisions that drive improvements in teaching, learning, and institutional effectiveness. We hope that this book serves as a valuable resource for developing your skills and confidence in working with data.

Thank you for joining us on this journey into quantitative methods in education. Let's begin!

1

FROM DATA TO DECISIONS: QUANTITATIVE RESEARCH METHODS AND MEASUREMENT

LEARNING OBJECTIVES

By the end of this chapter, you will be able to:

- Differentiate between the four scales of measurement—nominal, ordinal, interval, and ratio—and explain their impact on data analysis;
- Identify different types of variables, including qualitative versus quantitative, continuous versus discrete, and independent versus dependent variables, and understand their roles in research design; and
- Explain the importance of random sampling and random assignment, including how random sampling enhances external validity and how random assignment strengthens internal validity in research studies.

Research plays a fundamental role in expanding knowledge, solving problems, and making informed decisions across various fields. In education, research helps students, educators, administrators, and policymakers evaluate the effectiveness of instructional strategies, understand student learning processes, and improve educational outcomes. Whether investigating the impact of small class sizes on student achievement or assessing the effectiveness of technology-enhanced learning, research provides the foundation for evidence-based decision-making.

While this chapter does not cover all research methods used by educational researchers, it provides a solid foundation for understanding the quantitative methods introduced in this book. You will learn what research is, what data are, how to classify different types of data, and why quantitative methods matter in educational research. This foundational knowledge will support your understanding of the more advanced topics discussed in later chapters and equip you with the confidence to apply quantitative methods effectively in your own research.

WHAT ARE RESEARCH METHODS?

Research methods refer to the systematic approaches used to collect, analyze, and interpret data to answer research questions. These methods help researchers understand patterns, test theories, and make informed decisions based on evidence.

Let's consider a hypothetical scenario. A school principal wants to evaluate whether the instructional strategies used in teaching mathematics are effectively supporting student learning. In other words, "Are the current teaching methods in mathematics leading to improved student understanding and performance?"

To address this question, researchers collect data on student performance, analyze trends in how different teaching strategies impact learning processes and outcomes, and interpret these findings to assess effectiveness. The goal of the research is to provide evidence-based insights into whether current instructional methods are achieving their intended outcomes or require modification. By addressing research questions with systematic inquiry, educators and policymakers can make informed, data-driven decisions about potential adjustments, ultimately aiming to enhance student learning outcomes.

Additional examples of education-related research questions include:

- Does reducing class size in K–12 education produce significantly higher student achievement?
- Is a technology-enhanced learning approach more effective than traditional classroom instruction in improving student engagement and comprehension?
- How do different assessment methods impact students' motivation and learning outcomes?

TYPES OF RESEARCH METHODS

In educational research, methods are broadly classified into quantitative and qualitative approaches:

- **Quantitative research** focuses on numerical data and statistical analysis (e.g., test scores, Likert-type survey responses) and is useful for identifying patterns, testing hypotheses, and making predictions.
- **Qualitative research** explores nonnumerical data, such as interviews, observations, and open-ended responses, to understand deeper meanings and experiences and is commonly used for exploring complex social or behavioral issues.
- **Mixed-methods research** integrates both quantitative and qualitative approaches in a single study. Researchers may collect both numerical data (e.g., test scores) and nonnumerical data (e.g., interviews or observations) to gain deeper insights. This approach offers a more comprehensive understanding by capturing both statistical trends and qualitative perspectives.

Regardless of the type of research, a standard part of the empirical research process is to collect data on the phenomenon of interest and then analyze the data to derive meaningful conclusions. As we walk through this book, you will learn how to analyze *quantitative data*, interpret the results of analyses, and draw conclusions for your scientific inquiries.

OVERVIEW OF QUANTITATIVE METHODS

A key part of quantitative research is to apply statistical techniques to summarize and analyze data. A basic overview of the field of statistics is shown below. Statistics is broadly divided into:

- *Descriptive statistics*, used to summarize and describe data (e.g., means, standard deviations, frequency distributions); and
- *Inferential statistics*, used to make predictions or draw conclusions about a population based on a sample (e.g., hypothesis testing, regression analysis). There are further subdivisions under inferential statistics.

As we progress through this book, you will learn how to apply these methods to analyze quantitative data, interpret research findings, and draw meaningful conclusions for educational research.

UNDERSTANDING QUANTITATIVE DATA

As an educator and researcher, you are constantly surrounded by information. To conduct quantitative research that supports evidence-based practices in education, you must transform that information into data that can be analyzed using quantitative methods. This involves two basic steps:

- Operationalizing the concept by identifying and defining measurable variables that represent the idea you want to study; and
- Determining how to measure or collect data based on these variables (e.g., through survey responses, test scores, observational data).

These steps are closely connected and can easily be confused, as the way you define a variable often influences how it can be measured. However, keeping in mind that operationalization and measurement are related but distinct processes will help you approach research design more clearly and effectively.

SCALES OF MEASUREMENT

When we measure, we attempt to identify the dimensions, quantity, capacity, or degree of something. **Measurement** is the act of measuring by assigning symbols or numbers to something according to a specific set of rules.

Measurement can be categorized by the type of information communicated by the symbols or numbers assigned to variables of interest. In particular, there are four levels, or types, of information that are discussed. They are called the *four scales of measurement.*

A **nominal scale** is the simplest form of measurement scale, used for categorizing or labeling data without any order or numeric value, and classifies data into distinct groups based on characteristics or attributes, but there's no implied ranking or order between categories.

- A nominal variable is one that is used to categorize, label, classify, name, or identify data and group them into distinct categories or types.
- Numbers can be used to label the categories of a nominal variable, but the numbers serve only as markers, not as indicators of amount or quantity. For example, if you were studying a variable such as "mode of transportation," you might assign numerical codes to each category: 1 = car, 2 = bicycle, and 3 = bus.
- Some examples of nominal-level variables include country of birth, college major, personality type, experimental group (e.g., treatment or control), ethnicity, and race.

An **ordinal scale** is a type of measurement scale that categorizes and orders data based on a hierarchy or ranking and tells you the position of items relative to one another. However, the intervals between the ranks are not necessarily equal or consistent. This level of measurement enables one to make judgments about *rank order*.

- An ordinal variable is any variable whose categories can be ranked or ordered, but the intervals between the categories are not necessarily equal or known.
- Examples of ordinal-level variables include the order in which someone finishes a marathon, class rank, rating scales (e.g., strongly agree to strongly disagree), and course grades (A, B, C, D, F).

An **interval scale** is a type of measurement scale that categorizes and orders data points, with *equal distances* between each value. Unlike a ratio scale, an interval scale does not have an absolute zero point, meaning zero doesn't represent the total absence of what's being measured.

- Examples of interval-level variables include temperature measured in Celsius or Fahrenheit, IQ scores, and standardized test scores.
- For instance, the absence of a true zero is evident in temperature scales. Zero degrees Celsius does not mean there is no temperature; it simply represents a point on the scale.

A **ratio scale** is a type of measurement scale that has all the characteristics of an interval scale with the additional feature of an *absolute zero point*.

- A ratio scale has all characteristics of each lower-level scale: equal intervals (interval scale), rank order (ordinal scale), and the ability to mark a value with a name (nominal scale).
- Examples of ratio-level variables include number of correct answers, weight, height, response time, temperature in Kelvin, and annual income.
- For instance, the presence of a true zero is seen in income: if someone earns zero dollars in a year, this indicates a total absence of income. Zero truly means zero dollars.

VARIABLES

The basic building blocks of quantitative data analysis are variables. *Variables* (something that takes on different values or categories) are the opposite of *constants* (something that cannot vary, such as a single value). As you recall, the first step in data analysis is deciding how to operationalize and define your variables. Once you have done that, the second step is determining how to measure them to collect data for analysis.

Quantitative vs. Qualitative. The first step in working with data is determining whether a variable is qualitative or quantitative. Ask yourself: Can this variable be easily represented by a number? If not, it is likely qualitative.

Variables measured on a nominal scale are qualitative, while the other three scales (ordinal, interval, and ratio) are quantitative. Qualitative variables, also called categorical variables, represent characteristics that describe categories or group membership rather than numerical quantities (e.g., eye color, type of school, or preferred learning method).

Quantitative variables, on the other hand, represent measurable numerical values that express an amount or quantity (e.g., height, test scores, or number of study hours).

It is important to note that while qualitative (nominal) variables are not inherently numerical, they can still be incorporated into quantitative analyses by assigning numerical codes. This is often done using dichotomous variables, for example, coding group membership as 1 = treatment and 0 = control, which we will learn more about in later chapters.

Continuous vs. Discrete. Next, determine whether the variable is *continuous* or *discrete*. A helpful way to think about this distinction is to ask if you can have "half" of something. For example, you can have half a pound or score half a point on a test—these are continuous variables. But you can't have half a category, such as half of "blue" or half of "red"—these are discrete variables.

Qualitative (categorical) variables are always discrete because they represent distinct categories or groups. For quantitative variables, the distinction between continuous and discrete becomes a bit more nuanced.

Consider a Likert-type scale with responses such as strongly agree, agree, disagree, and strongly disagree. This is an ordinal scale, and if each response is coded numerically (e.g., 1 = strongly agree, 2 = agree, etc.), it becomes a discrete quantitative variable. However, if multiple Likert items are averaged or summed—for example, combining responses from five related questions—you can generate a variable that behaves more like a continuous one. In such cases, values may range across a spectrum, such as from 1.0 to 5.0, including decimals such as 3.7 or 4.2. These types of scores are typically treated as continuous in analysis because they provide more variability and can approximate interval-level measurement.

Why does this distinction matter? Because certain statistical techniques are only appropriate for continuous variables. For example, SAT scores are technically discrete, but because they have many possible values and tend to follow a normal distribution, they are often treated as continuous in educational research.

The main challenge arises with ordinal data that have only a few response categories (e.g., agree to disagree). Strictly speaking, such variables should not be treated as continuous. As you advance in your studies, you'll learn about specialized statistical methods designed for analyzing ordinal data appropriately.

Dependent vs. Independent. Finally, it is important to identify your *independent* and *dependent variables*. The **independent variable (IV)** is the variable you believe is the cause or influence in a relationship. The **dependent variable (DV)** is the outcome or effect—the variable that is influenced by the independent variable.

For example, consider the relationship between smoking and lung cancer. In this case, smoking is the independent variable because it is presumed to cause or influence lung cancer, which is the dependent variable.

Whenever you aim to make a claim about cause and effect—that is, that changes in one variable lead to changes in another—you must also be aware of extraneous or confounding variables. These are variables other than the independent variable that may influence the dependent variable. For instance, if lung cancer rates vary not because of smoking but instead due to exposure to air pollution, then air pollution would be a confounding variable.

You will learn more about how to control for these variables using statistical techniques such as multiple regression later in your studies. For now, it is most important to understand the basic distinction between independent and dependent variables and to be aware that other variables may influence the outcome.

RANDOM SAMPLING AND RANDOM ASSIGNMENT

In research, it is often impractical or impossible to collect data from an entire population, so we rely on sampling to study a smaller, manageable subset. *Sampling* refers to drawing a sample (the subset) from a population (the full set). The usual goal of sampling is to produce a *representative sample* (i.e., a sample that is similar to the population on all characteristics except that it includes fewer people rather than the complete population). A **statistic** is a numerical characteristic of a sample. A **parameter** is a numerical characteristic of a population. However, because a sample is only a subset, there will always be some *sampling error*—the difference between a sample statistic and the true population parameter.

To minimize bias and improve the accuracy of population estimates, researchers often use random sampling methods. Simple random sampling ensures that every individual in the population has an equal chance of being selected, making the error random rather than systematic. Unlike nonrandom sampling methods (e.g., convenience sampling), which may introduce bias when inferring population characteristics, random sampling enhances the representativeness of the sample, allowing the generalizability of findings to the broader population.

It is important to distinguish *random sampling* from *random assignment*, as they serve different purposes in research.

- Random sampling refers to how participants are selected from the population to be included in a study. It is used to ensure a representative sample.
- Random assignment, on the other hand, refers to how participants are assigned to different experimental conditions or groups (e.g., treatment vs. control group). It is a key feature of experimental research because it helps ensure differences between groups occur by chance, reducing potential confounding variables and allowing researchers to infer causality.

While *random sampling improves external validity* (the ability to generalize findings to a larger population), *random assignment improves internal validity* (helping to ensure that differences between groups are due to the independent variable and not other factors).

CAUSALITY IN EDUCATIONAL RESEARCH

Ideally, research results permit strong causal inferences such as "if intervention A is performed, then outcome B will occur." While such statements may be possible in some research domains, it is often difficult to establish them in social science fields such as education, psychology, and sociology. Let's explore why this is the case.

To establish causality, we first need to satisfy three key requirements: (1) temporal precedence, (2) covariation, and (3) elimination of alternative explanations.

- **Temporal precedence** means that the cause (independent variable, intervention A) must occur before the effect (dependent variable, outcome B). This ensures that observed changes in the dependent variable result from the changes in the independent variable, rather than the other way around.
- **Covariation** means that there must be a relationship between the independent variable (intervention A) and the dependent variable (outcome B). If no correlation exists, we cannot infer causation.
- **Elimination of alternative explanations** means that we should control for confounding variables that could influence the observed relationship between the independent and dependent variables. This is often done through random assignment in experiments.

In social science disciplines, causality is often hard to establish due to ethical and practical constraints. This is because we often cannot randomly assign participants to either experimental or treatment conditions to control for possible confounding variables as doing so is often impractical or unethical. For example, randomly assigning students to small or large class sizes could disrupt their education and raise fairness concerns. This limitation often leads educational researchers to use quasi-experimental designs or observational studies with statistical adjustments to approximate causal relationships.

To navigate these challenges, developing a strong foundation in *quantitative research methods* is essential. Learning basic techniques such as *descriptive statistics, correlation analysis*, and *simple regression* can help you analyze data, identify trends, and make informed conclusions. As you build your skills, you will gain confidence in interpreting research findings and applying quantitative methods to real-world educational questions. Whether you are designing a study, evaluating an intervention, or examining patterns in student learning, *quantitative methods provide valuable tools for drawing meaningful insights from data.*

CONCLUSION

Understanding the fundamental concepts of research methods, measurement scales, and variables is essential for effective data analysis and informed decision-making. By distinguishing between different scales of measurement, types of variables, and research approaches, researchers can ensure accurate data interpretation and robust conclusions. Mastering these foundational elements provides the tools necessary to navigate the complexities of quantitative and qualitative research, ultimately enabling evidence-based practices across various fields of study.

KEY TAKEAWAYS FOR EDUCATIONAL RESEARCHERS FROM CHAPTER 1

- Understanding measurement and variables is essential for conducting quantitative research in education. Accurately defining and measuring key concepts—such as student achievement, engagement, and instructional effectiveness—requires distinguishing between different scales of measurement (nominal, ordinal, interval, ratio) and types of variables (qualitative vs. quantitative, continuous vs. discrete, independent vs. dependent). This ensures that data collected in educational studies are meaningful and can be analyzed appropriately.

- Random sampling and random assignment serve different purposes. Random sampling improves external validity by ensuring a representative sample, while random assignment enhances internal validity by reducing confounding variables and strengthening causal claims in experimental research. For example, when studying broad educational trends, random sampling of students, teachers, or schools helps ensure that findings can be generalized to a larger population. In contrast, when evaluating the impact of a specific teaching strategy or intervention, random assignment to treatment and control conditions helps control for differences between students and between classrooms, allowing researchers to draw stronger conclusions about cause and effect.
- Establishing causality in educational research is challenging but important. Because ethical and practical constraints often prevent researchers from randomly assigning students to different learning conditions, educational research frequently relies on quasi-experimental designs or statistical methods to approximate causal relationships. Understanding these methods allows researchers to draw more accurate conclusions about the effectiveness of policies, instructional strategies, and other interventions in real-world educational settings.

KEY DEFINITIONS FROM CHAPTER 1

Covariation means that there must be a relationship between the independent variable (intervention A) and the dependent variable (outcome B). If no correlation exists, we cannot infer the causation.

A **dependent variable** is the outcome or result that is measured or observed in an experiment to see how it is affected by changes in another variable (the independent variable). The dependent variable is often called the "effect" or outcome variable.

Elimination of alternative explanations means that we should control for confounding variables that could influence the observed relationship between the independent and dependent variables. This is often done through random assignment in experiments.

An **independent variable** is a factor or condition that is intentionally manipulated or changed by a researcher in an experiment to observe its effect on another variable. The independent variable is often referred to as the "cause" or the predictor variable.

An **interval scale** is a type of measurement scale that categorizes and orders data points, with *equal distances* between each value. Unlike a ratio scale, an interval scale does not have an absolute zero point, meaning zero doesn't represent the total absence of what's being measured.

Measurement is formally defined as the act of measuring by assigning symbols or numbers to something according to a specific set of rules.

Mixed-methods research combines both quantitative and qualitative approaches in a single study. Researchers may collect numerical data (e.g., test scores) and also conduct interviews to gain deeper insights. This approach helps provide a more holistic understanding by capturing both statistical trends and personal experiences.

A **nominal scale** is the simplest form of measurement scale, used for categorizing or labeling data without any order or numeric value. The nominal scale classifies data into distinct groups based on characteristics or attributes, but there is no implied ranking or order between categories.

An **ordinal scale** is a type of measurement scale that categorizes and orders data based on a hierarchy or ranking. The ordinal scale tells you the position of items relative to one another, but the intervals between the ranks are not necessarily equal or consistent. This level of measurement enables one to make judgments about *rank order.*

A **parameter** is a numerical characteristic of a population.

A **ratio scale** is a type of measurement scale that has all the characteristics of an interval scale, with the additional feature of an absolute zero point.

A **statistic** is a numerical characteristic of a sample.

Temporal precedence means that the cause (independent variable; e.g., intervention A) must occur before the effect (dependent variable; outcome B). This ensures that observed changes in the dependent variable result from the changes in dependent variable rather than the other way around.

Qualitative research explores nonnumerical data, such as interviews, observations, and open-ended responses, to understand deeper meanings and experiences and is commonly used for exploring complex social or behavioral issues.

Quantitative research focuses on numerical data and statistical analysis (e.g., test scores, survey responses) and is useful for identifying patterns, testing hypotheses, and making predictions.

CHECK YOUR UNDERSTANDING

1. Which scale of measurement has an absolute zero point?
   a. Nominal
   b. Ordinal
   c. Interval
   d. Ratio

2. What type of variable can take on an infinite number of possible values within a given range?
   a. Discrete variable
   b. Continuous variable
   c. Nominal variable
   d. Ordinal variable

3. Which research method combines both quantitative and qualitative approaches in a single study?
   a. Quantitative research
   b. Qualitative research
   c. Mixed-methods research
   d. Observational research

4. In research, random assignment primarily enhances which type of validity?
   a. External validity
   b. Internal validity
   c. Construct validity
   d. Content validity

5. Which of the following is an example of a nominal scale?
   a. Temperature in Celsius
   b. Class ranking (e.g., first, second, third)
   c. College major (e.g., psychology, biology, history)
   d. Height in centimeters

2

UNDERSTANDING THE CENTER: MEASURES OF CENTRAL TENDENCY

LEARNING OBJECTIVES

By the end of this chapter, you will be able to:

- Define and distinguish among the three main measures of central tendency—mean, median, and mode—and describe situations in which each is most appropriate to use;
- Accurately calculate the mean, median, and mode for a given dataset and interpret the results in context; and
- Describe the concept of skewness and its relationship to measures of central tendency.

In this chapter and the next two chapters, we will discuss three ways of summarizing data. First, we look at how to generate a *typical score*—one value that can tell us something important about the data. In other words, how well does that *typical score* represent the entire dataset? In the next chapter we will talk about variability, which refers to how spread out the data are. In Chapter 4, we will learn how to visually represent data. For now, we are going to focus on different ways to summarize data with a single value, which is known as the *central tendency* of the data.

WHAT IS CENTRAL TENDENCY?

Central tendency refers to the statistical measure that identifies the single value or point that best represents a dataset's center, or typical, value. The most common measures of central tendency are the mean (average), median (middle value), and mode (most frequent value). These measures are used to provide a reasonably accurate description, or summary, of a group (think of a whole class, a grade level, a school, etc.). How do you know when to use the mean, median, or mode to describe a dataset? Let's walk through an example.

A teacher gave a quiz to eight students. Which measure of central tendency would be the most appropriate for summarizing the following scores?

$$13, 14, 10, 18, 11, 10, 16, 15$$

Let's use the three different measures of central tendency introduced above to summarize the scores and then decide which one provides the best description of the data. The first step is to reorder the scores from smallest to largest:

$$10, 10, 11, 13, 14, 15, 16, 18$$

The **mode** is simply the most frequently occurring number in a dataset. The most frequently occurring number in this example is 10 as it appears twice, and other numbers only appear once.

The **median** is the center point in a set of numbers listed in numerical order. It is also the 50th percentile. Sometimes the median is referred to in terms of depth (i.e., the value halfway into the ordered score distribution). The median provides a measure that is less influenced by extreme values compared to the mean.

- To calculate the median by hand, you first put your numbers in ascending or descending order, then check to see which of the following two rules apply:
  - *Rule 1. If you have an odd number of values, the median is the center number (e.g., 3 is the median for the numbers 1, 1, 3, 4, 9).*
  - *Rule 2. If you have an even number of numbers, the median is the average of the two innermost numbers (e.g., 2.5 is the median for the numbers 1, 2, 3, 7).*
- We can always define the median for N ordered raw scores with this formula: $\left(\frac{N+1}{2}\right)$. We have an even number of scores (Rule 2) for our score data, so the median is the $\frac{8+1}{2}$ = 4.5th ordered value.
  - The fourth value is 13. The fifth value is 14.
  - (13 + 14)/2 = 13.5
  - The median of our dataset is 13.5

The **mean** is the most commonly reported measure of central tendency in educational statistics. It is the arithmetic average of a dataset. More specifically, a sample mean, often denoted as $\bar{X}$, can be defined as $\bar{X} = \frac{\sum_{i=1}^{n} x_i}{n}$, where n represents the number of cases and x_i represents the ith value of variable x. In other words, $\sum_{i=1}^{n} x_i$ means the sum of first to nth values of x. The sample mean is said to estimate μ, which is the population mean. For our score data, the mean is:

$$\bar{X} = \frac{10 + 10 + 11 + 13 + 14 + 15 + 16 + 18}{8} = \frac{107}{8} = 13.4$$

The mean provides a central value that represents the overall distribution, but it can be heavily influenced by outliers or extreme values.

There are *five key characteristics of the mean* that are typically important when dealing with statistics in educational research:

1. Adding a new score or removing an existing score will change the mean unless the value is equal to the mean. For example, adding a score of 1 to our dataset would change the mean to 12.
2. Changing an existing score will change the mean. For example, changing the first 10 in our dataset to 6 would change the mean to 12.9.
3. Adding, subtracting, multiplying, or dividing each score by a constant will change the mean by that constant. For example, if we added 100 to each score in our sample, the new mean would be 113.4.
4. The sum of the differences of scores from their mean is equal to 0. Table 2.1 illustrates this for our example. Note that the exact mean of 13.375 is used to avoid rounding error.

TABLE 2.1 ***Sum of the Difference of Scores from the Mean***

x_i	$(x_i - \bar{X})$
10	10 - 13.375 = -3.375
10	10 - 13.375 = -3.375
11	11 - 13.375 = -2.375
13	13 - 13.375 = -.375
14	14 - 13.375 = .625
15	15 - 13.375 = 1.625
16	16 - 13.375 = 2.625
18	18 - 13.375 = 4.625
$\Sigma x_i = 107$	$\Sigma (x_i - \bar{X}) = 0$

5. The sum of squared distances of scores from the mean is minimal, meaning no other value would lead to a lower sum of squared distances (Table 2.2). Using squared distances will always produce a positive number. The larger the final product, the further scores deviate from the mean. The smaller the final product, the less scores deviate from the mean.

TABLE 2.2 ***Sum of Squared Difference of Scores from the Mean***

x_i	$(x_i - \bar{X})$
10	$(10 - 13.4)^2 = 11.56$
10	$(10 - 13.4)^2 = 11.56$
11	$(11 - 13.4)^2 = 5.76$
13	$(13 - 13.4)^2 = 0.16$
14	$(14 - 13.4)^2 = 0.36$
15	$(15 - 13.4)^2 = 2.56$
16	$(16 - 13.4)^2 = 6.76$
18	$(18 - 13.4)^2 = 21.16$
$\Sigma x_i = 107$	$\Sigma(x - \bar{X})^2 = 59.88$

Often, we only collect data from a subgroup of cases (a sample) from the entire collection of cases (population). The mean of a subsample of the data, $\bar{X}$, or the sample mean, is said to estimate the population mean. The formula for a population mean is $\mu_X = \frac{\sum_{i=1}^{N} x_i}{N}$. The difference between this and the formula for a sample mean as seen above is n and N. The N represents the entire collection of cases that belong to your target population (e.g., all undergraduate students at a university), while the n represents the sample size you obtained from the population (e.g., a random sample of 20 undergraduate students at the university). The difference between $\bar{X}$ and μ_X is typically not zero and is considered to be the sampling error.

We have three possible measures of central tendency for our data:

$$\text{Mode} = 10, \text{Median} = 13.5, \text{Mean} = 13.4$$

Which measure of central tendency do you think is the most appropriate to summarize how the students performed on the quiz? Why?

TABLE 2.3 *Advantages and Disadvantages of the Mode, Median, and Mean*

	MODE	MEDIAN	MEAN
Advantages	• Represents the most frequently occurring number • Is applicable to nominal data • Is not affected by outliers	• Is not affected by outliers • Does not assume the scale is interval	• Can be manipulated algebraically (i.e., inferential analyses are possible) • Can use all data to compute • Can make an inference about population characteristics • Will be more stable with large data
Disadvantages	• Doesn't represent the entire collection of data	• Is difficult to apply inferential analysis to the median	• Is influenced by outliers • Requires interval properties of the data

The word **outlier** comes up a few times in the prior table, and it means a score that is much higher than the next highest score or much lower than the next lowest score in your dataset. For example, imagine that instead of our original dataset, the scores looked like this:

$$10, 10, 11, 13, 14, 15, 16, 50$$

In this case, the mode and median stay the same, but the mean changes to 17.4. With the first dataset, either the median or the mean would have been an appropriate measure of central tendency. Why? We would probably use the mean due to the advantages listed in Table 2.3, but clearly the results were very similar. However, this second dataset has an outlier—one student who managed to score a 50. That single score has a major influence on the mean, so much so in this case that the mean is higher than the second-highest quiz score. You can imagine the outlier *pulling* the dataset up (or to the right), which makes the mean a less useful measure to summarize our dataset. That *pulling* is referred to as **skewness**, or lack of symmetry in the dataset. In the second example, we would use the median to describe the dataset.

SKEWNESS AND MEASURES OF CENTRAL TENDENCY

We will learn a lot more in the next few chapters about distribution and the normal curve. For now, just know that if you plot all the numeric data points from a dataset and draw a curved line to connect them, it would look like the examples in Figures 2.1, 2.2, and 2.3. The curve in the middle is *normal*, which has many important properties. For now, we care about the extent to which the curve is skewed, or pulled to one side or the other. A **negatively skewed** curve will be pulled to the left, meaning the longer tail of the curve is on the left side. A **positively skewed** curve will be pulled to the right, meaning the longer tail of the curve is on the right side. The dataset above (10, 10, 11, 13, 14, 15, 16, 50) would be positively skewed because the outlier of 50 would pull the curve to the right.

Note that when a variable is perfectly normally distributed, the mean, median, and mode are the same number. When the distribution is skewed to the left (i.e., negatively skewed), the mean shifts to the left the most, the median shifts to the left the second most, and the mode is the least affected by the presence of skew in the data. Therefore, when the data are negatively skewed, mean < median < mode.

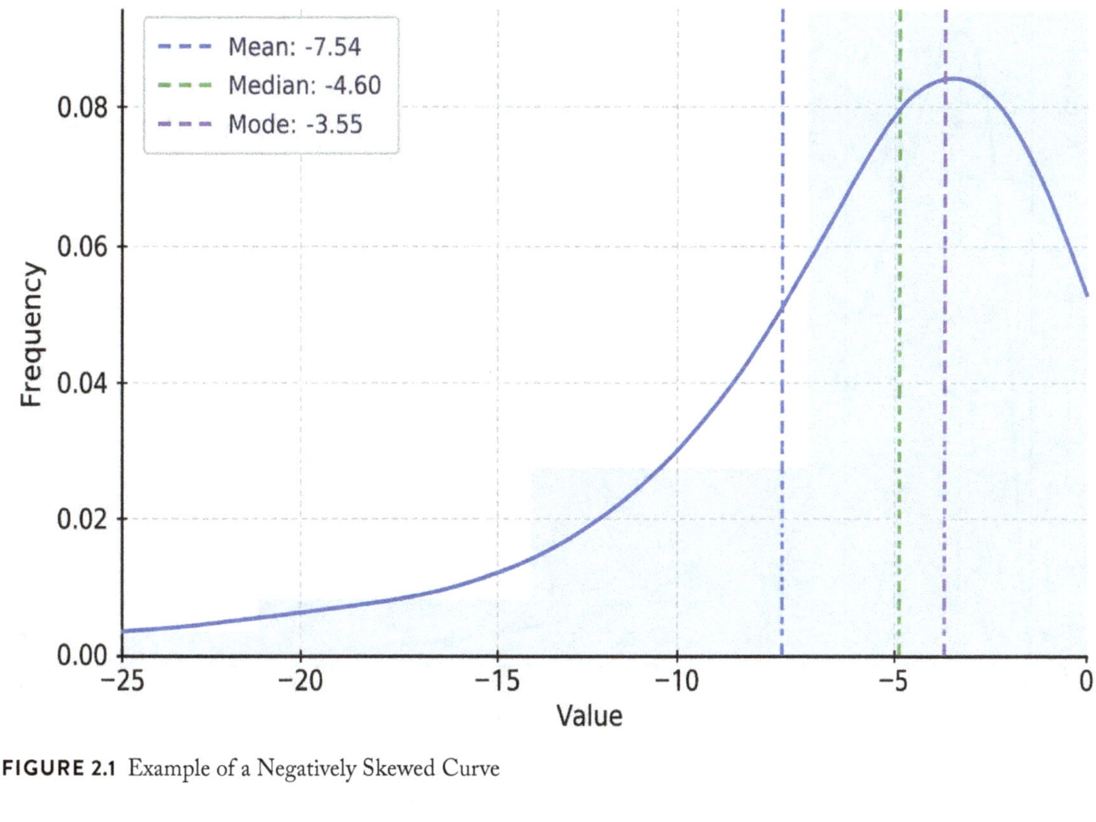

FIGURE 2.1 Example of a Negatively Skewed Curve

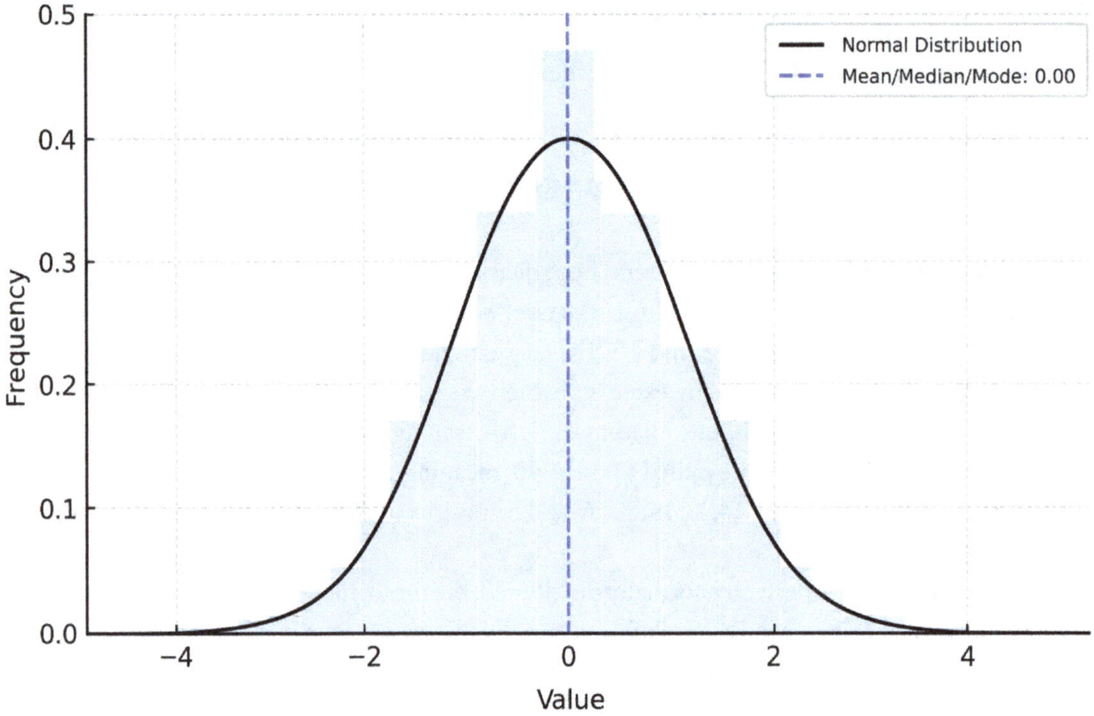

FIGURE 2.2 Example of a Normally Distributed Curve

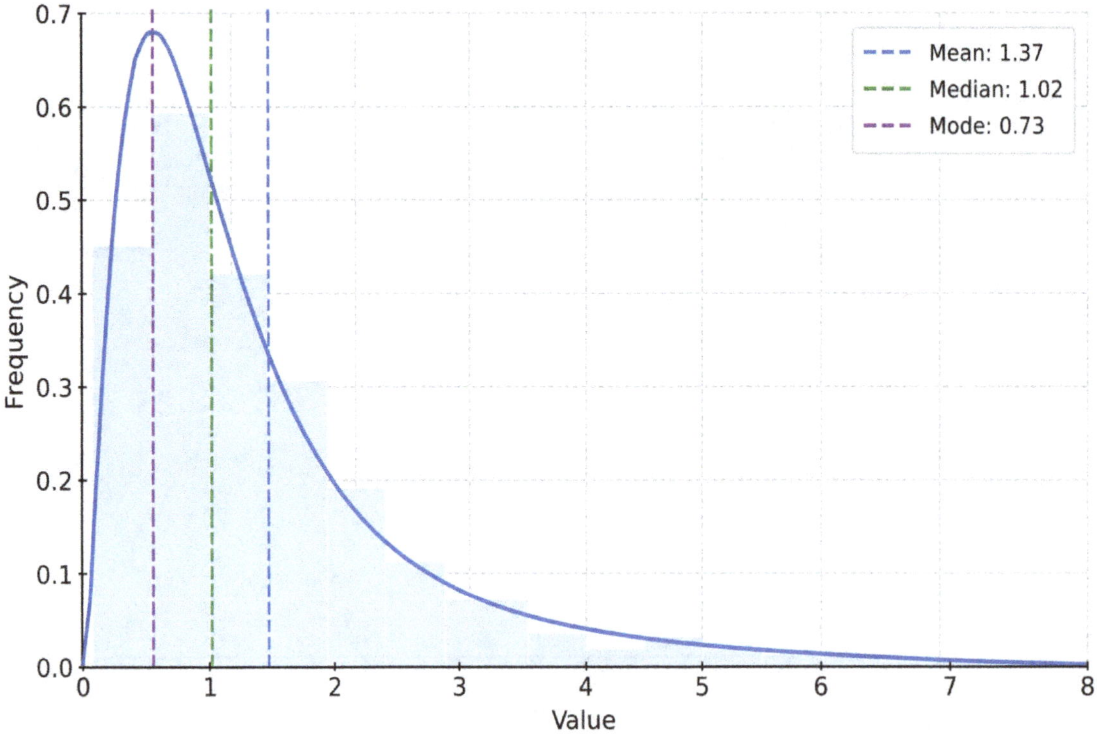

FIGURE 2.3 Example of a Positively Skewed Curve

When the variable is skewed to the right (i.e., positively skewed), the mean is shifted to the right the most, the median is shifted to the right the second most, and the mode is the least affected. Therefore, when the data are positively skewed, mean > median > mode. Again, this is what happened with our second dataset example (10, 10, 11, 13, 14, 15, 16, 50).

If you go to the end of the curve to where it is pulled out the most, you will see that the order goes mean, median, and mode as you "walk up the curve" for negatively and positively skewed curves.

You can use the following two rules to provide some information about skewness even when you cannot see a line graph on the data (i.e., all you need is the mean and the median):

Rule 1: If the mean is less than the median, the data are skewed to the left.

Rule 2: If the mean is greater than the median, the data are skewed to the right.

MODAL DISTRIBUTIONS

Any distribution in which one or more numbers occur most often can be referred to as a modal distribution. We typically describe these distributions in one of four ways. **Unimodal distributions** have a single mode, or single score that occurs most often, and take the shape of a curve. The normal distribution, or bell curve, is a good example. **Bimodal distributions** have two modes, and the mean and median are typically located between the humps. **Multimodal distributions** have more than two modes.

Nonmodal or uniform distributions have no unique mode because every value appears with equal frequency. As an example, 12 people take a test with scores ranging from 1 to 4. Three people each earn scores of 1, 2, 3, and 4. There is no mode, but the mean score is 2.5 and the median is 2.5—either would be an appropriate description of central tendency.

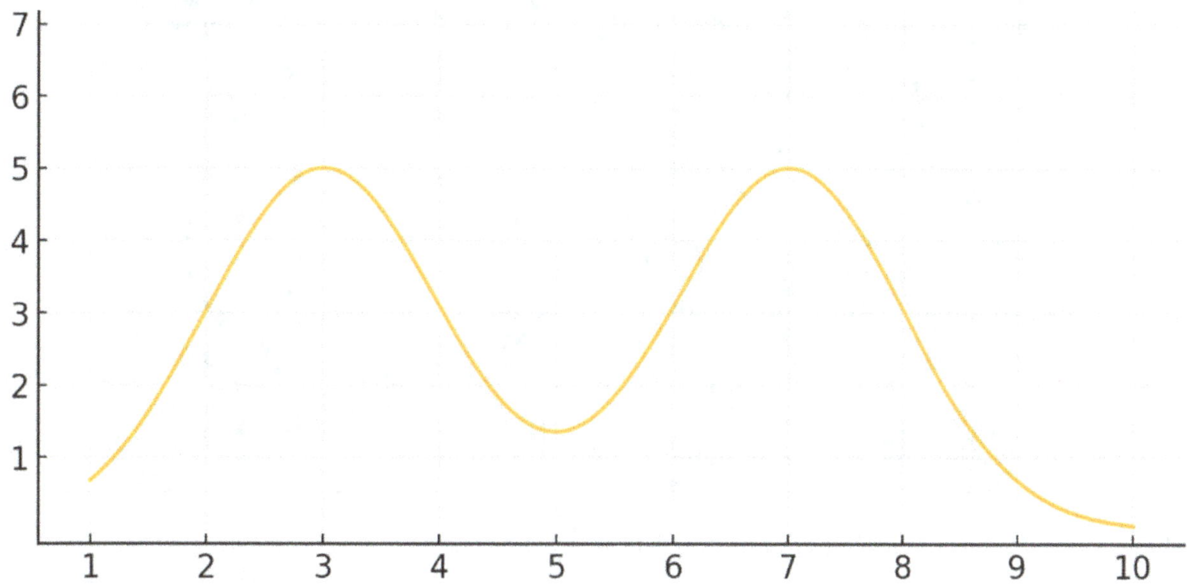

FIGURE 2.4 Example of a Bimodal Distribution

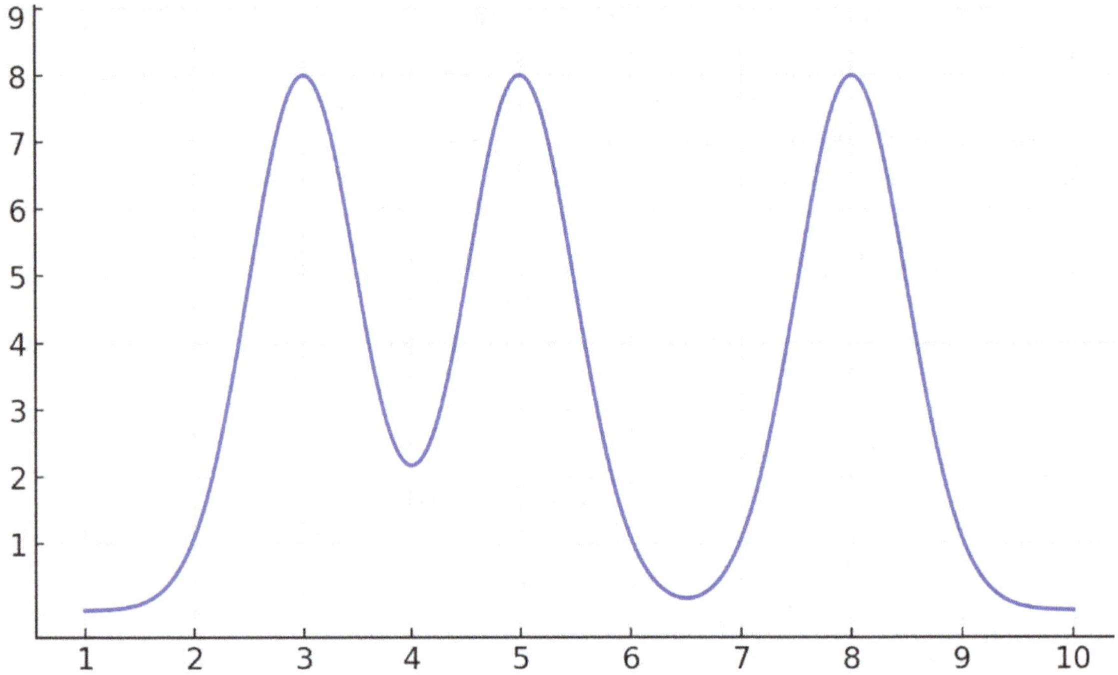

FIGURE 2.5 Example of a Multimodal Distribution

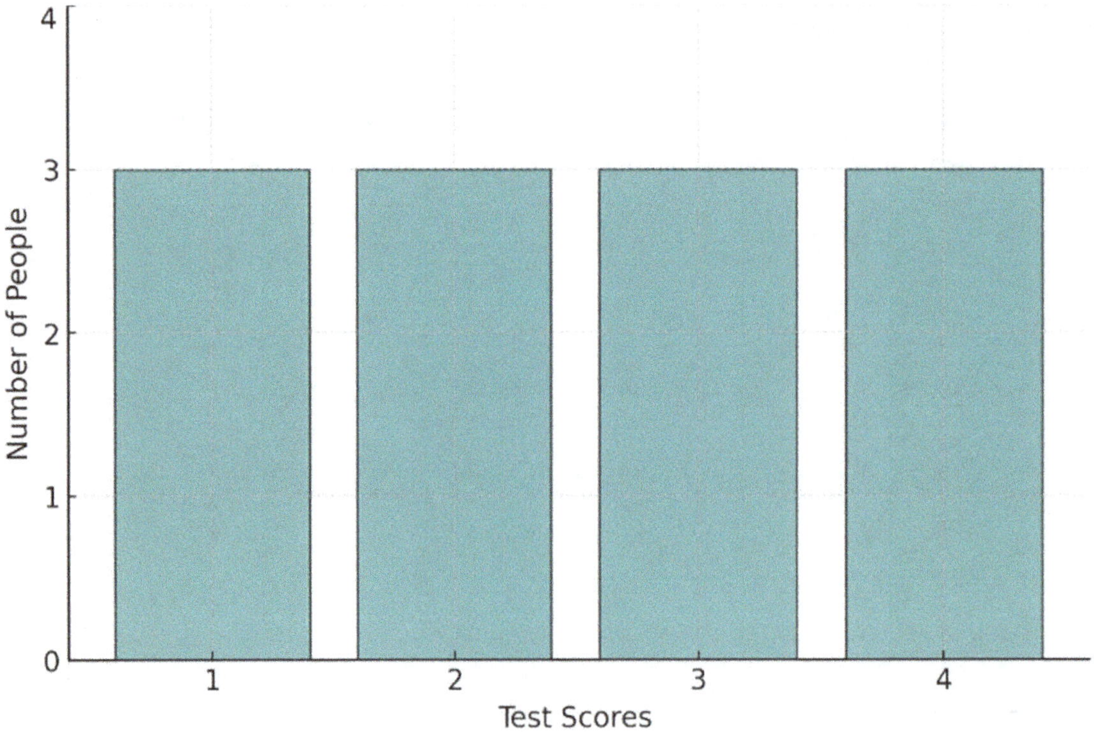

FIGURE 2.6 Example of Nonmodal or Uniform Distribution

CONCLUSION

This chapter has provided a comprehensive overview of measures of central tendency, emphasizing the significance of the mean, median, and mode in summarizing data. Understanding these measures is essential for accurately interpreting datasets and making informed decisions in educational research and statistical analysis. The chapter highlighted the impact of skewness on central tendency and the importance of selecting the most appropriate measure based on data distribution. By mastering these concepts, readers can develop a strong foundation in statistical reasoning, enabling them to analyze and communicate data effectively. Moving forward, Chapter 3 will explore variability, providing further insight into how data spread around these central measures.

KEY TAKEAWAYS FOR EDUCATIONAL RESEARCHERS FROM CHAPTER 2

- The mean, median, and mode are the three main measures of central tendency used to summarize a dataset. Mode represents the most frequently occurring value in a dataset and is useful for nominal data and is not affected by outliers. Median is the middle value in a dataset when ordered and less influenced by outliers than the mean. Mean is the arithmetic average of a dataset but is sensitive to outliers.
- Adding, removing, or altering value in the dataset will directly impact the mean but not necessarily the median or mode.
- Choosing an appropriate measure of central tendency depends on the shape of score distribution. As a general rule, use the mean for symmetric data, the median for skewed data, and the mode for nominal or categorical data.
- Understand the context of data and analyze the interested variables in the dataset to decide which measure of central tendency best summarize the variables of interest.

KEY DEFINITIONS FROM CHAPTER 2

Bimodal distributions are a distribution of scores where two scores occur most frequently.

Central tendency refers to the statistical measure that identifies the single value or point that best represents a dataset's center, or typical, value.

Mean is the arithmetic average of a dataset. The sum of a set of scores in a distribution, divided by the total number of scores.

Median is the center point in a set of numbers listed in numerical order.

Mode is simply the most frequently occurring number in a dataset.

Multimodal distributions have more than two modes.

Negatively skewed data have a mean that is less than the median. When graphed, the left tail is longer than the right tail.

Nonmodal, or uniform, distributions have no unique mode because every value appears with equal frequency.

An **outlier** is a score that is much higher than the next highest score or much lower than the next lowest score in your dataset.

Positively skewed data have a mean that is greater than the median. When graphed, the right tail is longer than the left tail.

Skewness means the extent to which a curve is pulled to one side or another.

Unimodal distributions have a single mode, or single score, that occurs most often.

CHECK YOUR UNDERSTANDING

1. A researcher is investigating if any relationship exists between the hours a student sleeps at night and the student's math performance across a school district with more than 1,000 students. However, the researcher can only survey 50 students. What notation below would be used to indicate the number of students surveyed?
   a. $N = 50$
   b. $n = 50$

2. The ______ is the value or values in a dataset that appear most frequently.
   a. Mean
   b. Median
   c. Mode

3. An educational researcher records the following sample of scores: 1, 2, 4, 4, 4, 5, 5, 7, 8, 10. Calculate the sample mean, median, and mode.
   a. Mean __________
   b. Median __________
   c. Mode __________

4. An educational researcher obtains the following sample scores: 1, 3, 3, 4, 6, 100. Using what method of central tendency to describe these sample scores might be problematic?
   a. Mean
   b. Median
   c. Mode

5. What direction of skewness is created by the following data: 1, 1, 2, 2, 10?
   a. Positive skewness
   b. Negative skewness
   c. No skewness

3

EXPLORING DATA SPREAD: MEASURES OF VARIABILITY

LEARNING OBJECTIVES

By the end of this chapter, you will be able to:

- Define and explain the concept of variability and its importance in statistical analysis, distinguishing between measures such as range, variance, and standard deviation; and
- Calculate measures of variability (range, interquartile range, variance, and standard deviation) for a given dataset and interpret what these values indicate about the spread of the data.

In this chapter, we are continuing the discussion about summarizing data. In Chapter 2, we discussed how we can use one statistic to represent the central tendency of a dataset. Most people are likely already familiar with the concept of using an average, or the mean, to summarize data. In this chapter, we will explore how we can use other statistics to represent the spread, or variability, in a dataset. **Variability** refers to the extent to which data points in a dataset differ from each other and/or from the central value (e.g., the mean) and is a measure of how spread out or dispersed the data are. High variability indicates that the data points are more spread out, while low variability means they are closer together. A measure of variability should be reported along with any measure of central tendency because they provide different but complementary information.

Variability is often a little tricky to understand, so let's look at two simple datasets to help with our learning.

- Set A: 47, 48, 49, 50, 51, 52, 53
- Set B: 20, 30, 40, 50, 60, 70, 80

The mean for each dataset is 50, but these sets of numbers are quite different! The numbers in Set B are more spread out from the center, which is another way of saying they have greater variability. If we only report the mean for each dataset, then we have really not summarized the data well.

Before we talk about the different measures of variability, Table 3.1 includes some examples of variables with high, low, and no variability.

TABLE 3.1 *Examples of High, Low, and No Variability*

HIGH VARIABILITY	LOW VARIABILITY	NO VARIABILITY
• Semester tuition rates across all public and private colleges and universities in the United States. • Salaries of college graduates entering the workforce. • Standardized math scores for all third grade students in a given state.	• Starting salaries of state employees in similar positions. • High school GPAs of students who are accepted to elite private colleges. • Standardized math scores for eighth grade students in a gifted education program.	• Grade level (for a program that only includes fifth graders). • Fixed-rate mortgage payments over the course of a loan. • Enrollment fee for a program in dollars.

MEASURES OF VARIABILITY

Just like with central tendency, there are several different ways to measure and report the variability of a dataset. We introduce four different measures of variability: range, interquartile range, standard deviation, and variance.

A relatively crude or simple indicator of variability is the **range.** The range is a measure of variability that represents the difference between the largest and smallest values in a dataset and provides a simple indication of how spread out the data are. For example, the range in Set A above is 6 (the range = the largest value - the smallest value; i.e., 53 - 47 = 6), and the range in Set B is 60 (80 - 20). You can see that even if we only knew the range, we would get a sense that these two datasets were different.

However, a major limitation of range is that it is affected by outliers. Let's insert a third dataset that looks like this: 0, 56, 57, 58, 59, 60, 60. We still have a mean of 50 and a range of 60, but the range seems misleading. That 0 in our third dataset is a complete outlier because it is far below the rest of the values. It could be an odd case or even a typo, but it changes the range of our third dataset from 4 to 60, completely changing our understanding of what the data look like.

Interquartile range (IQR) is the difference between the first and third quartiles, or the 25th and 75th percentiles. Much like how the median was better than the mean when the data included major outliers, the IQR is better than the range in the same circumstances because it is less likely to be affected by extreme values.

To find the IQR, start by organizing your dataset in ascending order. Next, divide the data into two halves to calculate the first quartile (Q_1) and third quartile (Q_3). The **first quartile (Q_1)** is the median of the lower half of the data, representing the 25th percentile, while the **third quartile (Q_3)** is the median of the upper half of the data, representing the 75th percentile. The IQR is then calculated as $Q_3 - Q_1$, which measures the range of the middle 50% of the data, providing a robust indicator of variability that is less sensitive to outliers. Let's compute the IQR with a sample set of data.

Suppose you have the dataset 3, 7, 8, 12, 13, 15, 16, 20, 21.

- We first find the median.
  - Since we have nine values, the median is the fifth value: (9 + 1)/2 = 5. The fifth value is 13. The median of this dataset is 13.

- We next find the median of the lower half of the dataset. Since we have four values, we will need to add the two middle values (7 + 8) and then divide by 2 to get 7.5. The first quartile is equal to 7.5.
- We then find the median of the upper half of the dataset. Since we again have four values, we will need to add the two middle values (16 + 20) and divide by 2 to get 18. The third quartile is equal to 18.
- Last, we find the difference between the third quartile and the first quartile (18 – 7.5), which is equal to 10.5. Thus, the IQR for this example is 10.5.

Standard deviation (*SD*) tells us how far away, on average, each individual data point is from the mean. In other words, is this a dataset where most of the individual points are clustered close together so that each point is very close to the mean (small standard deviation)? Or are the individual points spread much farther apart, so any given value could be quite different from the mean (large standard deviation)?

Think about how you might figure out the standard deviation if you had to invent statistics yourself. It was stated earlier that it is how far away each data point is, on average. We know that we find an average by adding up a bunch of values and dividing by the number of values, right? So, in theory, we could find the mean, and then take every single individual data point and figure out how far away they are from the mean. Add them all up, divide by the number of data points, and then we have the average distance away. Right?

Not quite, due to one problem: negative numbers. By definition, some data points are going to be above the mean, and some will be below. Say you have a mean of 50, and one data point is 45 while another is 55. If we take the differences and add them up, we get -5 + 5 = 0. This is not helpful at all. So, what do we do? Think back to Chapter 2 when we were learning about the mean. The mathematical way to solve this problem is to square the differences, which results in only positive numbers.

This brings us to **variance**, which is a measure of variability that indicates how far the data points in a dataset are spread out from the mean. Variance is calculated as the average of the squared differences between each data point and the mean; the equation is listed in Table 3.2. By squaring the differences, variance gives greater weight to larger deviations, making it sensitive to extreme values in the dataset. While variance effectively quantifies dispersion, it is expressed in squared units of the original data, which can make it less intuitive for interpretation compared to the standard deviation. Variance is a fundamental concept in statistics, as it forms the basis for other measures of variability, such as the standard deviation, and is crucial in fields like probability, regression, and hypothesis testing.

Let's calculate the variance for a very simple dataset containing just four numbers: 5, 7, 9, 11.

- The mean for our dataset is 8:
  - $(5 + 7 + 9 + 11)/4 = 8$
- Subtract the mean from each of our values and square the result:
  - $5 - 8 = -3, -3^2 = 9$
  - $7 - 8 = -1, -1^2 = 1$
  - $9 - 8 = 1, 1^2 = 1$
  - $11 - 8 = 3, 3^2 = 9$
- Sum the squared differences and divide by the number of data points:
  - $9 + 1 + 1 + 9 = 20, 20/4 = 5$

The variance for our dataset is 5.

TABLE 3.2 *Population and Sample Variance and Standard Deviation*

	POPULATION (σ^2, σ)	SAMPLE (s^2, s)
Variance	$\sigma_x^2 = \frac{\sum (x_i - \mu)^2}{N}$	$s_x^2 = \frac{\sum (x_i - \bar{x})^2}{n-1}$
Standard Deviation	$\sigma_x = \sqrt{\frac{\sum (x_i - \mu)^2}{N}}$	$s_x = \sqrt{\frac{\sum (x_i - \bar{x})^2}{n-1}}$

This concept of the sum of squares is going to become very important when we learn topics such as analysis of variance (ANOVA) in later chapters. If you find yourself confused later about what "sum of squares" means, just remember that it has to do with variance.

Back to the current topic, we now have a value of 5, the variance, that tells us the average squared distance each point is from the mean. However, it is hard to interpret squared values, so what is our last step? We take the square root of the variance, and we call that the standard deviation, which is on the same scale as the original variable. Essentially, we square the differences to deal with negative numbers, and then we unsquare the result so that the statistic is easier to work with for interpretation. This means that the standard deviation for our example is approximately 2.24 (i.e., $\sqrt{5}$). Anytime you see variance and standard deviation reported, the second value will be the square root of the first.

A couple more notes before we get to the formulas. First, we can find the variance and standard deviation for a population or a sample, and the notation is different for each (see Table 3.2). Most of the time we're working with samples, but it is good to know both. Second, if we only have a sample, then we have to divide by $n - 1$ rather than N, where N represents the population size and n represents sample size, respectively. The quantity of $n - 1$ for the sample variance and standard deviation estimators is known as the "degrees of freedom" (*df*). We use the *df* to obtain an unbiased, or more accurate, estimate of the population parameter when we only have a sample to analyze.

Table 3.2 shows how to calculate the variance and standard deviation by hand, with the correct notations for populations and samples: σ_x^2 and σ_x represent population variance and standard deviation of variable x, respectively; s_x^2 and s_x represent sample variance and standard deviation of variable x, respectively; μ is a population mean; and $\bar{x}$ is a sample mean of variable x.

CHARACTERISTICS OF VARIABILITY

Here is a summary of the characteristics of data variability measures.

- Higher values indicate a larger amount of variability than lower numbers.
- When you have no variability, the numbers are constant (e.g., for the dataset 3,3,3,3,3,3, the variance and standard deviation equal 0).
- The variance tells you the average squared distance from the mean.
- The standard deviation is just the square root of the variance (i.e., it brings the squared units back to regular units).

- The standard deviation tells you approximately how far the numbers tend to vary from the mean. Making a judgment about how large or small the standard deviation depends entirely on the unit of measurement for your variable.
  - If the entire range of scores is 1 to 100, then you can say that a dataset with a standard deviation of 20 has more variability than a dataset with a standard deviation of 10.
  - However, a standard deviation of 10 on a dataset that ranges from 1 to 100 could indicate much greater variability than a standard deviation of 20 on a dataset that ranges from 1 to 1,000.

STANDARD DEVIATION AND THE NORMAL CURVE

We are going to learn a lot more about the normal curve, or bell curve, later, but many current and future educators already have at least some knowledge of the concept. Thus, let's focus on how the normal curve relates to standard deviation.

If data are *normally distributed*, then an easy rule to apply to the data is the **empirical rule,** or the *68-95-99.7 percent rule*:

- Approximately 68%, or just over two-thirds, of all individual cases will fall within one standard deviation of the mean, either above or below.
- Approximately 95% of cases will fall within two standard deviations of the mean.
- Approximately 99.7% of cases will fall within three standard deviations of the mean.

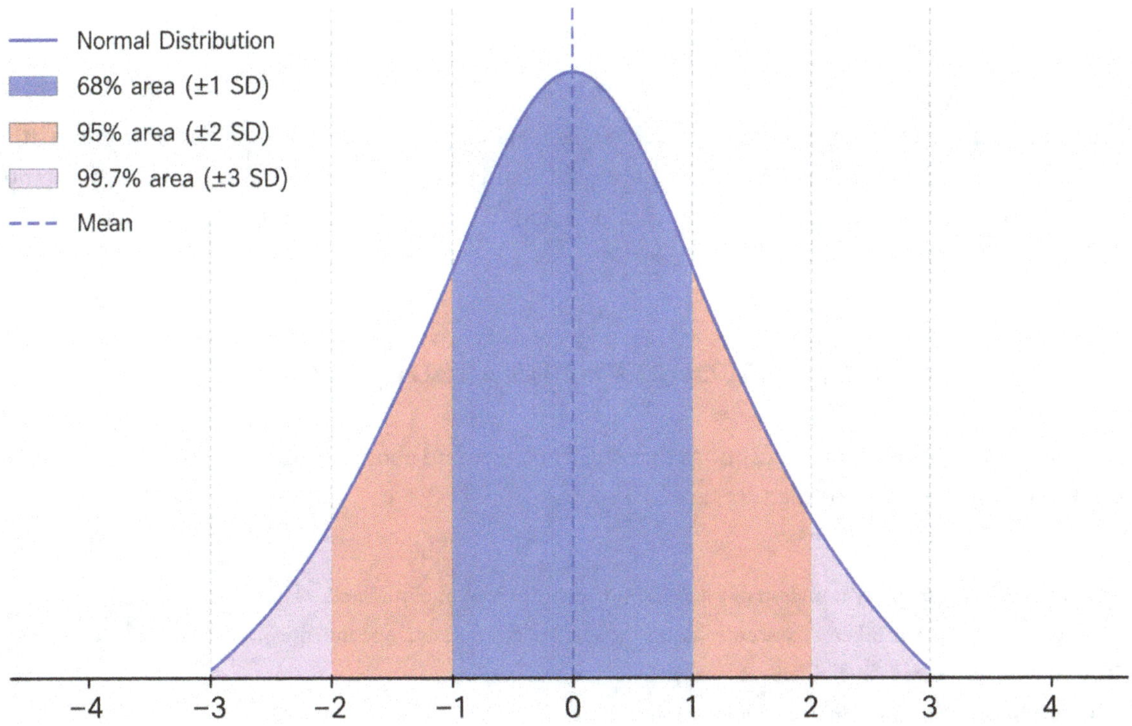

FIGURE 3.1 Normal Distribution Showing the Empirical Rule

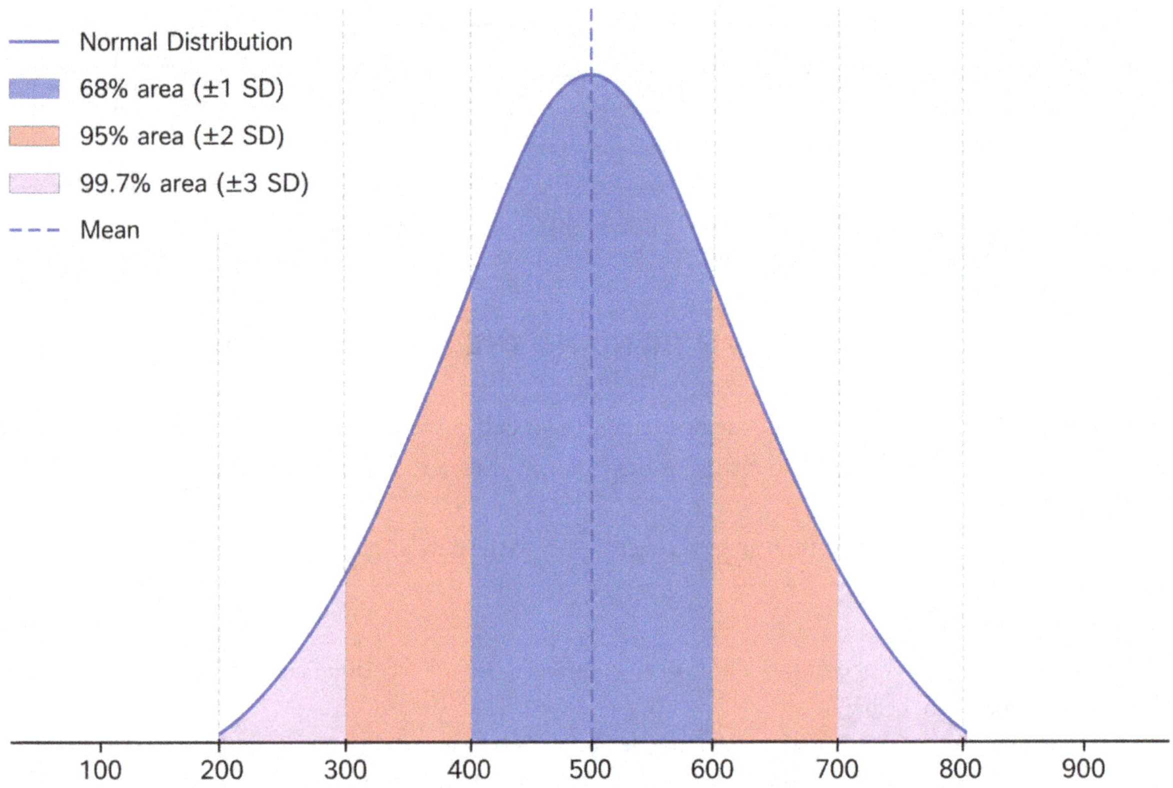

FIGURE 3.2 Normal Distribution of Standardized Test Scores

When you hear the term *standardized tests*, what that usually means is that the test has been standardized to fit the normal distribution. If you hear that a national standardized test has a mean of 500 and a standard deviation of 100, then you can apply the empirical rule and know that 68% of test takers score between 400 and 600 points.

COMMON MISCONCEPTIONS ABOUT VARIABILITY

Here are some common past misconceptions that students have had about variability, with brief explanations of the correct understanding for each concept.

1. **Misconception**: Variability is undesirable and should be minimized or eliminated.
   Reality: Variability is an inherent and necessary component of any dataset. In many cases, it offers crucial insights and is essential for understanding differences within a dataset and patterns.
2. **Misconception**: Variability in data occurs due to measurement errors.
   Reality: Although variability may include measurement error, it also reflects genuine differences between observations, which are essential for understanding patterns and drawing meaningful conclusions.

3. **Misconception**: Low variability is often interpreted as a sign of greater accuracy and consistency in the data and considered a desired characteristic of data.
   Reality: While low variability may indicate precision, it does not necessarily ensure accuracy or representativeness. For example, a thermometer that consistently provides incorrect readings has low variability but isn't accurate.
4. **Misconception**: Increasing the sample size (e.g., from 10 cases to 100 cases) eliminates variability in the data.
   Reality: Larger sample sizes reduce sampling variability (the differences between sample statistics and the population parameters, which is sampling error), but the natural variability in the population remains unchanged. This means that in theory, the sample size doesn't affect the inherent variability in data.
5. **Misconception**: It is commonly believed that all measures of variability, such as range, variance, standard deviation, and interquartile range, tell the same story.
   Reality: Various measures of variability emphasize different characteristics of a dataset. For example, the range focuses solely on the extremes, whereas the standard deviation accounts for the distances of all data points from the mean.
6. **Misconception**: Variability measures, such as standard deviation, are unrelated to measures of central tendency, such as the mean.
   Reality: Variability is closely linked to the central value used in its calculation. For instance, variance is calculated relative to the mean, highlighting the importance of the chosen measures of central tendency in interpreting variability.
7. **Misconception**: Mean and standard deviation values alone give enough insights into a dataset.
   Reality: Mean and standard deviation offer valuable statistical summaries, but they may miss crucial patterns such as extreme values (outliers), asymmetric distributions (skewed distributions), and/or multiple peaks in the data. Additional analysis beyond these basic measures is often needed for a complete understanding.
8. **Misconception**: High variability indicates that the data may be unreliable.
   Reality: Actually, high variability in data doesn't necessarily mean the data are untrustworthy. High variability simply means there's a wide spread or dispersion in the values. The data could be perfectly accurate and reliable while still showing large variations.
9. **Misconception**: Knowing standard deviation of a dataset is sufficient to understand how data are spread.
   Reality: While standard deviation measures typical distance from the mean, it fails to capture the full picture of how data are distributed, including the shape, symmetry, and concentration of values.

CONCLUSION

This chapter has provided a thorough exploration of variability, a crucial aspect of statistical analysis that complements measures of central tendency. By understanding range, interquartile range, variance, and standard deviation, readers can better assess how data are spread and how individual data points deviate from the mean. This chapter emphasized the importance of selecting the appropriate measure of variability based on data characteristics, ensuring a more accurate interpretation of statistical findings. Additionally, the chapter highlighted common misconceptions about variability, reinforcing the idea that variability is not an error but instead is an essential component of data analysis. As we move forward, Chapter 4 will introduce visual representations of data, providing further insights into how data distributions can be effectively communicated and understood.

KEY TAKEAWAYS FOR EDUCATIONAL RESEARCHERS FROM CHAPTER 3

- Variability highlights the extent of differences of each data point within a dataset beyond the central value. Two datasets may have the same mean values but differ drastically in their spread, impacting the interpretation of results.
- High variability in datasets (e.g., eighth graders' standardized reading test scores across states) may suggest diverse population characteristics. On the other hand, low variability (e.g., standardized test scores within a specialized group or subgroup) indicates uniformity, useful for specific interventions.
- Range can be largely impacted by outliers, while IQR or standard deviation will be less impacted by outliers.
- Central tendency, such as mean, median, and mode, and variability are complementary metrics for understanding how data look. Reporting *both* metrics (e.g., mean and standard deviation) ensures a more complete summary.
- Educational researchers should choose measures of central tendency and variability that align with their data characteristics and research questions.

KEY DEFINITIONS FROM CHAPTER 3

The **empirical rule,** also known as the 68-95-99.7 percent rule, is a guideline stating that 68% of the data falls within one standard deviation of the mean, 95% of the data falls within two standard deviations of the mean, and 99.7% of the data falls within three standard deviations of the mean.

First quartile is another name for the median of the lower half of the values in a dataset (25th percentile).

Interquartile range (IQR) is a measure of variability that indicates the spread of the middle 50% of a dataset. It is the difference between the third quartile and the first quartile.

Range is a measure of variability that represents the difference between the largest and smallest values in a dataset and provides a simple indication of how spread out the data are.

Standard deviation (*SD*) is the square root of the variance and is expressed in the same units as the data. *SD* quantifies how much the individual data points deviate, on average, from the mean of the dataset.

Third quartile is another name for the median of the upper half of the values in a dataset (75th percentile).

Variability refers to the extent to which data points in a dataset differ from each other and from the central value (e.g., the mean or median). Variability is a measure of how spread out or dispersed the data are. High variability indicates that the data points are more spread out, while low variability means they are closer together.

Variance is a measure of variability that indicates how far the data points in a dataset are spread out from the mean.

CHECK YOUR UNDERSTANDING

1. What is the range of the following dataset: 1, 2, 3, 4, 5, 6, 7, 8, 9, 10?
   a. 10
   b. 9
   c. 1
   d. 0

2. The ______ is the square root of the variance?
    a. Interquartile range
    b. Range
    c. Standard deviation
    d. Mean

3. What is the population variance of the following dataset: 1, 2, 3, 4, 5?
    a. 1
    b. 2
    c. 2.5
    d. 3

4. What is the population standard deviation of the following dataset: 1, 2, 3, 4, 5?
    a. 1.41
    b. 1.73
    c. 4.00
    d. 9.00

5. What percentage of values fall within one *SD* of the mean if data are normally distributed?
    a. 50%
    b. 68%
    c. 95%
    d. 99.7%

4

DISPLAYING DATA GRAPHICALLY

LEARNING OBJECTIVES

By the end of this chapter, you will be able to:

- Identify appropriate graphical methods (e.g., bar graphs, histograms, box plots, scatter plots) to summarize and communicate patterns, trends, and variability in categorical and continuous data; and
- Construct clear, accurate, and meaningful data visualizations that effectively communicate key insights while avoiding common pitfalls such as distortion, poor labeling, or overcomplicated designs.

Exploratory data analysis is an approach that involves examining datasets to summarize their main characteristics, identify patterns, and uncover insights, often using visual methods. This technique is widely used in education-related fields to gain insights from educational data, improve decision-making, and inform policy and practice. You could take an entire course about data visualization, but for now we will only summarize some of the key concepts.

Data visualization refers to the graphical representation or structured representation of information and data to facilitate understanding and interpretation. By using charts, graphs, figures and tables, we present information in a way that highlights trends, outliers, and patterns in raw data. While graphs and charts are commonly associated with data visualization in educational research, tables also serve as a fundamental tool by systematically organizing numerical and categorical data for easy comparison. Unlike graphical displays, which emphasize patterns and relationships, tables provide precise values and detailed breakdowns, making them particularly useful when exact numerical reporting is required. Understanding how to display data is important for multiple steps in the research process. You may notice an interesting pattern that you want to investigate more closely, so creating well-designed tables and graphs might be the start of deeper investigation into your research.

Frequently, the first step in data analysis is visualizing the data to confirm that it meets the assumptions of your planned analysis method. For example:

- If you want to know whether your dependent variable is approximately normally distributed, you create a histogram.
- If you want to see if the amount of variation is similar between two groups, you create a box plot.
- If you want to see if there is a linear relationship between two variables, you create a scatter plot.

A good graph should

- Show the data with the appropriate visualization (e.g., bar chart, scatter plot, histogram) based on the type of data and relationships being displayed;
- Induce the reader to think about the data being presented;
- Get as much information across as possible within limited space (use a clean and simple design that enhances clarity without excessive grid lines or redundant elements);
- Make large data sets coherent;
- Encourage the reader to ask additional questions about the data;
- Communicate the findings or your key message about the data; and
- Maintain uniform styles, colors, and labeling conventions across multiple graphs.

A good graph should NOT

- Distort the data (intentionally or unintentionally);
- Use 3D effects for a graph plotting two variables;
- Go overboard with patterns and effects that distract from the key message (overloading a graph with too many data series can make it unreadable); and
- Have badly labeled axes (number of what ???).

Displaying your data graphically can also be a critical part of sharing the results of any analysis, and you want to do so in a way that communicates your message effectively.

While graphical representations are powerful for identifying patterns and trends, there are instances when a structured, numerical representation is more effective. In cases where precision, exact comparisons, or detailed breakdowns are needed, tables serve as an essential tool for organizing and presenting data.

A **table** is an organized arrangement of data in rows and columns and is designed to systematically display values, frequencies, or relationships for easy comparison and interpretation. Tables are essential tools for summarizing data in educational contexts because they provide a clear, structured way to organize and present information. Tables allow educators, administrators, and policymakers to quickly interpret numerical and categorical data, making it easier to identify trends, compare variables, and draw meaningful conclusions. For instance, tables can display student performance across subjects, attendance rates, or resource allocation across schools, enabling stakeholders to make data-driven decisions. Unlike graphical displays, tables provide precise values, making them particularly useful when exact numbers are necessary for analysis or reporting. Moreover, tables are versatile, as they can accommodate large amounts of data and can be customized to highlight specific details, such as subgroup comparisons and longitudinal trends, fostering better understanding and informed actions in educational settings.

Let's walk through a scenario to illustrate how tables and graphical displays can be used. A school administrator wants to compare the average test scores of students across grade levels in one middle school. Table 4.1 shows average scores. It is very easy to see that the average scores for grades 6, 7, and 8 are 78, 85, and 92, respectively.

Another way we could visualize this data is through the use of a **bar graph**. Bar graphs are particularly effective for displaying data *across discrete categories*, such as grade levels, school districts, or subject areas, making complex information easier to understand. For example, bar charts can be used to compare average test scores

TABLE 4.1 *Average Test Score by Grade Level*

GRADE LEVEL	AVERAGE TEST SCORE
6	78
7	85
8	92

across grade levels, participation rates in extracurricular activities, or funding allocations across departments. Their straightforward design allows educators and administrators to quickly identify patterns, trends, and disparities, such as gaps in performance or resource distribution. Bar graphs also help effectively communicate findings to diverse audiences, including students, parents, and policymakers, as the visual format is easy to interpret. By simplifying the presentation of categorical data, bar graphs support data-driven decision-making and facilitate meaningful discussions about improving educational outcomes.

Bar graphs use vertical bars to represent the data. The height of the bars often represents the frequencies of some dependent variable for the categories that are displayed along the X-axis. Note that by tradition, the X-axis (independent variable) is the horizontal axis, and the Y-axis (dependent variable) is the vertical axis. The following bar graph shows the same information as Table 4.1.

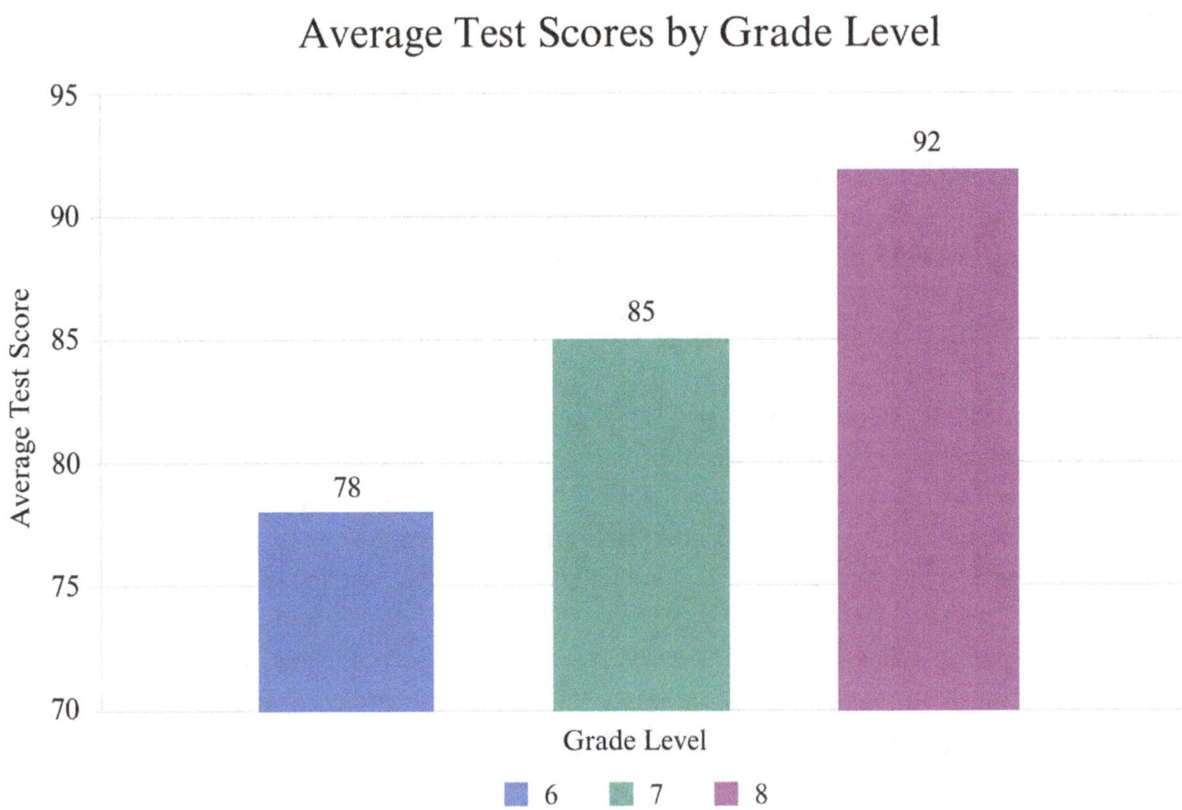

FIGURE 4.1 Bar Graph of Average Test Scores by Grade Level

Histograms summarize the frequency of *continuous data* that are grouped. Values of the independent variable are along the X-axis, grouped by equal intervals. Typically, the height of the rectangle, measured by the Y-axis, equals the frequency recorded for each interval. The base of each rectangle begins and ends at the lower and upper boundaries, seen from left to right, of each interval, and each rectangle touches adjacent rectangles at the boundaries of each interval.

It may seem that a bar graph and a histogram are the same, but in the world of statistics, it matters whether the bars touch each other. A bar graph, with separated bars, is used for categorical or discrete variables or sometimes for variables based on ordinal scales as well. The separation shows that these are discrete categories rather than groupings of a continuous variable where the group size could be adjusted to whatever you want. Unlike bar charts, histograms group data into intervals, or bins, to show the distribution of a dataset, making them useful for identifying skewness, outliers, and/or overall patterns in continuous variables.

Our school administrator now wants to look at the distribution of test scores for just seventh graders, but they decided to group scores in the following categories: 50 to <60, 60 to <70, 70 to <80, 80 to <90, 90 to 100. The histogram showing a summary of the dataset is included in Figure 4.2.

A **stem-and-leaf plot** is a graphic display in which each individual score from the original data set is listed in a relatively compact manner. Common digits shared by all scores are listed to the left, and the remaining digits for each score are listed to the right. So, the *stems* are the first digit(s), and the *leaves* are the last digit(s). Rather than giving a frequency count, this is a way of summarizing the data visually, because you can see

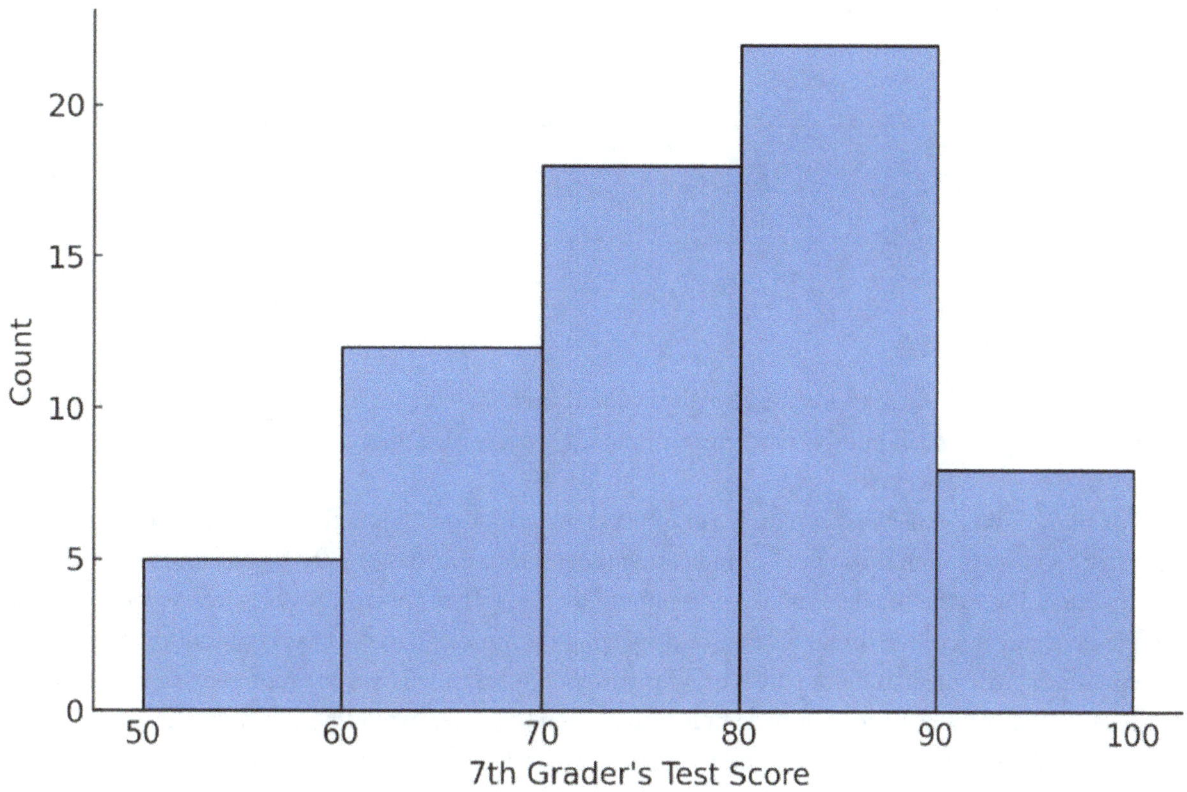

FIGURE 4.2 Histogram of Seventh Grade Test Scores

TABLE 4.2 ***Stem-and-Leaf Plot of Test Scores***

5	4 5 5 5 6
6	3 4 4 5 5 5 5 6 6 6 6 7 7
7	3 4 4 4 5 5 5 5 5 5 6 6 6 6 7 7
8	3 3 4 4 4 4 5 5 5 5 5 5 5 5 5 6 6 6 6 6
9	4 4 5 5 5 5 6 6

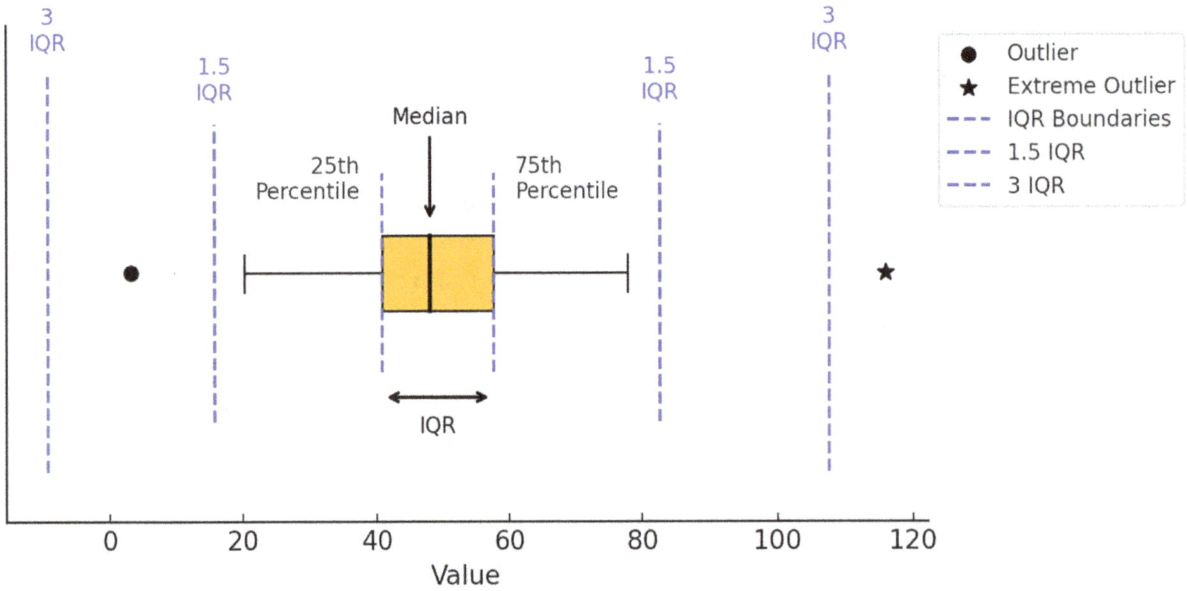

FIGURE 4.3 Box Plot and Interquartile Range

which stems have the most leaves while also displaying the actual scores. Stem-and-leaf plots are best suited for small datasets where all numbers are greater than 0 (Moore, McCabe, & Craig, 2021). Table 4.2. is one example of a stem-and-leaf plot.

A **box plot,** or box-and-whisker plot, is a graphical representation of data that shows its central tendency, variability, and distribution through a box indicating the interquartile range (IQR), a line for the median, and *whiskers* extending to the smallest and largest nonoutlier values. They are typically oriented vertically, but the example in Figure 4.3 is horizontal for ease of displaying the various parts. At the center of the box plot is a line representing the median (50th percentile) for the data, surrounded by a box representing the interquartile range (IQR, or the 25th to 75th percentile). Then there are lines extending out from the box (these are the *whiskers*). Most often these will extend to the smallest and largest observed points in the dataset that are within 1.5 times the IQR. If there are outliers in the dataset that fall beyond the whiskers, they are typically represented by individual circles.

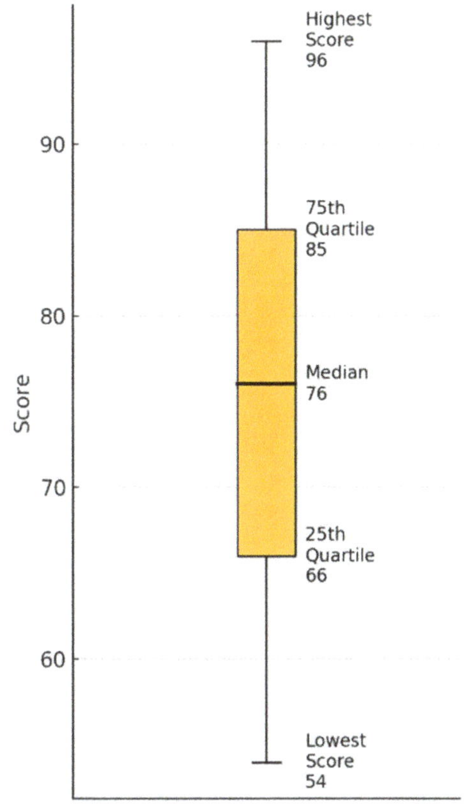

FIGURE 4.4 Box Plot of Seventh Grade Test Scores

Figure 4.4 shows a box plot of our seventh grade test scores. The median score is 76, the 75th quartile is 85, the 25th quarter is 66, the lowest score is 54, and the highest score is 96.

A **scatterplot** is a graphical representation that displays individual data points based on two variables, with one variable plotted along the X-axis and the other along the Y-axis, allowing researchers to visualize relationships between them. In educational research, scatterplots are essential for identifying correlations and trends, such as the relationship between socioeconomic status and academic performance or between hours spent studying and test scores. By highlighting patterns, such as positive, negative, or no correlation, scatterplots provide insights into potential causal or influential factors that can guide further investigation. Additionally, scatterplots help researchers detect outliers or anomalies, such as students performing significantly above or below the expected trend, which can inform targeted interventions or support. This ability to visually explore complex data relationships makes scatterplots a valuable tool in understanding and improving educational outcomes. Examples of scatterplots showing each individual test score are included in Figure 4.5.

A **quantile-quantile (*Q-Q*) plot** is a graphical tool used to assess whether a dataset follows a specific theoretical distribution, most commonly the normal distribution. In a *Q-Q* plot, the quantiles of the observed data are plotted against the quantiles of the chosen distribution (e.g., normal distribution). If the points in the plot fall approximately along a 45-degree reference line, this suggests that the data conform to that distribution. Deviations from this line indicate departures from the expected distribution, such as skewness, kurtosis, and outliers. *Q-Q* plots are particularly valuable in educational research when determining if data, such as student

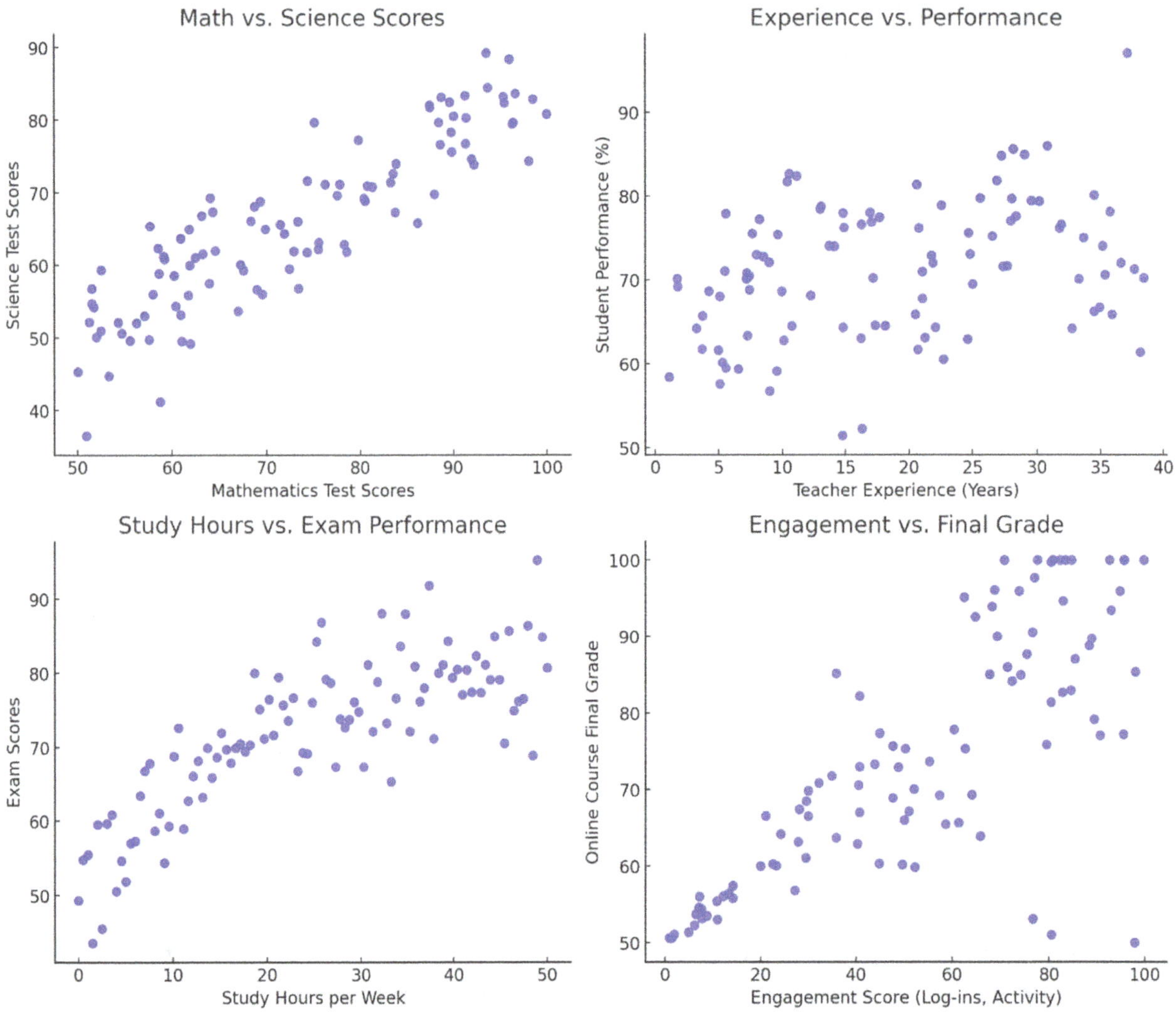

FIGURE 4.5 Examples of Scatterplots

test scores or survey responses, meet the assumptions required for parametric statistical tests. For instance, a researcher analyzing standardized test scores might use a *Q-Q* plot to check for normality before conducting a *t* test or an analysis of variance (ANOVA). While histograms provide a general sense of distribution, *Q-Q* plots offer a more precise visual check by directly comparing observed data with a theoretical model.

Imagine that an educational researcher wants to assess whether a sample of students' math scores follows a normal distribution. The *Q-Q* plot in Figure 4.6 demonstrates how well the data align with a normal distribution. Most points fall along the 45-degree line, indicating normality, although a few points deviate slightly at the ends, suggesting potential outliers.

Including *Q-Q* plots in your data visualization tool kit enhances your ability to diagnose data distribution issues, ensuring that your chosen analytical methods are appropriate and reliable. This tool, combined with histograms, box plots, and other visualizations, provides a comprehensive approach to exploratory data analysis.

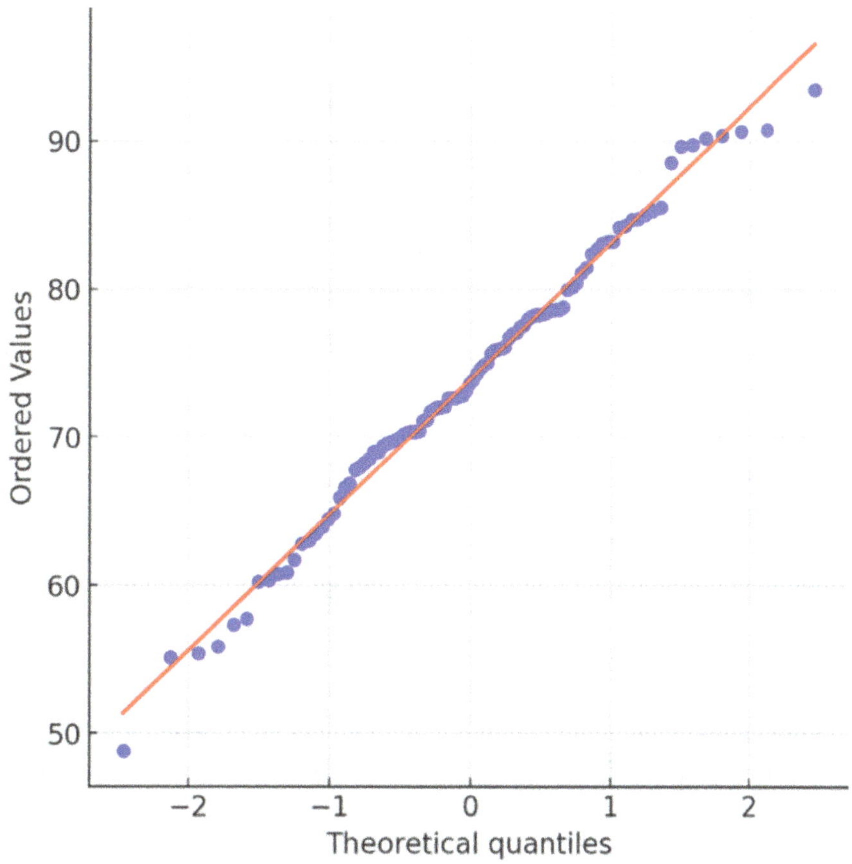

FIGURE 4.6 *Q-Q* Plot of Student Math Scores

A **pie chart** is a circular graph divided into slices where each slice represents a proportion of a whole, making it ideal for visualizing percentages or relative frequencies. In educational research, pie charts are often used to illustrate the composition of groups or categories, such as the distribution of students by socioeconomic status, the percentage of students enrolled in various extracurricular activities, or the allocation of a school's budget across different programs. For example, a pie chart could show the proportion of students achieving different performance levels on a standardized test, allowing researchers to identify the largest and smallest groups at a quick glance. Pie charts are particularly effective for presenting summary data to audiences who may not have extensive statistical training, as they provide an intuitive and visually engaging way to communicate the breakdown of categories. However, pie charts are most useful when there are relatively few categories, as too many slices can make interpretation challenging. An example showing student enrollment across grades six, seven, and eight from the administrator's middle school is included in Figure 4.7.

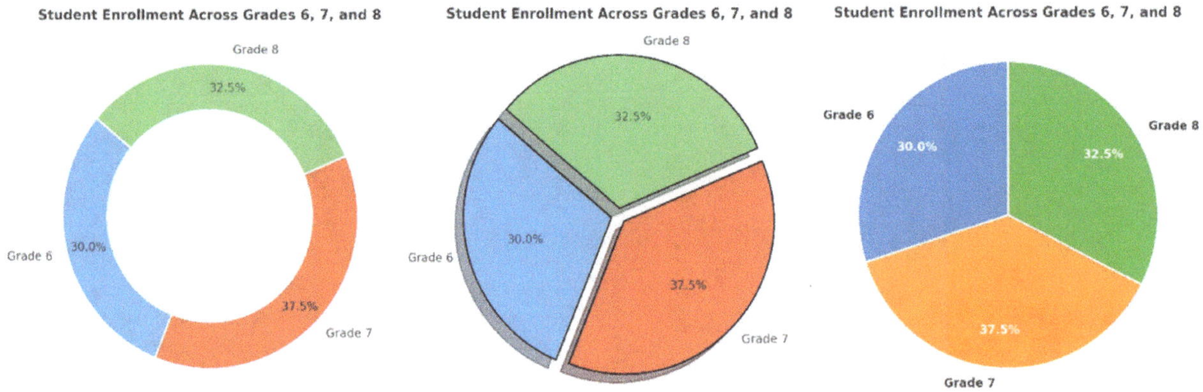

FIGURE 4.7 Pie Charts for Student Enrollment

CONCLUSION

Chapter 4 emphasizes the importance of data visualization in educational research, providing an overview of various graphical methods and their applications. From bar charts and histograms to box plots and scatterplots, each visualization tool serves a unique purpose in summarizing and communicating data. These tools help researchers and educators uncover patterns, trends, and outliers, offering valuable insights that inform decision-making and policy development. The chapter also highlights the distinction between categorical and continuous data representations and underscores the importance of designing graphs that are clear, accurate, and effective in delivering key messages.

Through practical examples, such as comparing test scores across grade levels and visualizing the relationship between two continuous variables such as study hours and academic performance, this chapter demonstrates how graphical displays can reveal meaningful patterns and insights from raw data. By integrating tables, stem-and-leaf plots, and pie charts alongside more complex visualizations, educators and researchers are equipped with the tools to present findings in ways that are both clear and informative. This foundational understanding of data visualization sets the stage for exploring deeper analytical techniques and leveraging data to improve educational outcomes. As we move forward, Chapter 5 will explore the normal distribution, an essential statistical concept that will further refine our ability to interpret data and make meaningful inferences.

KEY TAKEAWAYS FOR EDUCATIONAL RESEARCHERS FROM CHAPTER 4

- Incorporate data visualization to summarize patterns and trends. In educational research, data visualization helps uncover meaningful insights from student performance metrics, survey responses, and institutional data. Bar graphs, histograms, and scatterplots allow researchers to identify trends in student achievement, compare subgroup differences, and detect outliers that may indicate unique challenges or exceptional performance. By transforming raw numbers into graphical representations, educators and policymakers can more easily interpret and act on findings.

- Choose the right visualization for your data and analysis needs. Different types of data require different graphical techniques to accurately convey insights. Choose the correct visualization (e.g., box plots for variability, *Q-Q* plots for checking normality) based on the type of data and analysis requirements to ensure clarity and enhance interpretability.
- Ensure clarity and accuracy in graph design. Well-designed graphs improve comprehension and reduce the risk of misinterpretation. In educational research, this means clearly labeling axes, ensuring consistent scales when comparing multiple groups, and avoiding misleading visual distortions (e.g., truncating axes to exaggerate differences). Additionally, color choices and legends should enhance readability, especially when presenting findings to diverse audiences, including educators, administrators, and policymakers. A clear, accurate graph allows stakeholders to draw valid conclusions and make informed decisions.
- Use data visualization for exploratory analyses. Before applying statistical models, researchers can use graphical methods to check assumptions. For instance, histograms and box plots help assess the distribution of student test scores before conducting an ANOVA, while scatterplots with trend lines can reveal nonlinear relationships between study time and grade point average. Detecting patterns early through visualization can prevent errors in model selection and improve the robustness of conclusions in educational research.
- Incorporate visualizations in research reporting for effective communication. Effective visualizations bridge the gap between complex analyses and actionable insights. When reporting findings on student achievement or the effectiveness of instructional interventions, incorporating well-crafted charts, tables, and dashboards can make data accessible to decision-makers. Visuals enhance storytelling in research, making data-driven recommendations more compelling and easier to act upon.

KEY DEFINITIONS FROM CHAPTER 4

A **bar graph** is a graphical representation of data using rectangular bars, where the length or height of each bar corresponds to the value or frequency of a specific category or group. There is a gap between each bar.

A **box plot** (or box-and-whisker plot) is a graphical representation of data that shows its central tendency, variability, and distribution through a box indicating the IQR, a line for the median, and "whiskers" extending to the smallest and largest nonoutlier values.

Exploratory data analysis is an approach that involves examining datasets to summarize their main characteristics, identify patterns, and uncover insights, often using visual methods.

A **histogram** summarizes the frequency of continuous data that are grouped. The bars on a histogram touch one another.

A **pie chart** is a circular graph divided into slices, where each slice represents a proportion of a whole, making it ideal for visualizing percentages and relative frequencies.

A **scatterplot** is a graphical representation that displays individual data points based on two variables, with one variable plotted along the X-axis and the other along the Y-axis, allowing researchers to visualize relationships between them.

A **stem-and-leaf plot** is a graphic display where each individual score from the original data set is listed in a relatively compact manner. Common digits shared by all scores are listed to the left, and the remaining digits for each score are listed to the right.

A **quantile-quantile (*Q*-*Q*) plot** is a graphical tool used to assess whether a dataset follows a specific theoretical distribution, most commonly the normal distribution.

A **table** is an organized arrangement of data in rows and columns designed to systematically display values, frequencies, or relationships for easy comparison and interpretation.

CHECK YOUR UNDERSTANDING

1. What is the primary purpose of using a histogram in educational research?
   a. To compare categories of data.
   b. To display the relationship between two variables.
   c. To summarize the frequency of continuous data grouped into intervals.
   d. To show proportions of a whole.

2. Which visualization method is best suited for identifying relationships or correlations between two variables?
   a. Pie chart
   b. Box plot
   c. Scatterplot
   d. Stem-and-leaf plot

3. In a box-and-whisker plot, the IQR is represented by:
   a. The lines extending from the box to the smallest and largest values.
   b. The box itself, showing the range between the 25th and 75th percentiles.
   c. The line in the center of the box representing the median.
   d. The individual dots outside the whiskers.

4. What is a key advantage of tables in presenting data?
   a. They simplify relationships between variables visually.
   b. They display trends and patterns intuitively.
   c. They highlight outliers in large datasets.
   d. They summarize data with precise values for comparison.

5. What is the key difference between a bar chart and a histogram?
   a. Bar charts use touching bars to represent continuous data, while histograms use separated bars for discrete data.
   b. Bar charts use separated bars for discrete data, while histograms use touching bars for continuous data.
   c. Bar charts are only used for percentages, while histograms are used for counts.
   d. There is no difference; the terms are interchangeable.

5

THE NORMAL DISTRIBUTION

LEARNING OBJECTIVES

By the end of this chapter, you will be able to:

- Identify the defining characteristics of a normal distribution;
- Compute z scores from raw data and interpret their meaning in terms of standard deviations from the mean; and
- Use the properties of a normal distribution to calculate probabilities and interpret results for real-world scenarios using z tables and statistical software or applets.

The objectives for this chapter are to learn about the normal distribution, sometimes called the bell curve due to its shape, and to understand the concepts of probability associated with the normal distribution. An understanding of the normal distribution is very important because we use what we know about the normal distribution to make inferences about populations based on samples.

WHAT IS A NORMAL DISTRIBUTION?

A **normal distribution**, or bell-shaped distribution, has the following characteristics:

- The shape is symmetric, unimodal, and bell-shaped.
- The mean, median, and mode are the same—all are right in the center of the distribution.
- Following the empirical rule, approximately 68% of the values in a normal distribution fall within one standard deviation of the mean, approximately 95% of the values fall within two standard deviations of the mean, and approximately 99.7% of the values fall within three standard deviations of the mean.

For example, suppose a class took a science test and the students' scores were normally distributed. The distribution of students' scores had a mean of 50 points (the center of the normal distribution) and a standard deviation (*SD*) of 15 points. Based on the empirical rule, you know that about 68% of the students scored between 35 (one *SD* below the mean) and 65 (one *SD* above the mean) points.

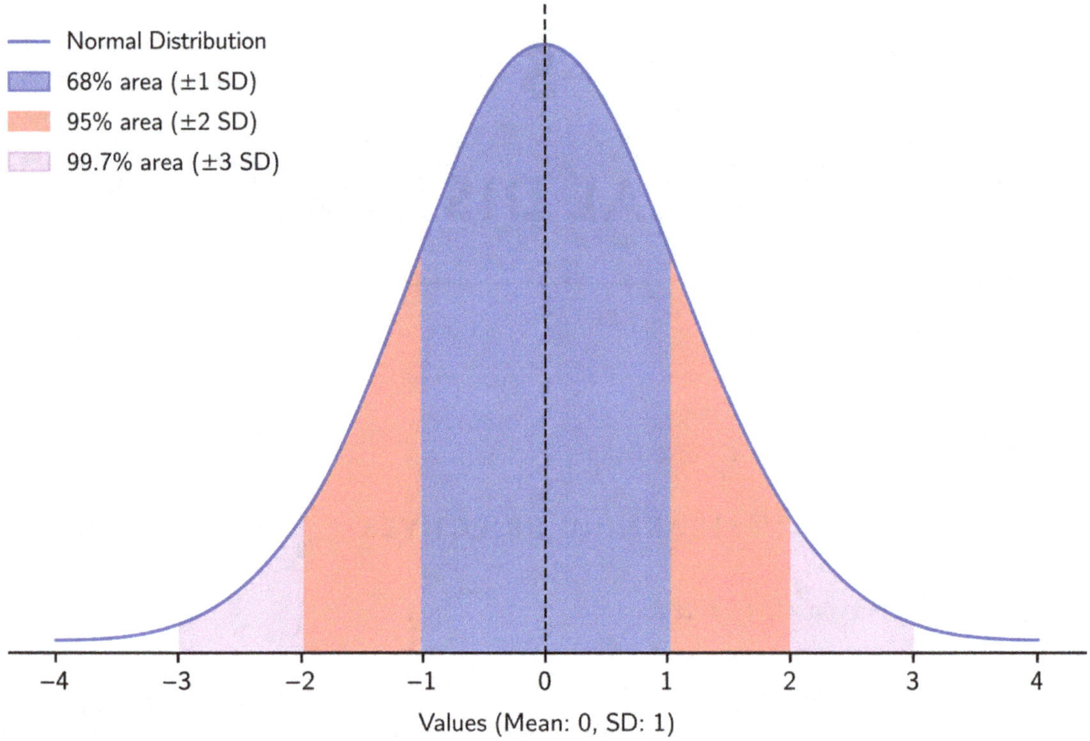

FIGURE 5.1 The Empirical Rule for a Normal Distribution

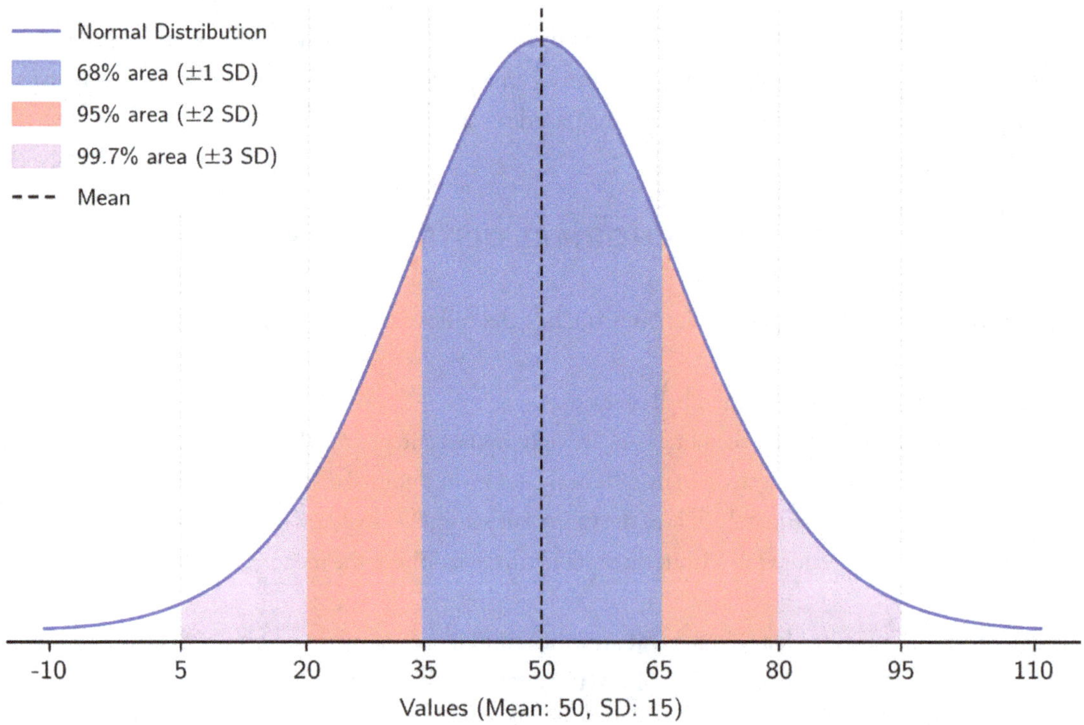

FIGURE 5.2 Normal Distribution with Mean of 50 and Standard Deviation of 15

It is important to understand the characteristics of the normal distribution not only because certain variables (e.g., weight, height, test scores) are likely to have normal distributions but also because many of the statistical techniques used to make inferences about populations rely on the properties of the normal distribution.

This rule is something you can use to interpret results from normally distributed data, but it is also part of what defines a normal distribution. Not all tests that students take are normally distributed. In those cases, the empirical rule would not apply.

Software programs such as SPSS® and JASP can provide tests of normality that tell you how close your distribution is to the normal distribution. You can assess how closely your data approximate normality by examining two key statistics: skewness and kurtosis.

- As we learned in Chapter 2, **skewness** means the extent to which a curve is pulled to one side or the other. Skewness is about symmetry or, really, the lack of symmetry. In a true normal distribution, half of the data points are above the mean, and half are below. A skewed distribution has more outliers on one side of the mean.
- **Kurtosis** measures the *tailedness* of a distribution, or how heavy or light the tails are compared to a normal distribution. The strict definition can be confusing. For now, think about it in terms of how likely it is that there are outliers, or extreme values, in your data. A *heavy-tailed* distribution will have a higher kurtosis value, and a greater percentage of scores in the distribution will be out in the tails rather than on the *shoulders*.

Figure 5.3 demonstrates how kurtosis affects distribution shape:

Positive kurtosis ($K > 0$): The blue< plot represents a distribution with a narrow peak and thick tails.
Zero kurtosis ($K = 0$): The green plot represents a standard normal distribution with neither exaggerated tails nor an unusually sharp peak.
Negative kurtosis ($K < 0$): The red plot represents a distribution with a flatter peak and thinner tails

A true normal distribution will have skewness and kurtosis values of 0. The closer the reported values are to 0, the more normal the distribution. A common rule of thumb is that a distribution with skewness and kurtosis values between -1 and 1 can be considered approximately normal. Figure 5.4 shows several examples of how skewness and kurtosis are displayed across several curves.

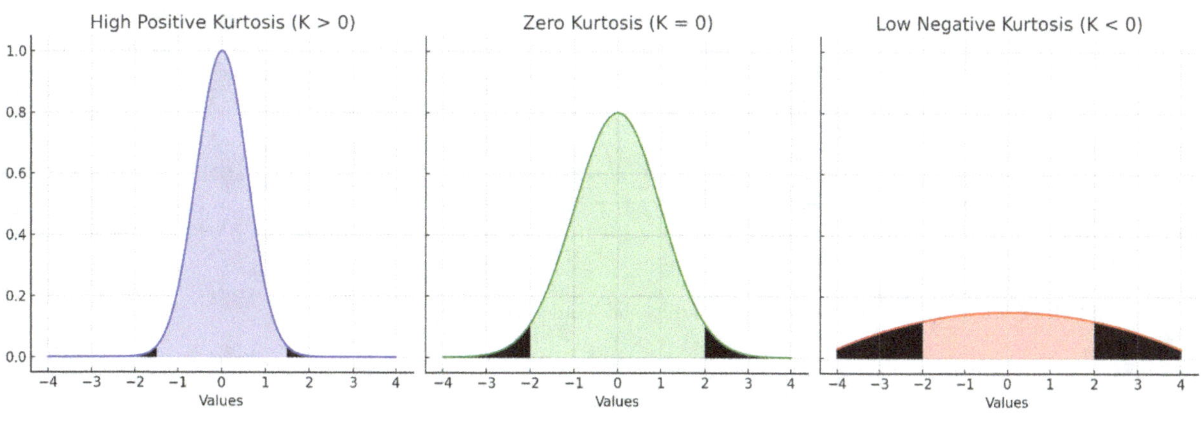

FIGURE 5.3 How Kurtosis Affects Distribution Shape

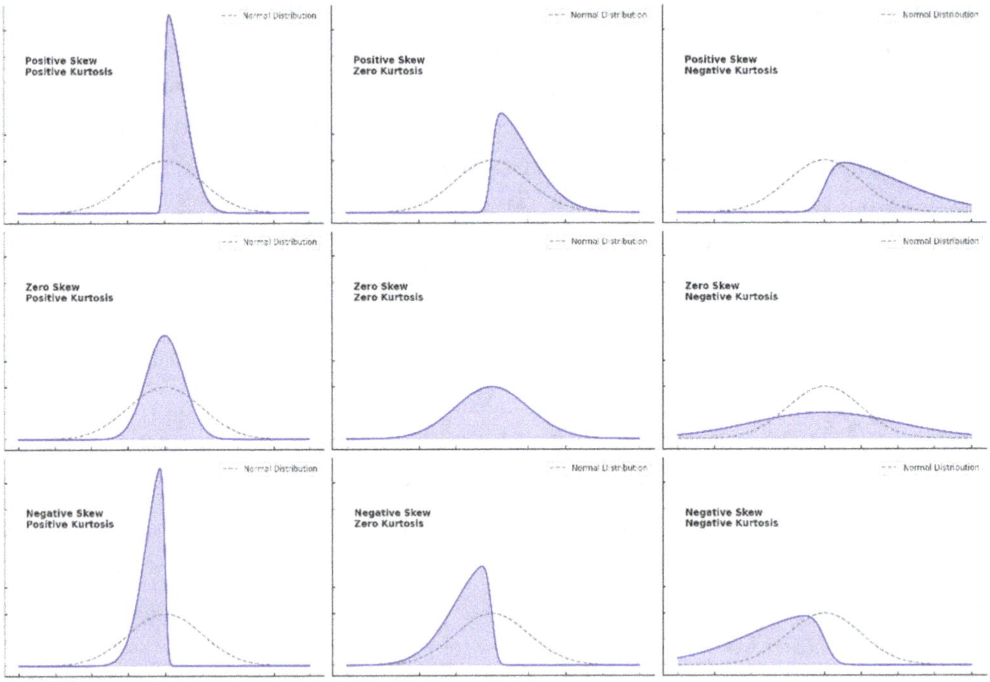

FIGURE 5.4 Effects of Skewness and Kurtosis on Distribution Shape

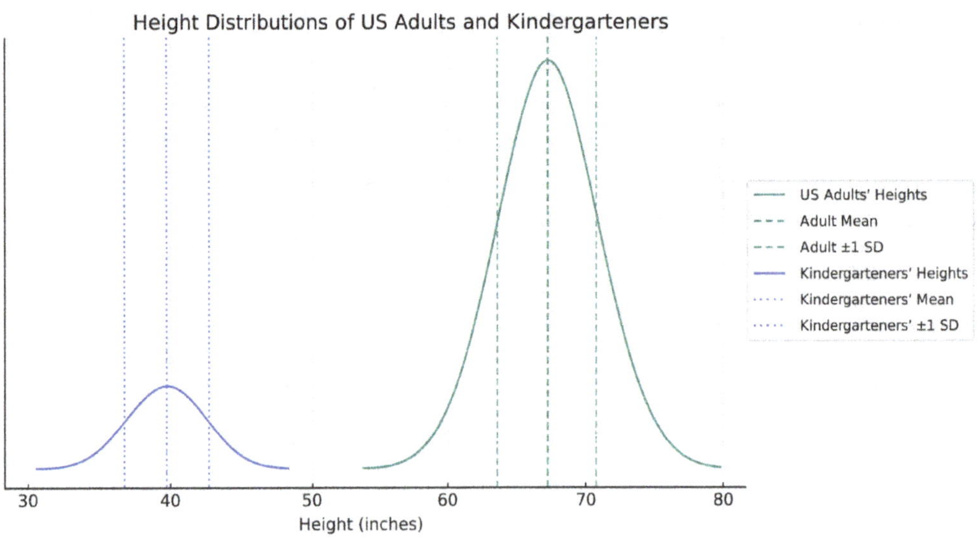

FIGURE 5.5 Height Distributions of US Adults and Kindergarteners

A normal distribution can have any mean and any standard deviation—it is about the shape of the distribution, not the specific numbers. For example, if you are looking at the heights of people (Figure 5.5), the mean and standard deviation will be quite different depending on whether your sample includes all people in the United States or all kindergarten students in one elementary school. Both height distributions will likely be normal and have the same basic shape, but the average height and variation in heights will probably both be lower if we are just looking at kindergartners.

Z SCORES AND THE STANDARD NORMAL DISTRIBUTION

We often convert raw data values in a normal distribution to *z* scores, or standardized scores. A ***z* score** is a statistical measure that describes how far a particular data point is from the mean, center, of its distribution. Instead of describing the difference between one score and the mean in terms of points or inches or whatever unit was used for the original distribution, we are using standard deviation units. This allows for the comparison of scores from different distributions or scales. We do this because simply knowing the difference between one score and the overall mean doesn't tell us much. For example, you learn that you scored 5 points above the mean. Is your performance a lot better than average? Is your performance barely above average? The answer depends on the standard deviation. If the standard deviation of this distribution was 20, your score of 5 points above the mean would be considered close to average. However, if the standard deviation was only 2, your score would be considered outstanding.

Using the example above, if you scored 5 points above the mean and the standard deviation was 20, your *z* score would be 0.25. Your score would be a quarter of a standard deviation above the mean. However, if the standard deviation was 2, your *z* score would be 2.5. In this case, you would have scored 2.5 standard deviations above the mean. Figure 5.6 helps visualize how far different *z* scores are from the mean.

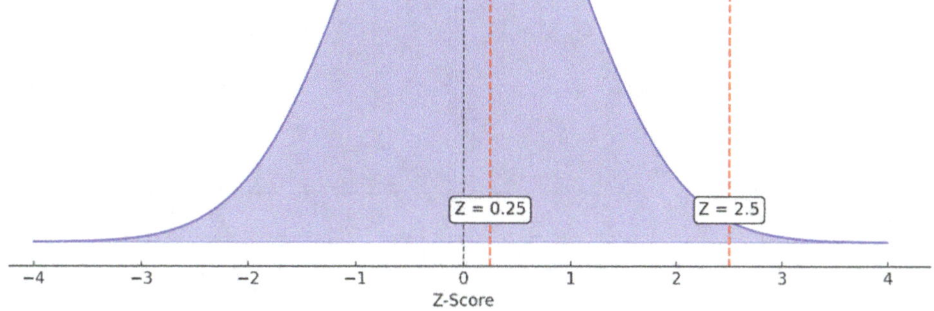

FIGURE 5.6 Visualizing Z Scores on the Standard Normal Distribution

If we convert an entire distribution from raw scores to z scores, that means we are standardizing the dataset. The mean of the distribution then becomes 0, the standard deviation becomes 1, and we refer to the distribution as the **standard normal distribution**.

What we have discussed so far are likely the most important concepts for you to understand. The rest of this chapter will mostly involve finding areas under the normal curve by using z scores and the z table. You will also learn how you can get to the same answer using online applets.

FINDING AREAS UNDER THE NORMAL CURVE

Let's say we have a continuous random variable: body temperature. Imagine that we have access to the body temperatures of a particular population (e.g., all residents of the United States), and when we graph this data, we see a normal distribution with a mean of 98.2 and a standard deviation of 0.73. Remember that we collected data from individuals who belong to a particular population (i.e., all Americans). So, these numbers (μ = 98.2 and σ = 0.73) are not sample statistics; they are population parameters. Our normal distribution is illustrated in Figure 5.7.

We now have three questions we want to answer based on this distribution.

Question #1: What is the probability that a randomly sampled individual from this population has a body temperature of 98.6 degrees or higher?

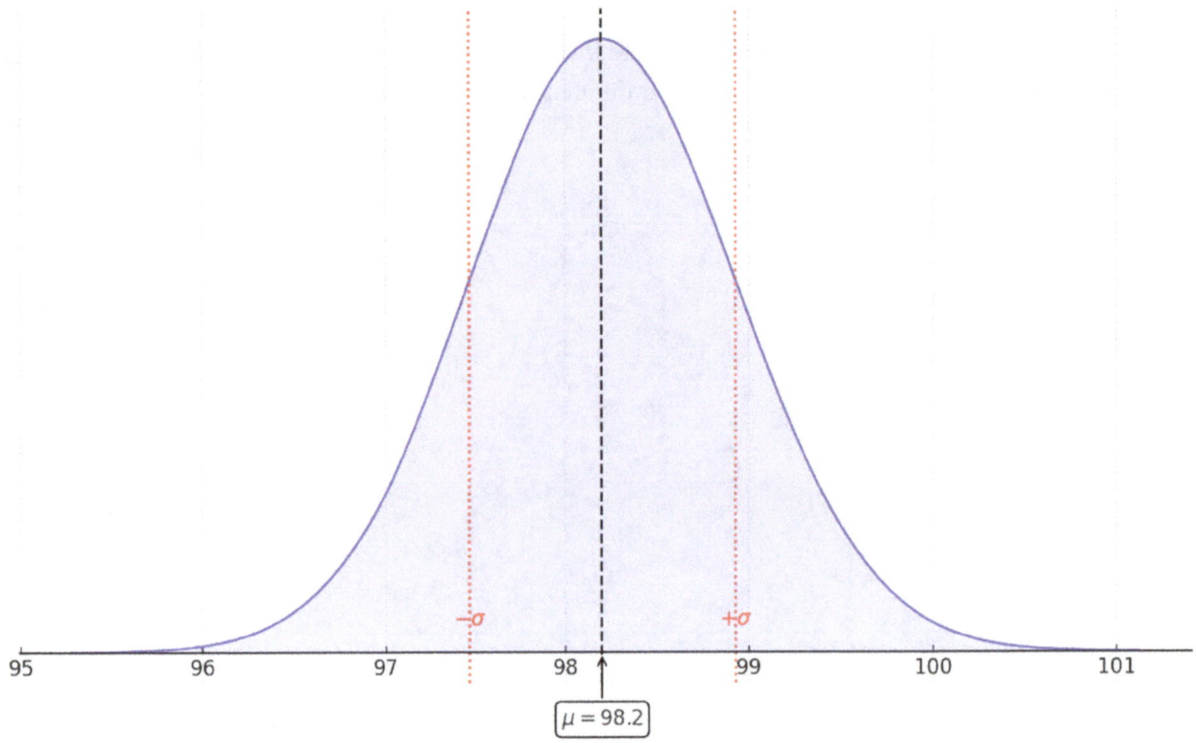

FIGURE 5.7 Normal Distribution with μ = 98.2 and σ = 0.73

If we let X denote body temperature, we are basically asking what *the probability is that X is greater than or equal* to 98.6 degrees. We can equivalently express this as follows: $P(X \geq 98.6) = ?$ In shorthand, P represents probability.

To answer the question, we first need to turn our random variable X into a z score. We then need to use this z score to find the area under the normal curve.

We convert X to a z score using the *z transformation formula*. The formula differs for a population of scores versus a sample of scores.

$$\text{Population of scores: } z = \frac{x - \mu}{\sigma}$$

$$\text{Sample of scores: } z = \frac{x - M}{SD}$$

Since we obtained information for the entire population of the United States, $z = \frac{98.6 - 98.2}{.73} = .55$, our z-score is 0.55. So, we can rewrite the above shorthand notation like this: $P(X \geq 98.6) = P(z \geq 0.55) = ?$.

Therefore, the probability that we would obtain a temperature that is 98.6 degrees or higher is the same as the probability of obtaining a z score of 0.55 or higher. The probability is the area under the curve, to the right of a z score of 0.55, or to the right of the body temperature of 98.6. Figure 5.8 shows you this area—it is the area to the right of the red line.

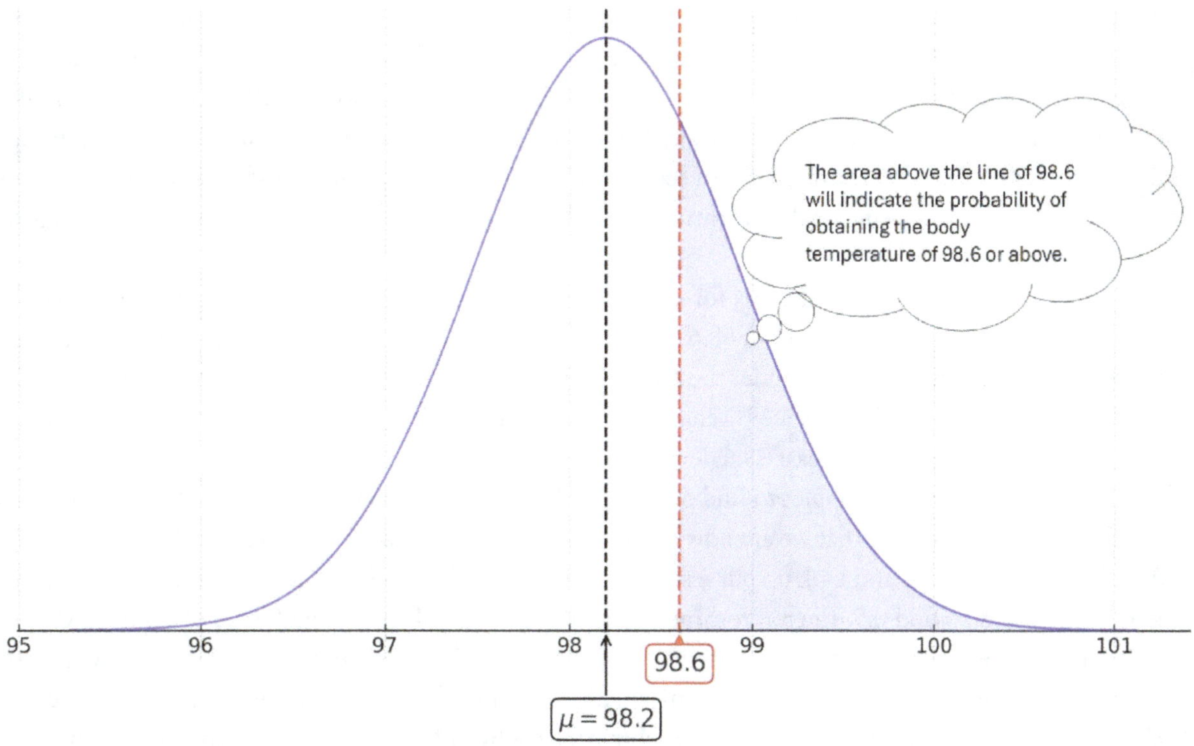

FIGURE 5.8 Shaded Area Showing $P(X \geq 98.6)$ in the Normal Distribution

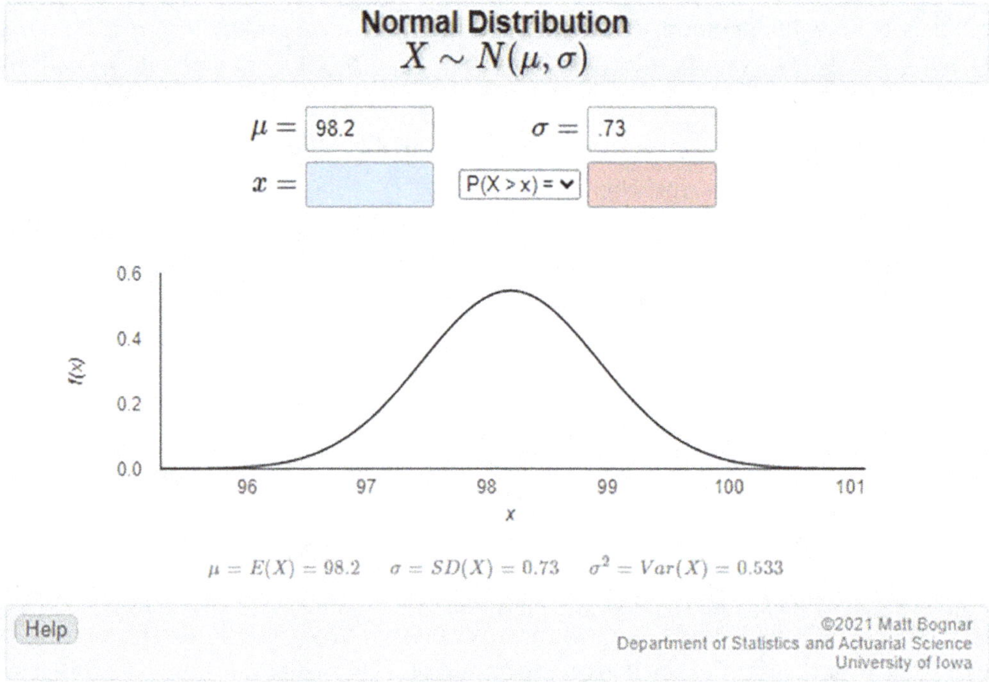

FIGURE 5.9 Probability of Observing a Value or Greater under the Normal Distribution: Applet screenshot with a Specific Mean and Standard Deviation Entered

Now, let's determine the probability of obtaining a body temperature of 98.6 or higher, which corresponds to a *z* score of 0.55 or above. Here, we want to introduce a free applet (https://homepage.divms.uiowa.edu/~mbognar/applets/normal.html) to help find the probability (Bognar, 2021). Enter the population mean, μ, of 98.2 and the standard deviation, σ, of .73 to define the normal distribution, as shown in Figure 5.9. You will see that the X-axis has been scaled using the specified mean and standard deviation. This is a good example of how the normal distribution can take on different values for μ and σ without changing shape.

Next, enter the value of 98.6 in the box for *x*. Figure 5.10 shows the output. According to the applet, the probability of obtaining a temperature of 98.6, or a *z* score of 0.55, is roughly 0.29. Recall that probabilities range from 0 to 1, with the total area under the normal curve equaling 1 (or 100%). Thus, a probability of 0.29 indicates that there is approximately a 29% chance of randomly selecting an individual from this population who has a body temperature of at least 98.6 degrees.

Although it is easy to use an applet to find our answer, let's learn how to use the Table A1 in the appendix to find the answer. Remember that we are now expressing our question as follows: P(z≥*0.55*) = ?.

We must go to Table A1 and look up our *z* score of 0.55. The column on the far left indicates the first two values of the *z* score, so find 0.5. The top row indicates the next two values, so find 0.05. Follow the 0.05 column down to where it intersects the 0.5 column to find a value of .7088. This indicates the cumulative probability up to 0.55 standard deviations above the mean. Since we are interested in the area on the opposite side of 0.55 (a body temperature of at least 98.6), we subtract .7088 from 1 to get .2912, roughly the same number given by the applet. This means that the probability of selecting an individual at random who has a body temperature of 98.6 degrees or greater is .2912, or 29.12%.

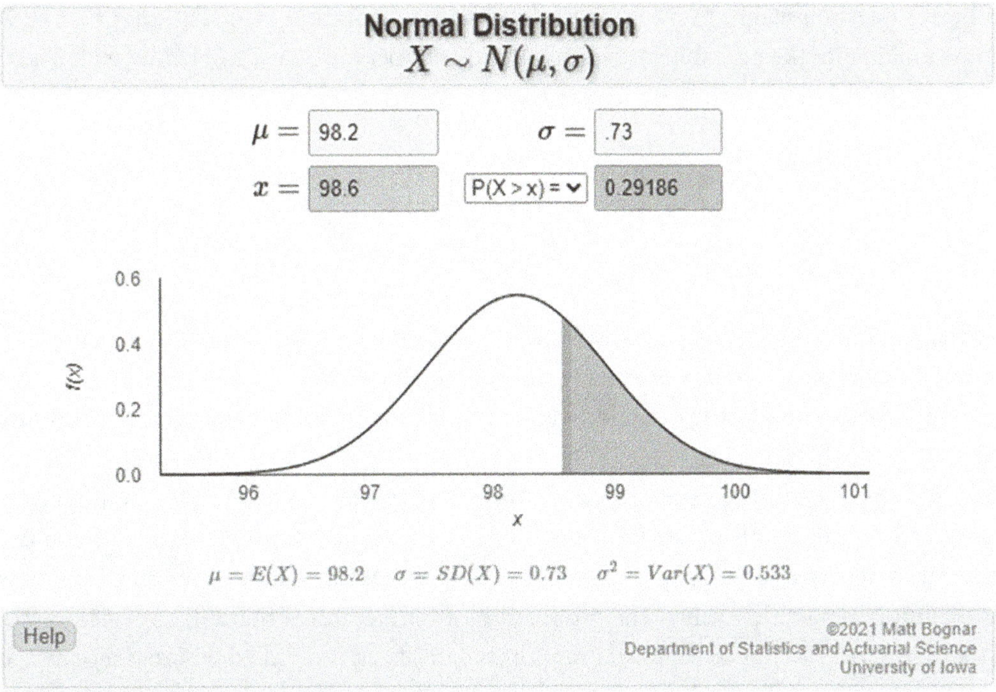

FIGURE 5.10 Probability of Observing a Value or Greater under the Normal Distribution: Applet screenshot showing the probability of $P(X \geq x)$

Question #2: What is the probability that a randomly sampled individual from this population has a body temperature of 97.5 degrees or lower?

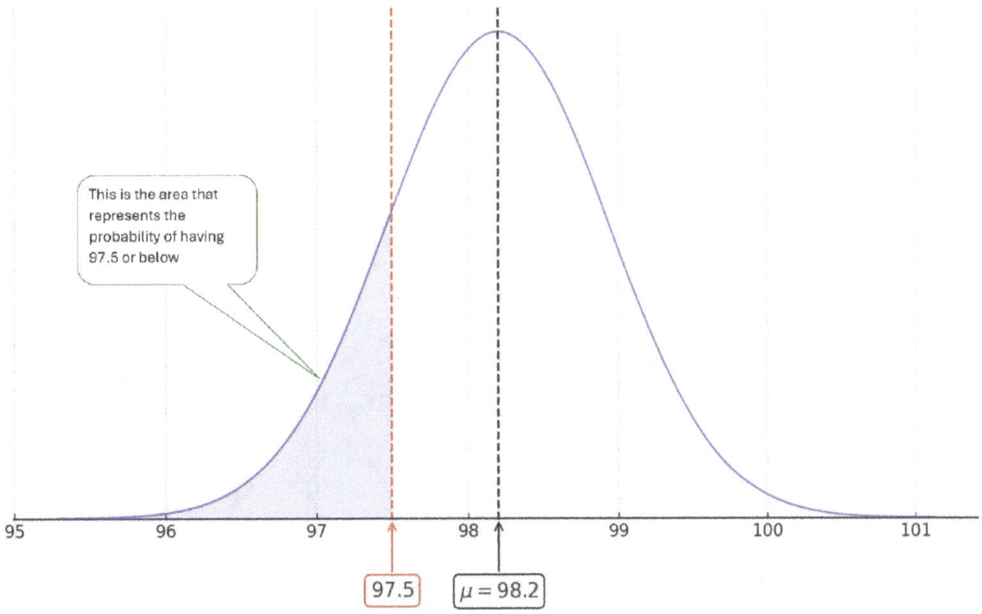

FIGURE 5.11 Shaded Area Showing $P(X \leq 97.5)$ in the Normal Distribution

Hopefully, this question will be easy to answer. In shorthand notation, we are asking: $P(X < 97.5) = ?$.

Now, we are looking for the probability of having a body temperature of 97.5 or below, which is represented by the left side of the red line. To answer this question, we first turn 97.5 into a *z* score:

$$z = \frac{97.5 - 98.2}{0.73} = -0.96$$

$$z = \frac{X - \mu}{\sigma}$$

Let's use the applet (https://homepage.divms.uiowa.edu/~mbognar/applets/normal.html) to find $P(x \leq 97.5) = P(z \leq -0.96)$, or the probability of *z* is less than or equal to -0.96.

The probability that a randomly selected individual from the population has a body temperature less than 97.5 is 0.1688, or 16.9% (Figure 5.12).

Now, let's find this probability using Table A1. Since our *z* score is –0.96, find the intersection of the –0.9 row and the 0.06 column. The stated value is .1685. Since we are interested in the probability of having a body temperature less than or equal to 97.5 and the table represents cumulative probability up until that point, we can simply report that value. The proportion of temperatures that fall at or below 97.5 is about 0.1685, or there is a 16.85% chance we would randomly sample an individual who has a body temperature of 97.5 degrees or less. Again, the value is slightly off from the one obtained from the applet due to rounding differences.

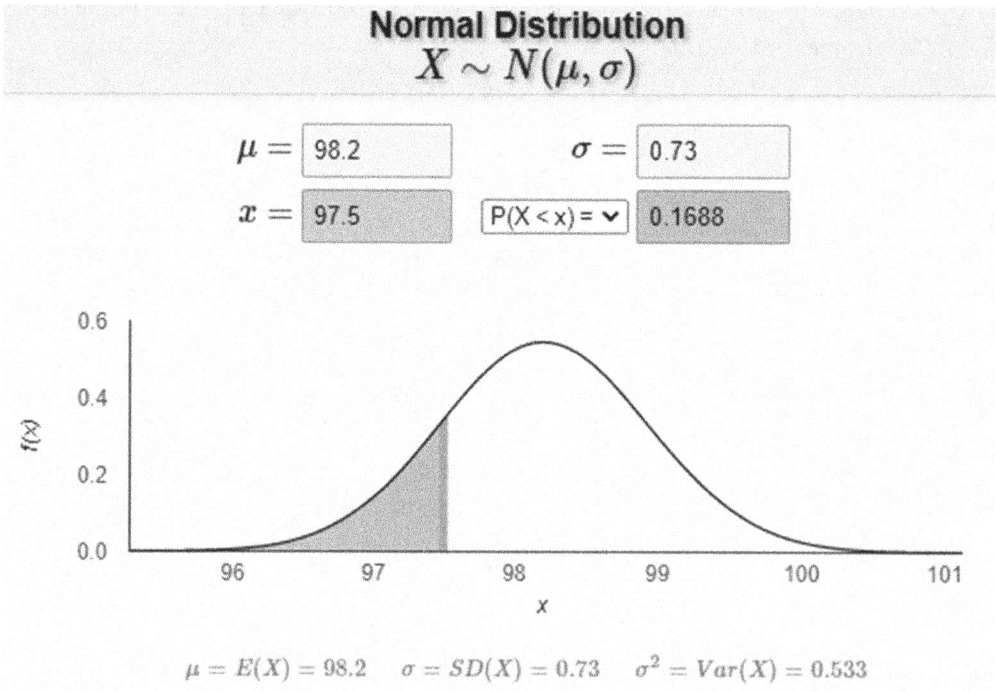

FIGURE 5.12 Probability of Observing a Value or Smaller under the Normal Distribution: Applet Screenshot Showing the Probability of $P(X \leq x)$

Question #3: What is the probability that a randomly sampled individual from this population has a body temperature between 97.5 and 98.6 degrees?

With this final question, we are asking $P(97.5 \leq X \leq 98.6) = ?$ Although this might seem like a more difficult question, it is actually the same basic concept. The first step is to convert both values to z scores, which we have already done. We know that 98.6 expressed as a z score is about 0.55 and that 97.5 expressed as a z score is about -0.96. We have also already figured out the probability of having a temperature above 98.6, $z = 0.55$, and the probability of having a temperature below 97.5, $z = -0.96$. We know that the entire probability under the normal distribution is 1.0, so we can get to the probability of scoring between these two z scores by subtracting the probabilities we already know.

Here's how the math works out:

$$P(97.5 \leq X \leq 98.6) = 1 - P(X \leq 97.5) + P(X \geq 98.6) = 1 - (0.1685 + 0.2912) = 1 - 0.4597 = 0.5403$$

This means that there is a probability of about 0.54, or 54%, that we would randomly sample an individual from this population who has a body temperature between 97.5 and 98.6 degrees (Figure 5.13).

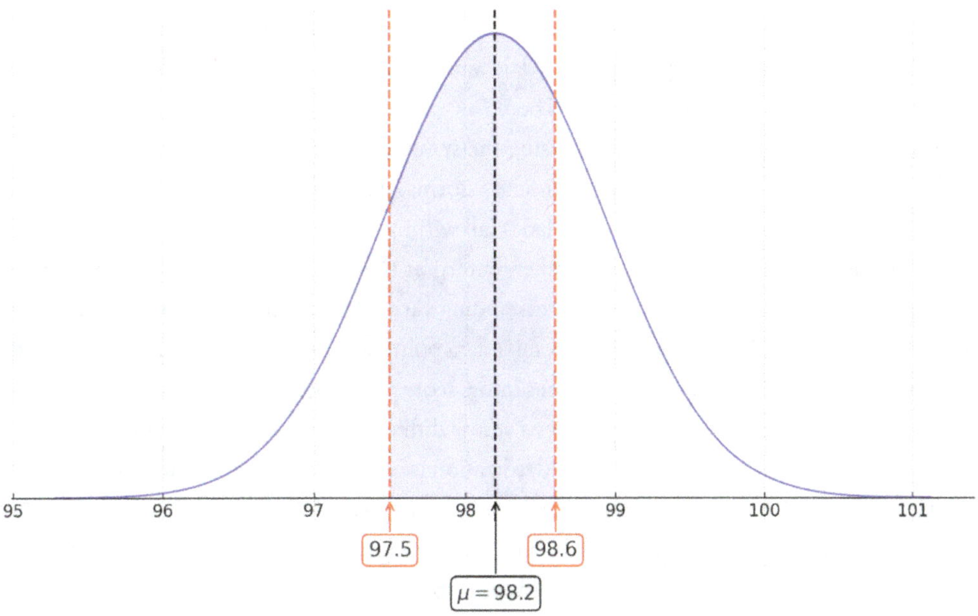

FIGURE 5.13 Shaded Area Showing $P(97.5 \leq X \leq 98.6)$ in the Normal Distribution

CONCLUSION

Chapter 5 has delved into the fundamental concepts of the normal distribution, a cornerstone of statistical analysis. The chapter outlined the key features of this distribution, including its symmetric, bell-shaped curve; its unimodal nature; and the alignment of its mean, median, and mode at the center. The chapter revisited the empirical rule (68-95-99.7%), explaining how it provides a framework for interpreting the proportion of data within one, two, or three standard deviations from the mean. Additionally, the chapter emphasized the importance of understanding normal distribution properties for making population inferences and highlighted how skewness and kurtosis helped assess a distribution's symmetry and the likelihood of outliers.

This chapter also introduced *z* scores as a way to standardize data, enabling meaningful comparisons across different datasets or scales. The chapter also explained how to calculate *z* scores, interpret their significance as standard deviations from the mean, and apply them in the context of the standard normal distribution, where the mean was 0 and the standard deviation was 1. Key applications, such as using *z* scores to compute probabilities and find areas under the normal curve, were explored through practical examples involving real-world scenarios such as body temperature analysis. Tools such as *z* tables and statistical software were presented as essential for calculating probabilities and interpreting results, building a strong foundation for further statistical methods such as sampling distributions. Together, the chapter equipped readers with essential skills for understanding and applying the principles of the normal distribution in various contexts.

KEY TAKEAWAYS FOR EDUCATIONAL RESEARCHERS FROM CHAPTER 5

- Understanding the characteristics of a normal distribution (symmetry, unimodal shape, and alignment of mean, median, and mode) is important for interpreting data and essential for assessing its appropriateness for applying statistical analyses you will learn later in this book.
- Skewness and kurtosis are key metrics for assessing the symmetry and *tailedness* of a distribution, helping researchers evaluate how closely their data approximate a normal distribution. A distribution is generally considered approximately normal when these two statistics fall within the range of -1 to 1.
- The empirical rule (68-95-99.7%) helps researchers interpret the portion of data within standard deviations from the means, making it a critical tool for understanding variability of data in education research contexts.
- *Z* scores provide a standardized measure of how far a data point deviates from the mean, expressed in terms of standard deviation units. By converting raw scores into *z* scores—a process known as standardization—we enable meaningful comparisons of relative positions across different datasets even when the original scores are measured on varying scales. Understanding standardization is essential for cross-study comparisons and evaluating individual performance within a population or reference group.
- Although probabilities are rarely calculated manually using *z* scores in practice, understanding the process provides researchers with a foundational knowledge of probability, enabling them to make population inferences and conduct hypothesis testing—essential components of quantitative analysis in educational research.

KEY DEFINITIONS FROM CHAPTER 5

Kurtosis measures the *tailedness* of a distribution, or how heavy or light the tails are compared to a normal distribution.

Normal distribution (bell curve) is a statistical distribution that is symmetric about its mean. All three measures of central tendency (mean, median, and mode) are equal and located at the center of the distribution, and the left and right halves of the curve are mirror images of each other.

Skewness means the extent to which a curve is pulled to one side or the other.

The **standard normal distribution** is a specific type of normal distribution that has been standardized so that the mean is 0, the standard deviation is 1, and the distribution is symmetric.

A **z score** (standardized score) is a statistical measure that describes how far a particular data point is from the mean of its distribution, measured in terms of standard deviations.

CHECK YOUR UNDERSTANDING

1. A distribution with high kurtosis has lighter tails than a normal distribution.
   a. True
   b. False

2. Which of the following values for skewness and kurtosis indicates that a distribution is approximately normal?
   a. Skewness = 0.5, kurtosis = 2.5
   b. Skewness = 0.25, kurtosis = 0.5
   c. Skewness = 1.0, kurtosis = 3.5
   d. Skewness = -2.0, kurtosis = 4.0

3. What is the probability of a data point falling within one standard deviation of the mean in a normal distribution?
   a. 68%
   b. 95%
   c. 99.7%
   d. 50%

4. A standardized test has a mean score of 500 and a standard deviation of 100. What is the probability that a randomly selected student scores higher than 650? Use the z score formula and the standard normal distribution table to find the answer.
   a. 0.1056 (10.56%)
   b. 0.1587 (15.87%)
   c. 0.0668 (6.68%)
   d. 0.0228 (2.28%)

5. What is the purpose of converting raw data to z scores?
   a. To reduce the range of the data.
   b. To standardize data for comparison across distributions.
   c. To eliminate outliers.
   d. To calculate the mean and median more accurately.

6

SAMPLING DISTRIBUTIONS

LEARNING OBJECTIVES

By the end of this chapter, you will be able to:

- Identify and describe various sampling methods (e.g., random sampling, convenience sampling, purposive sampling);
- Understand the concept of a sampling distribution, including its relationship to sample means, and describe how the central limit theorem ensures that sampling distributions approach normality as sample size increases; and
- Estimate and interpret the value of the standard error of the mean.

This chapter's main concept, sampling distributions, is challenging but fundamental to inferential statistics. Please stay engaged as we explore this complex topic.

WHAT IS A SAMPLE?

A **sample** is a small portion of the full population being studied. Since gathering information from everyone in the targeted population is often not realistic, collecting information from a smaller sample allows us to estimate population parameters. For example, if you wanted to know the average SAT math score of first-year students at a given institution, you could likely get that information from the institution's website. But what if you didn't have access to the website and decided to estimate the average by asking a handful of students what their SAT math score was? The students you talked to would represent your sample.

A few examples of sampling techniques include random sampling, convenience sampling, purposive sampling, and snowball sampling. Here we only cover methods fundamental to social science research.

Random sampling, also known as simple random sampling, is a sampling method where every individual in a population has an equal chance of being selected to participate in a study. This approach is widely used in research because it minimizes bias and helps ensure that the sample is representative of the larger population being studied. This can be achieved by using random number generators or drawing names from a hat. For example, all student identification numbers from an institution are entered in a random number generator, and 100 students are selected at random.

Some researchers utilize *systematic random sampling*. When utilizing this technique, a sample is obtained by selecting one element on a regular interval, such as every 10th name from a list, starting at a random point. For example, all student identification numbers from an institution are ordered on a spreadsheet, and every 50th student is selected for the study.

Convenience sampling is another common technique used in the social sciences. When using this technique, a sample is taken from a group of people who are easy to contact. This method is often used in pilot studies and during preliminary research efforts to get a gross estimate of the results without incurring the costs or time required to select a random sample. For example, a researcher has an office in a university building, so the researcher sits in the lobby and speaks with students who pass by to obtain a sample.

Participants are sometimes chosen based on the purpose of the study and the researcher's judgment. This type of sampling is called **purposive sampling,** and it is often used when a limited number of people have expertise in the area being studied. For example, a researcher wants to focus on SAT scores for engineering students, so the researcher surveys students who are studying engineering at their university. The researcher also decides to focus on juniors and seniors, since they have spent more time at the university, and selects students with different academic standings in an attempt to create a more representative sample.

Less frequently, existing study participants recruit future participants from their acquaintances. This is called **snowball sampling** because the sample group appears to grow like a rolling snowball. This method is often used in social science research when the focus is on specific characteristics that are hard to find. For example, a researcher is interested in studying the experiences of parents who homeschool children with special needs, focusing on the unique challenges and strategies they use. The researcher begins by reaching out to a few parents through online homeschooling communities or local support groups for families of children with special needs. These initial participants are asked to refer other homeschooling parents in their networks who might be willing to participate in the study.

Random sampling is arguably the most important sampling technique in educational research. You should not just ask the five students sitting near you when trying to answer a research question because there might be something about that group that makes them systematically different from the overall population. For example, maybe you want to investigate study habits among all high school students, but you teach Advanced Placement (AP) classes. This particular group of students might spend more time studying due to factors such as a more rigorous curriculum or higher academic motivation, meaning their study habits may not reflect those of the broader high school population.

The kind of sample you choose has implications for what you eventually say about your results. Your goal is to obtain a sample that is an accurate representation of the population. If your sample is not an accurate representation, your inferences will not be valid. In other words, your statistical calculations might be correct, but the inferences you make from the obtained statistics may not provide a valid description of population characteristics. For example, if your entire sample is drawn from urban private schools in the Chicago area, then you cannot claim that your results say much about the experiences of students in all schools across the United States.

Let's imagine that we draw a sample of students that is both random and representative of our population. We now have a distribution of scores for the students in our sample, and from that we can calculate the mean and standard deviation. In order to use that sample to draw inferences about our population, however, we must trust that the sample mean and standard deviation are accurate estimates of population parameters.

How do we know that they are? Technically we don't, but we can assume that they are close enough, and we can quantify how far off our estimate is likely to be.

THEORETICAL SAMPLING DISTRIBUTIONS

Everything we have talked about so far is something you will do in your own research. But to explain further, we need to step outside of real-world applications and into the theoretical statistics world.

Imagine that you had unlimited time and money. Instead of taking just one sample, imagine that you randomly selected 100 different samples from the same population to measure math SAT scores. Each time, you create a distribution of student SAT scores, and you calculate the mean of those scores. In every sample, the mean SAT score will be slightly different because of random chance. Some sample means might be unusually high or low just because you randomly selected individuals with unusually high or low SAT scores.

Now, take each of those sample means and list each of them out. Instead of a distribution of SAT scores, you have a new distribution of sample means. This is called a **sampling distribution**, which is the probability distribution of a statistic (such as the mean, proportion, or standard deviation) calculated from all possible samples of a given size taken from a population. Note that you can have a sampling distribution of sample variances or any other statistic, but for this purpose we are going to focus on the sampling distribution of sample means.

CENTRAL LIMIT THEOREM

Quick reminder that we are not in the real world right now, and you would never actually create a sampling distribution. We are still in theoretical statistics world trying to understand why we can trust that our sample mean is a good estimate of the population mean. The reason why is something called the **central limit theorem**, which states that when a sufficiently large random sample is drawn from any population, regardless of the population's distribution, the sampling distribution of the sample mean will approximate a normal distribution, provided that the sample size is large enough.

Think back to the real world where we had one sample of SAT math scores. It is possible that the distribution of scores was approximately normal, but maybe it was not. Imagine that you drew this sample from one of the top institutions in the United States. There will be plenty of students who earned near-perfect scores. The original distribution was probably negatively skewed, with a few lower scores pulling the distribution left.

The magic of the central limit theorem is that no matter what your original distribution of SAT scores looks like, the sampling distribution will be approximately normal, especially as the size of our sample exceeds 30. We will explore how that works out in a minute, but first, think about what it means. If our sampling distribution is normal, then we know from the 68-95-99.7 rule that 95% of our sample means will be within two standard deviations of the true population mean.

The prior statement is very important, so we will say it again in a different way. Remember, we are in theoretical statistics world, and we just took 100 individual samples and created a distribution of sample means. The center of this distribution is equal to the population mean. Due to the central limit theorem, most of the 100 sample means will be close to the true population mean—close enough to be considered a good estimate of the population mean.

Figure 6.1 shows that regardless of the shape of the original distribution (top row), the sampling distribution of sample means becomes approximately normal, especially as the sample size (n) increases.

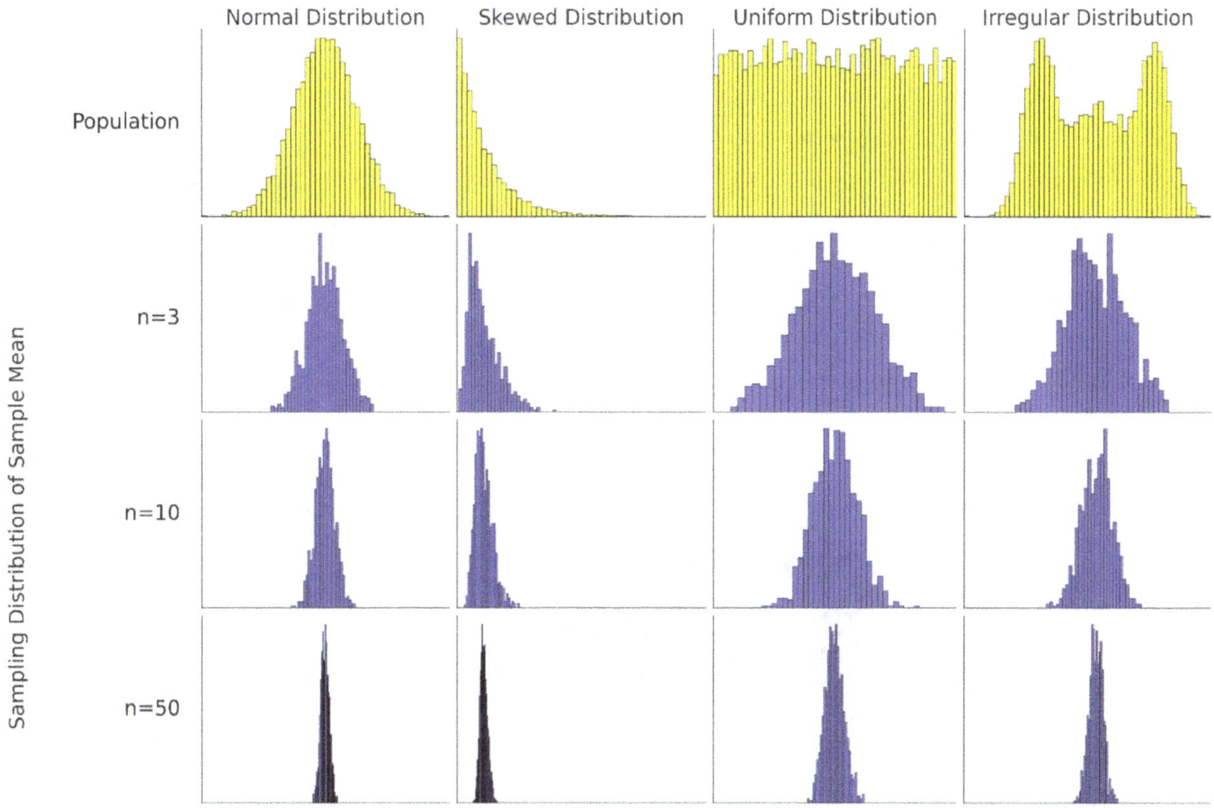

FIGURE 6.1 Example Sampling Distributions of Sample Means

CENTRAL LIMIT THEOREM IN ACTION

The concept can be hard to understand, so let's use a sampling distributions applet (https://onlinestatbook.com/stat_sim/sampling_dist/index.html) to walk through some scenarios. The starting point is a normal distribution with mean = 16 and standard deviation (*SD*) = 5. We are still in theoretical statistics world, where we magically know the population parameters—remember that in the real world we usually do not know those. To make it more interesting, change the starting distribution to a skewed distribution (Figure 6.2).

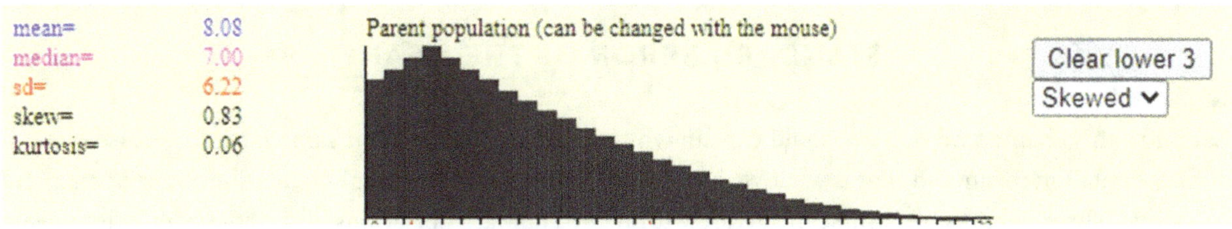

FIGURE 6.2 Sampling Distributions (onlinestatbook.com): Screenshot from Sampling Distribution Applet

On the left in Figure 6.2, we can now see a new population mean of 8.08 and standard deviation of 6.22. Our next step is to take a sample of this data. In the applet, click on the word "Animated" and you will see what happens when you take a random sample of 5 from the population. Click it four more times, and you will have three distributions: the population at the top, which in the real world we do not have; your most recent five-person sample; and then your sampling distribution. Do you notice how the sample mean is slightly different each time you take a sample? Then, each of those sample means are added to the third distribution.

Rather than clicking "Animated" 100 times, let's speed this up. Click on 10,000 to see what would happen if you pulled 10,000 samples from the original population. Do you notice how the mean of your sampling distribution is now very close to the true population mean? Your example will look a little different since each sample is random, but you should see something like Figure 6.3.

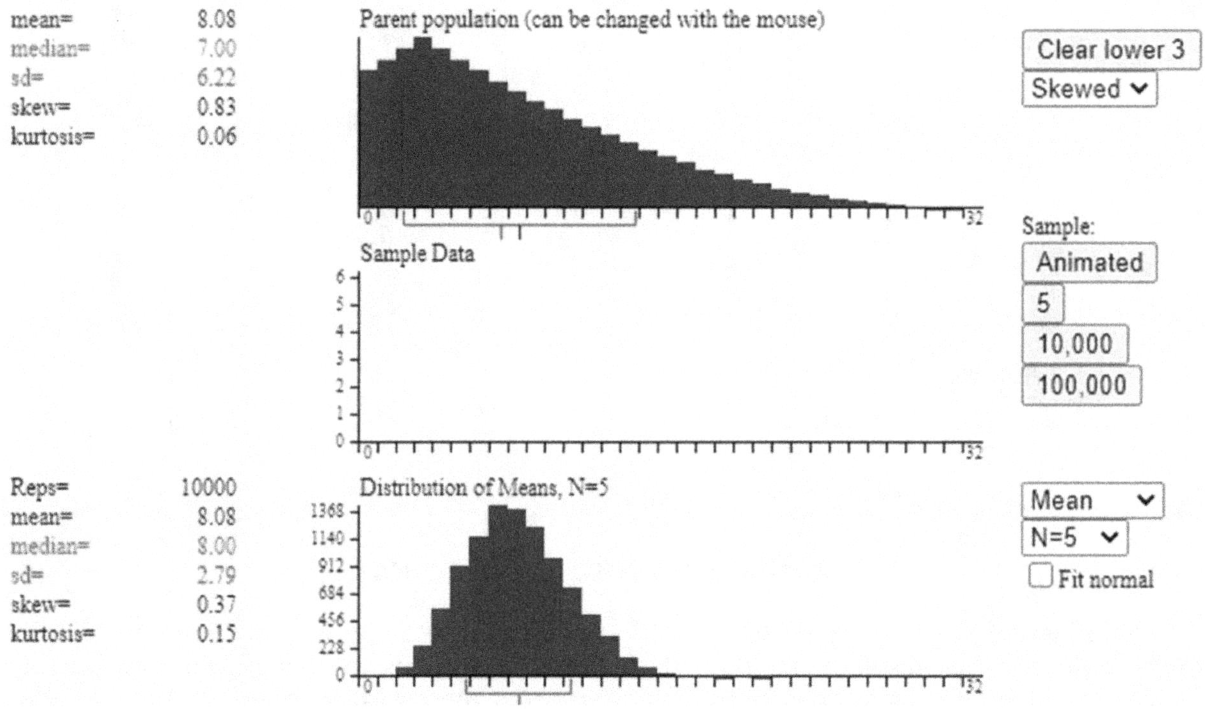

FIGURE 6.3 Sampling Distributions (onlinestatbook.com): Screenshot showing 10,000 sample simulation

Even though the distribution of scores in the original population was skewed, the distribution of sample means looks a lot more like a normal distribution.

STANDARD ERROR OF THE MEAN

Earlier in this chapter, we said we would explain why we can assume that our sample mean is close enough to the population mean, and that is what we just discussed. Because the sampling distribution is normal, we know that 95% of all possible sample means are within two standard deviations of the true population mean,

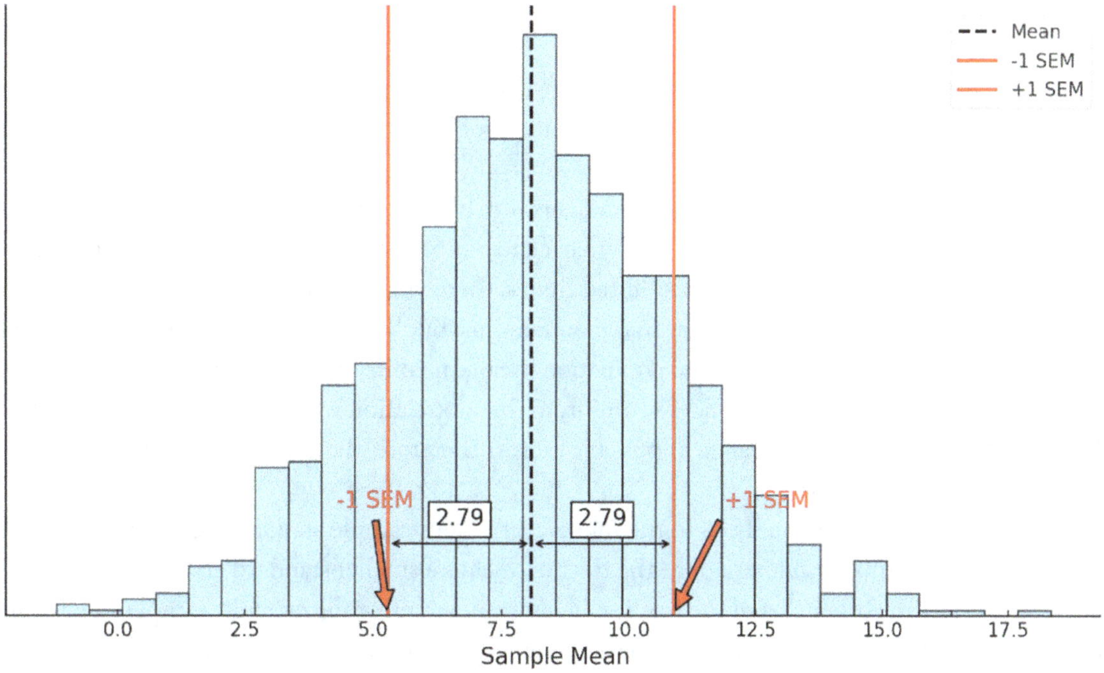

FIGURE 6.4 Sampling Distribution with Standard Error of The Mean (SEM) of 2.79 Points

and that is good enough. When we get into hypothesis testing, you will see how we manage the possibility that we randomly ended up with an oddball sample that is not a great estimate of the population, but for now, we are assuming that our sample mean is close.

How close is it, though? We can't say for sure, because remember, in the real world we only took one sample, and we don't actually know the population parameters. But now we get to the second promise made earlier in the chapter: we can estimate how close our sample mean is likely to be to the population mean. We do this by calculating the standard error of the mean, which you will often see referred to as simply standard error. The **standard error of the mean (**S_M**, or SEM)** quantifies how much a sample mean is likely to differ from the true population mean.

Before getting into the calculations, think about this conceptually. What exactly are we doing here? We have a distribution of sample means, and we want to know how far the average sample mean is likely to be from the true population mean. This should sound very similar to what we learned in Chapter 3: standard deviation. Just like we can find the standard deviation of our distribution of SAT scores, we can find the standard deviation of our sampling distribution of sample means. Look back to Figure 6.3, and you can see that the standard deviation of the sampling distribution is 2.79. This tells us that the average sample mean is 2.79 points away from the true population mean. Figure 6.4 illustrates this.

Do you see a slight problem? We can quantify how far our sample mean is likely to be from the true population mean, but to get the standard deviation of the sampling distribution, we first must create a sampling distribution, and that doesn't happen in the real world. So, what do we do? We can estimate it. Here is how you calculate an estimate of standard error:

$$S_M = \frac{s}{\sqrt{n}}$$

What that says is that the standard error of the mean (S_M) is equal to the sample standard deviation (s) divided by the square root of the sample size (n). You should probably memorize that, but of course you should also understand why it makes sense.

If you think about what would cause your standard error to be high or low, there are two main factors. First is the variability of the original distribution. Let's go back to our SAT math score example. If your population is all students who took the SAT across the United States, there will be quite a bit of variation in scores. A random sample could end up including some major outliers, just by chance, and when that happens your sample mean will be farther away from the true population mean. In contrast, say your population was students admitted to an elite honors college. The SAT scores from that population will likely be more tightly clustered together near the higher end of the range of possible scores. Therefore, there are fewer outliers that could be included in your random sample.

Look back at the formula above. The s in the numerator is your sample standard deviation. As variability decreases, s decreases; a smaller numerator in the fraction means a smaller standard error.

Next, let's consider the other part of the fraction in our standard error formula: the sample size. If you take a very small sample and you end up with one outlier, the outlier will have a greater influence on the sample mean than if it was just one of many scores. Let's go back to the sampling distribution applet (https://onlinestatbook.com/stat_sim/sampling_dist/index.html) so you can see how this works. Figure 6.3 provides a screenshot of the sampling distribution after taking 10,000 samples of size $n = 5$. Go ahead and change that to $n = 25$ and then click on the 10,000 button again. Figure 6.5 shows what this adjustment looked like in our example.

You should notice two things. First, the mean did not change very much. Even with a very small sample size, after we take 10,000 samples, we will have a very accurate estimate of the population mean. Changing the sample size to 25 did not get us much closer. But what changed? When our sample size was only 5, the sampling distribution was much wider. By random chance, there were some really oddball samples included that would have been pretty poor estimates of the population mean. The statistic showed that the standard deviation of the sampling distribution (i.e., standard error of the mean) was 2.79 when $n = 5$. But in this second version, when $n = 25$, the standard error is much lower, at 1.24.

Go back to the formula and you will see how this works out. The denominator of the fraction is the square root of the sample size. As sample size increases, the denominator gets bigger, and the standard error gets smaller.

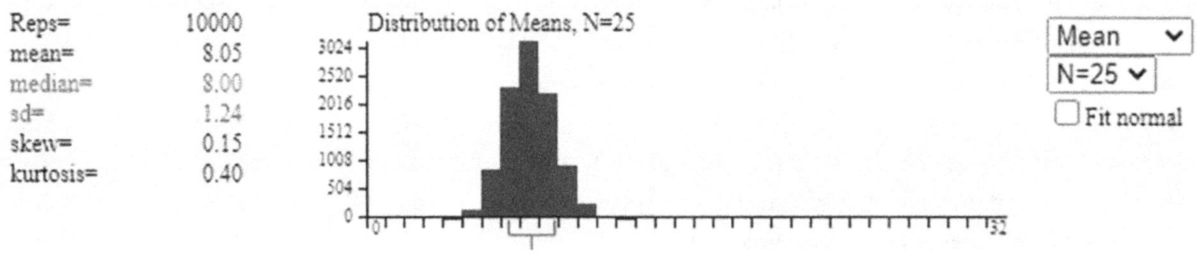

FIGURE 6.5 Sampling Distributions (onlinestatbook.com): Screenshot showing 10,000 Samples of Size $n = 25$

CONCLUSION

We spent a lot of time in the theoretical statistics world in this chapter. Let's go back to the real world for a quick summary before moving to hypothesis testing in Chapter 7.

- If your sample is both random and a good representation of the population, we can safely assume that the sample mean is a useful estimate of the population mean, especially when your sample size is at least 30.
- We know that your sample may be different from the population simply due to sampling error, but we can estimate the size of that error by calculating the standard error of the mean.
- Standard error of the mean (S_M, sometimes referred to SEM) is related to our sample standard deviation, but it is not the same thing. We divide our sample standard deviation (*SD*) by the square root of the sample size to calculate standard error. Thus, standard error, which is the standard deviation of the theoretical sampling distribution, will always be smaller than our actual sample standard deviation.
- The standard error is measured in the same units as your original dependent variable. For example, if you took a sample of 100 students and found that their average SAT math score was 600 with a standard deviation of 50, the standard error would be 5, calculated as 50 divided by the square root of 100.
  - Relating to standard deviation, this means that individual scores in the sample deviate from the sample mean by an average of 50 points. However, when considering the sample mean as an estimate of the population mean, the standard error tells us how much the sample mean is likely to differ from the true population mean.
  - According to the 68-95-99.7 rule, we can be 95% confident that the true population mean lies within two standard errors of our sample mean. In this example, that means we are 95% confident that the true population mean SAT math score is within 10 points (2×5) of 600 (590 to 610).

KEY TAKEAWAYS FOR EDUCATIONAL RESEARCHERS FROM CHAPTER 6

- Representativeness and random sampling matter. To ensure that your sample mean is a valid estimate of the population mean, your sample should be both random and representative of the population. This is particularly important in educational research, where studying diverse groups (e.g., students from different schools, districts, or demographics) helps ensure meaningful and generalizable results.
- The standard error of mean (S_M, or SEM) quantifies how much a sample mean is likely to differ from the true population mean. For instance, when comparing average test scores or teacher evaluations across schools, SEM helps determine the reliability of your estimates. Calculating SEM allows researchers to account for variability and acknowledge potential sampling error in their conclusions.
- Larger sample sizes reduce standard error, leading to more precise estimates of population parameters. Education researchers should prioritize collecting sufficient and representative participants for research, especially when dealing with diverse populations, to minimize the influence of outliers and improve the reliability of their findings (e.g., assessing standardized test performance across districts).
- When the central limit theorem is applicable with a sufficiently large sample size (i.e., $n > 30$), the 68-95-99.7 rule can be applied to the sampling distribution of means. This rule states that approximately 95% of sample means will fall within two standard errors of the true population mean. Education researchers can use this principle to

evaluate how closely their sample mean approximates the population mean, gaining insights into the reliability of their sample statistic for making inferences about the population mean.

KEY DEFINITIONS FROM CHAPTER 6

The **central limit theorem** states that when a sufficiently large random sample is drawn from any population (regardless of the population's distribution), the sampling distribution of the sample mean will approximate a normal distribution, provided the sample size is large enough.

Convenience sampling utilizes a sample taken from a group of people who are easy to contact.

Purposive sampling utilizes a sample chosen based on the purpose of the study and the researcher's judgment.

Random sampling is a sampling technique whereby every member of the population has an equal chance of being selected.

Sample refers to a small portion of the full population being studied.

A **sampling distribution** is the probability distribution of a statistic (such as the mean, proportion, or standard deviation) calculated from all possible samples of a given size taken from a population.

Snowball sampling utilizes a technique whereby existing study subjects recruit future subjects from among their acquaintances.

The **standard error of the mean (**S_M, or *SEM***)** quantifies how much a sample mean is likely to differ from the true population mean.

CHECK YOUR UNDERSTANDING

1. What is a sampling distribution?
   a. A distribution of a population's characteristics
   b. A distribution of a statistic (like a mean or proportion) calculated from all possible samples of a specific size
   c. A list of all individual data points in a population
   d. A random collection of data from a single sample

2. According to the central limit theorem, what happens to the sampling distribution of the sample mean as the sample size increases?
   a. It becomes more skewed.
   b. It approximates a normal distribution, regardless of the population distribution.
   c. It moves closer to the smallest value in the sample.
   d. It diverges from the population mean.

3. Which sampling technique ensures that every member of the population has an equal chance of being selected?
   a. Convenience sampling
   b. Purposive sampling
   c. Random sampling
   d. Snowball sampling

4. What is the standard error of the mean (SEM), and how is it calculated?
   a. It is the variability of individual data points and is equal to the population standard deviation.
   b. It is the standard deviation of the sampling distribution of sample means, calculated by dividing the sample standard deviation by the square root of the sample size.
   c. It measures the deviation of a single data point from the sample mean, using the formula *SD/n*.
   d. It is a fixed value representing the population variance.

5. What happens to the sampling distribution of the sample mean if the population distribution is normal?
   a. The sampling distribution will always be skewed, regardless of sample size.
   b. The sampling distribution will only be normal if the sample size is large.
   c. The sampling distribution will match the population distribution for any sample size.
   d. The sampling distribution will be normal, regardless of sample size.

7

FOUNDATIONS OF STATISTICAL INFERENCE: UNDERSTANDING HYPOTHESIS TESTING

LEARNING OBJECTIVES

By the end of this chapter, you will be able to:

- Identify and formulate null and alternative hypotheses based on research questions and determine their relevance in statistical analysis;
- Understand when and how a one-sample *z* test is used to evaluate the difference between a sample mean and a known population mean;
- Accurately interpret a *p* value to draw meaningful conclusions from the result of a one-sample *z* test and explain the finding; and
- Demonstrate the ability to execute hypothesis testing, including setting decision criteria, computing test statistics, and making evidence-based conclusions.

In this chapter, we will begin learning methods for inferential statistical analysis of educational data to address specific research questions. Stating a research question identifies the specific reason for, or purpose of, your investigation. A research question can help educators differentiate between real patterns in the data and patterns that may happen simply by chance. In addition, focusing on your research question helps ensure that your study stays on track and avoids distractions from looking at unrelated variables or issues.

WHAT IS A HYPOTHESIS?

The American Psychological Association (2018) defines a **hypothesis** as "an empirically testable proposition about some fact, behavior, relationship, or the like, usually based on theory, that states an expected outcome resulting from specific conditions or assumptions." Research hypothesis is also defined as "a statement describing the investigator's expectation about the pattern of data that may result from a given study." By stating specific expectations before the data are collected, the investigator makes a commitment about the direction (e.g., Method A will yield higher final exam scores than Method B) and magnitude (e.g., participants' income will increase with more education) of potential relationships based on the study's theoretical framework and related prior studies.

Let's consider a specific scenario. A third grade teacher had reason to suspect (hypothesize) that the district's current math curriculum was not benefiting both high- and low-achieving students. The teacher was concerned that using this curriculum might result in low third grade math performance across the district on the state assessment compared to the overall state average.

You could imagine all kinds of interesting questions to ask about this situation, but only some could be turned into a quantitative research question with a testable hypothesis. For example:

- Is the average third grade math performance in the district significantly different from the overall state third grade average?
- Do students in the teacher's class who rank above average in math show the same level of growth over the course of the year as students who rank below average in math?

HYPOTHESIS TESTING

Hypothesis testing is a statistical procedure used to make decisions or draw conclusions about a population based on sample data. This type of testing provides a structured framework to determine whether there is enough evidence in the observed data to support a specific hypothesis about a population parameter, such as the mean.

There are four major steps in the process of hypothesis testing:

1. State the research question and determine the null and alternative hypotheses.
2. Set the criteria for a decision. How much of a difference is significant enough to draw a conclusion?
3. Verify that data meet necessary assumptions for the statistical method and calculate the test statistic.
4. Draw a conclusion for your research question.

We will use the teacher's first research question as an example to review the steps in hypothesis testing. We will assume that the teacher had access to the district-wide average math score computed based on a sample of 400 third graders ($\bar{X} = 95$) and the state population results ($\mu = 100, \sigma = 15$).

Before proceeding, we need to introduce the concept of a **statistical test**. A statistical test is a mathematical tool used to analyze data and evaluate a hypothesis. Specifically, a statistical test determines the likelihood that an observed pattern in the sample data (e.g., the sample mean) occurred due to random chance or represents a real characteristic of the population (e.g., the population mean). In essence, a statistical test operates within the process of hypothesis testing.

In this example, an appropriate statistical test is the **one-sample z test**, the first statistical test we will introduce. The one-sample z test is a method used to determine whether a specific mean ($\bar{X} = 95$) from a single sample (e.g., the sample of 400 third graders) is significantly different from a known or hypothesized population mean ($\mu = 100$), assuming that the population standard deviation is known ($\sigma = 15$). Now, let's review the steps of hypothesis testing to understand how to conduct a one-sample z test.

STEP 1: STATE THE RESEARCH QUESTION AND DETERMINE THE NULL AND ALTERNATIVE HYPOTHESES

As a reminder, we are using the following research question: Is the average third grade math performance in the teacher's district significantly different from the overall state third grade average? At first it may seem odd to ask this, since we already know they are different: $95 \neq 100$, but there are some uncertainties to contend with:

- The district average is based on a sample. We know from Chapter 6 that our sample mean could vary slightly or to a large extent, depending on which students were absent on the day of the test. Before we make a conclusion about the difference, we need to account for possible sampling error.
- We also need to account for how much natural variation there is in test scores among students. The standard deviation of the state population is 15 points, which means that some students score well above and below the mean. We use this information to judge whether the observed average difference of 5 points should be considered statistically significant or whether it could likely be only due to chance.

Whenever we begin a hypothesis test, we need to state two different hypotheses. The **null hypothesis** (H_0) is the statement that is being tested and most commonly means some version of *nothing to see here* (e.g., no difference, no effect, no relationship). We assume that the null is true unless we have evidence that it is not. We also state this hypothesis in terms of the population parameter (in this case the population mean, μ). In this example, the proper format for the null hypothesis is:

$$H_0: \mu = 100$$

We must also make a statement about what we suspect is true instead of the null hypothesis. In our example, the teacher believes that the district's students will have an average score that is different than the state average. This is known as the **alternative hypothesis,** and it is commonly noted as H_1 or H_a:

$$H_1: \mu \neq 100$$

Before we move to Step 2, think about what the sampling distribution of sample means would look like for this scenario. We know the population mean (100) and the standard deviation (15) in this case. Thus, we know that the sampling distribution would be centered at $\mu = 100$ and that the standard error would be:

$$\frac{\sigma}{\sqrt{n}} = \frac{15}{\sqrt{400}} = \frac{15}{20} = 0.75$$

The sampling distribution would look like Figure 7.1.

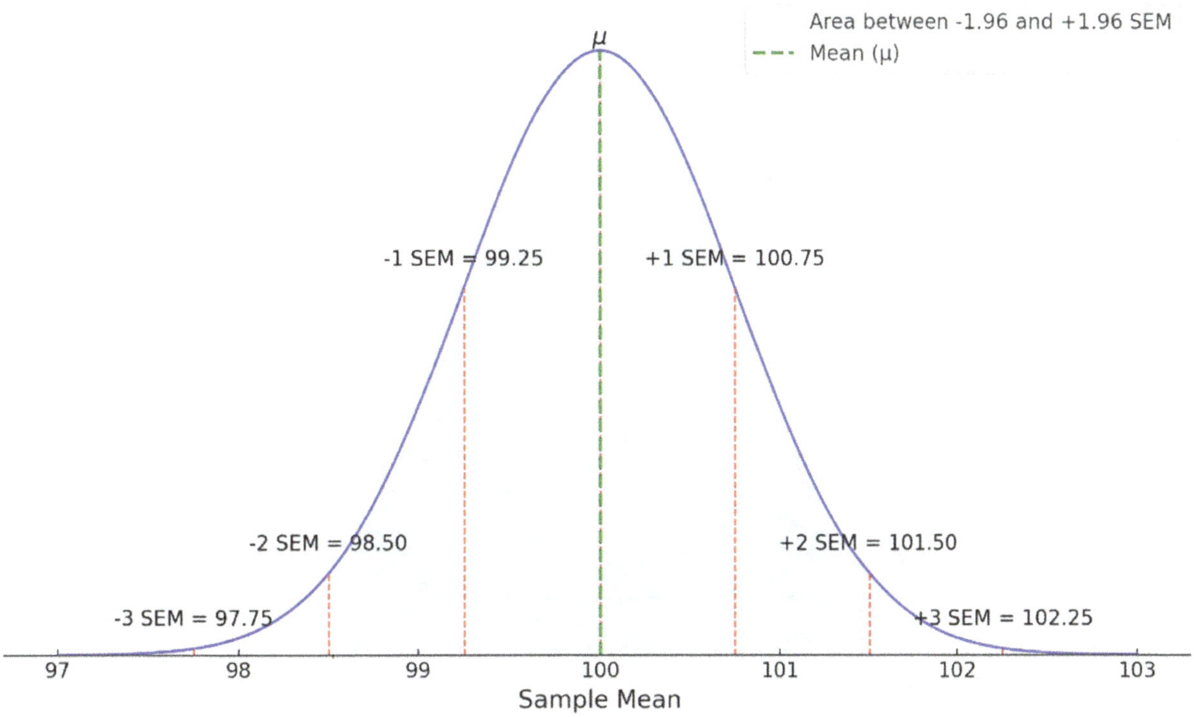

FIGURE 7.1 Sampling Distribution for $\mu = 100$ and SEM = .75

STEP 2: SET THE CRITERIA FOR A DECISION. HOW MUCH OF A DIFFERENCE IS SIGNIFICANT ENOUGH TO DRAW A CONCLUSION?

Earlier, we talked about whether there was a "significant" difference. We must decide what counts as significant to answer that question. In statistics, we set something called an **alpha level** (α) to allow us to interpret our results. In social science research, the alpha level is often set at $\alpha = 0.05$. The alpha level gives us the exact probability of committing a **Type I error**, rejecting the null hypothesis when the null hypothesis is really true. By setting $\alpha = 0.05$, we are basically saying that we want there to be no more than a 5% chance that we reject the null hypothesis in error. This threshold has become standard because it balances the risk of errors (i.e., Type I errors mentioned above and Type II errors that we will cover later this chapter) with practical considerations such as sample size and complexity while ensuring consistency across studies.

After establishing our decision criteria, we translate them into **critical values**, which serve as thresholds or cutoff points that define the rejection region in hypothesis testing. Critical values indicate how far a sample mean must fall from the center of a sampling distribution to confidently conclude that a result is statistically significant (i.e., unlikely to be due to random chance). As a starting place we will work with the familiar *z* distribution (standard normal distribution with a mean of 0 and a standard deviation of 1).

Before we know the critical values, we must decide what type of test we will be conducting. Most of the time when we pose a research question like the example above, we are just asking if our sample mean is different from the known population mean, hence the alternative hypothesis of $\mu \neq 100$. But if you recall, in our example the teacher's actual hypothesis was that the district's average would be lower than the state population

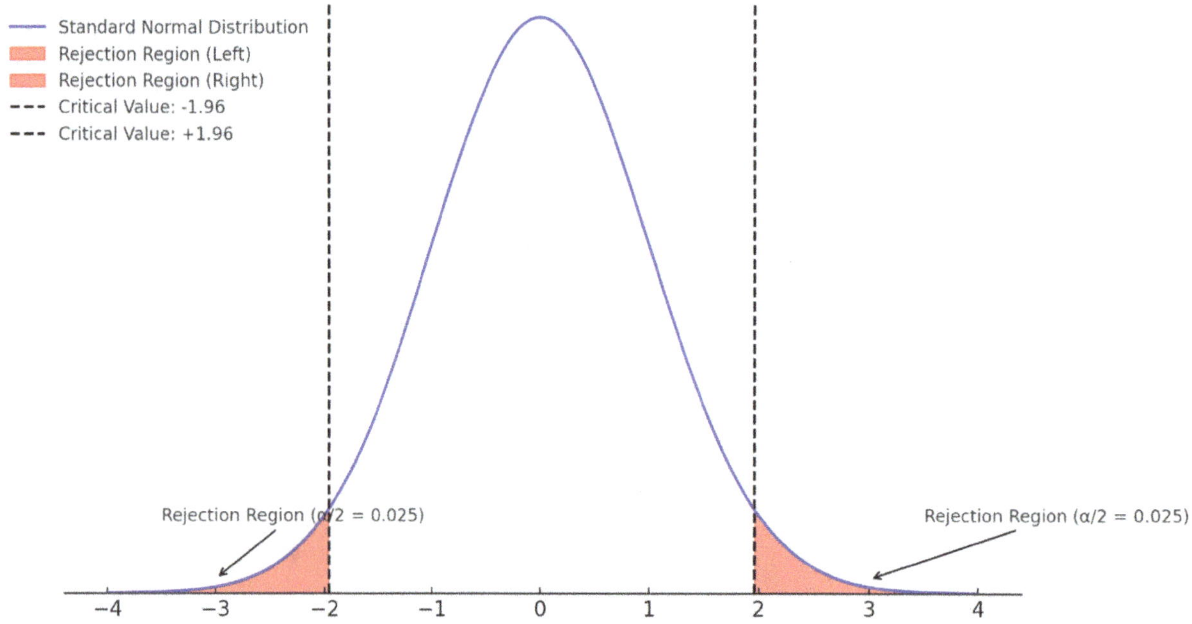

FIGURE 7.2 Critical Values and Rejection Regions for Two-Sided Test with $\alpha = 0.05$

mean. To test this kind of hypothesis, you need to use a **directional test**, also called a *one-sided test,* such as $\mu < 100$ (note that even with a directional test, the null hypothesis is still $\mu = 100$).

You will typically use a **nondirectional test**, also called a *two-sided test*, that accounts for both extremes of the z distribution. This approach sets critical values at the far ends, allowing for the possibility that the sample mean might significantly differ from the population mean—either much lower or much higher. When accepting a 5% error risk in incorrectly rejecting the null hypothesis, this probability is split evenly between the distribution's left and right tails (Figure 7.2). Additionally, consulting a z table shows that 2.5% of values are beyond this point toward the tail on the X-axis. The corresponding z score is approximately 1.96, which is nearly two standard deviations away from the mean.

If you had a strong theoretical justification to expect that your sample mean would be less than or greater than the population mean, you could conduct a one-sided test. In that case, the entire 5% chance of rejecting the null in error is located in one tail (Figures 7.3 and 7.4). If you use a z table to identify that z score, you will see that there is not an exact match to 0.05 for the area in the tail. We would use 1.645, since it is the z score in between 0.0505 and 0.0495.

Look back at Figure 7.1. The dashed lines represent one and two standard deviations above and below the mean (in this case SEMs, as the distribution is a sampling distribution of means), but the shaded region is just inside the lines for SD = ±2; it is meant to represent $z = \pm 1.96$. In this example, it is actually the unshaded part of the distribution that represents the rejection region for a two-sided test. Try to estimate where $z = 1.645$ and $z = -1.645$ along the X-axis. Can you see how a sample mean could be closer to the population mean but still fall in the rejection region if we used a one-directional test to define the rejection region? Smaller critical values (by which we mean closer to the center) make it easier to reject the null hypothesis. This will matter more when we get to the one-sample t test, which we will learn in Chapter 8.

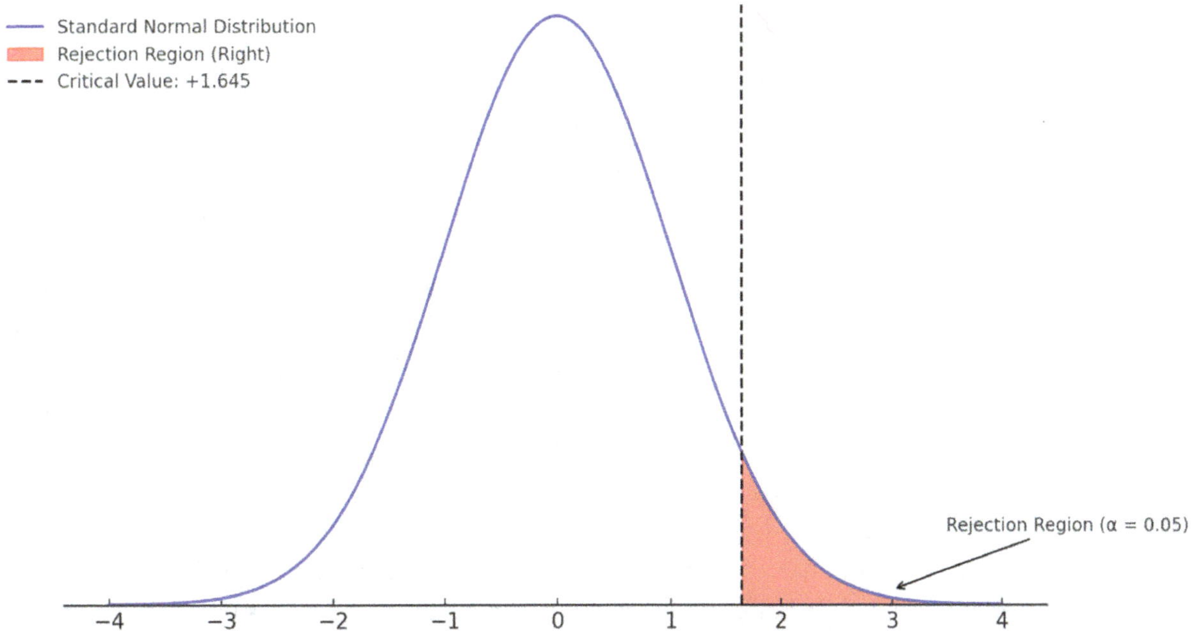

FIGURE 7.3 Critical Value and Rejection Region for One-Sided Test with $\alpha = 0.05$

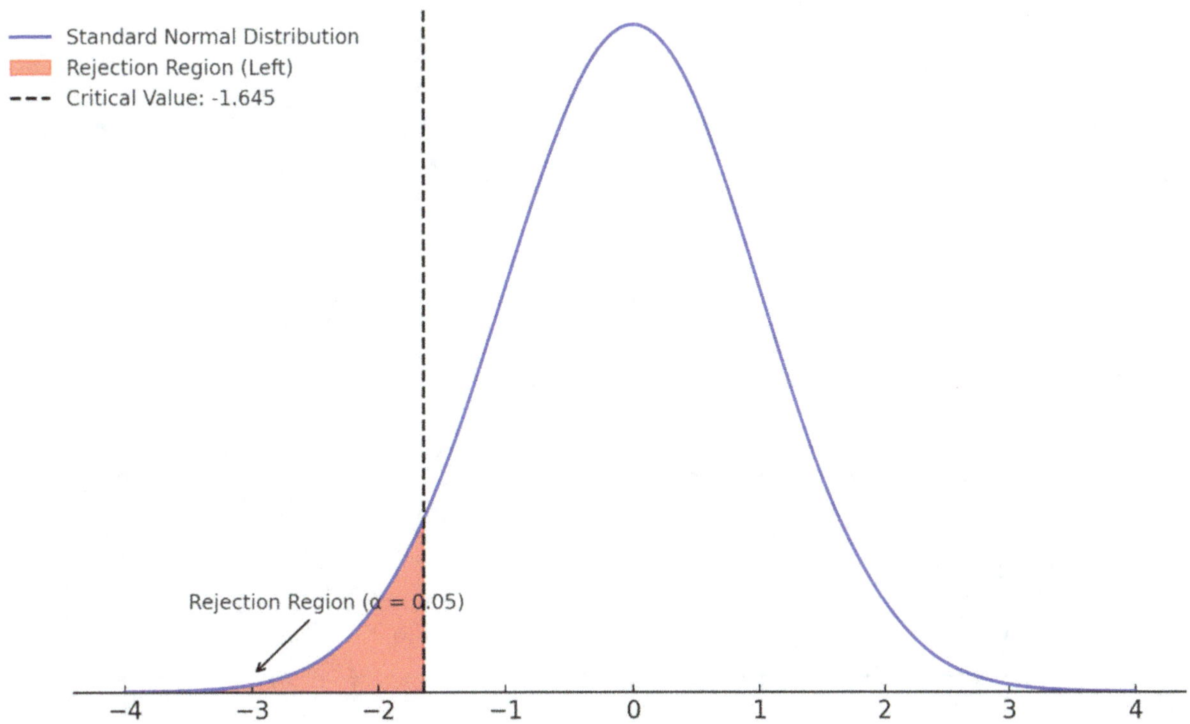

FIGURE 7.4 Critical Value and Rejection Region for One-Sided Test with $\alpha = 0.05$

STEP 3: VERIFY THAT DATA MEET NECESSARY ASSUMPTIONS FOR THE STATISTICAL METHOD AND CALCULATE THE TEST STATISTIC

Before you conduct your analysis by computing a test statistic or asking a software system to compute it for you, you should always check the assumptions required for whatever analysis method you are using.

With the one-sample z test, we assume that we collected a random, representative sample and that our variable is distributed normally in the population. We don't always know whether the population distribution is normal, but we can examine the distribution of our sample (create a histogram, check skewness and kurtosis statistics, etc.) because the representative sample should reflect the population characteristics in its distribution. We also know that even if the dependent variable is not normally distributed, our old friend the central limit theorem assures us that as long as our sample is larger than 30, the sampling distribution will be approximately normal. We can still proceed if our sample is smaller than 30, but we need to ensure that it does not contain extreme outliers and is not extremely skewed, since either of those issues would affect our sample mean.

For now, let's assume the teacher's data meet the conditions for the one-sample z test, so we move on to compute the test statistic. For this test, the test statistic is called the z statistic. The denominator accounts for the standard error (the standard deviation of the sampling distribution of the means under the assumption that the null hypothesis is true), and the numerator is the difference between the observed mean from empirical data and the hypothetical population value that you specify for the test. It is calculated using:

$$z = \frac{\bar{X} - \mu}{\frac{\sigma}{\sqrt{n}}}$$

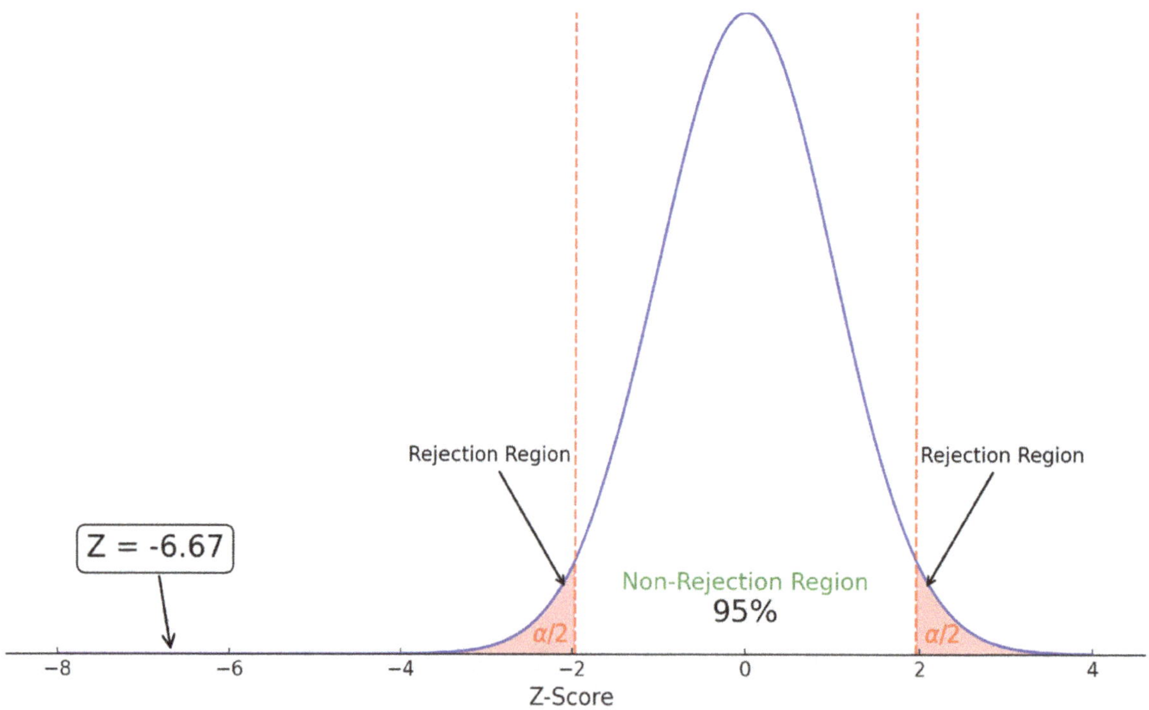

FIGURE 7.5 Location of Obtained z Score Relative to Rejection Regions

We will now plug in the numbers from the teacher's research question:

$$z = \frac{95 - 100}{\frac{15}{\sqrt{400}}} = \frac{-5}{0.75} = -6.667$$

This z statistic tells you that the sample mean of 95 is more than six standard deviations below the center of the sampling distribution of the mean for our data. Look back at Figure 7.2 and follow the X-axis as far left as you would need to go to reach -6.667. We think it is clear that this test statistic is in the rejection region (Figure 7.5).

Of course, it is not always this obvious, which brings us to the last few steps in the process of hypothesis testing.

STEP 4: DRAW A CONCLUSION FOR YOUR RESEARCH QUESTION

This step has two subparts, and the first is finding the p value. The p value tells us how probable it is to have obtained the sample mean we obtained, or something even more extreme (meaning, even farther away from μ), *if the null hypothesis is true*. We will share a few misconceptions about p values because this statement is more complicated than you might think, but for now we will move on to how you find them.

Normally, this is something that statistical software packages will calculate for you, but you can also use a z table, or standard normal cumulative probability table, to find p values for a one-sample z test. As it turns out, the lowest value in Table A1 in the appendix is –3.49. This means that we do not know the exact p value, but we do know it is even less than 0.0002, which is the p value associated with a z score of –3.49. Statistical software packages will often not report exact p values less than 0.001, and you will sometimes see p values equal to 0.000 as part of an output table. Please do not ever report that your p value is 0.000! Statisticians never say that there is a zero probability that our sample mean was different only due to chance even if the probability is very, very, small. If you see 0.000 when using statistical software, please report your p value as < 0.001.

The second subpart to this step is making and reporting your conclusion. Because the z statistic of -6.667 is much farther out into the tails of the distribution than -1.96, we reject the null hypothesis and conclude that the math performance of third graders in the teacher's district is significantly different from the state average. Note that if our results were different (say our z statistic was -1.5), our conclusion would be that we failed to reject the null hypothesis or sometimes that we retained the null hypothesis. We should not say that we "accept" the null or that the null is true. We just reject the null or fail to reject the null.

WHAT *P* VALUES TELL US

Remember that a p value indicates the likelihood (or probability) of obtaining your test statistic *if the null were true*. We are always starting from the hypothesis that the null is true and then testing that hypothesis. Some common misconceptions about p values are the following:

- A p value is the probability that the null hypothesis is true. Not quite. P values are about your test statistic and allow you to draw conclusions about the null hypothesis.

- A very small p value means the null hypothesis must be false. No. There's always a *chance* of committing a Type I error even if it is an extremely small chance.
- A p value indicates the size of an effect (i.e., a small p value indicates a large effect size). No. We will get to that in Chapter 8 when we review the concept of effect size, but you can have a highly significant result that is pretty much meaningless in terms of real-world application.

If your p value is very small, then either the null hypothesis is false or you somehow obtained a very unlikely sample from the population. On the other hand, your p value could be larger than 0.05 even if the null hypothesis is false. Sometimes your planned analysis method just does not have enough power to detect a significant difference, usually because the sample size is too small. **Statistical power** is the probability that a test will correctly reject a false null hypothesis, essentially measuring a test's ability to detect an effect when one truly exists. This depends on several factors including the sample size, the effect size, the significance level, and the variability within the data. High statistical power means there is a greater likelihood of identifying true effects, reducing the risk of **Type II errors** where significant effects are mistakenly deemed nonsignificant.

CONCLUSION

In this chapter, we introduced the foundational concepts of hypothesis testing, focusing on its importance in making data-driven decisions and drawing conclusions about a population based on sample data. We examined the process of hypothesis testing through the example of a one-sample z test, where we determined whether a sample mean significantly differs from a known population mean. The four key steps of hypothesis testing—stating hypotheses, setting decision criteria, verifying assumptions and calculating test statistics, and drawing conclusions—serve as a structured framework for analyzing data.

STEPS IN HYPOTHESIS TESTING

1. *State the research question and hypotheses.* Researchers begin by clearly defining their research question and establishing the null and alternative hypotheses. In one example, the district average math score was compared to the state average to determine if the difference was statistically significant.
2. *Set the decision criteria.* This involves selecting an alpha level (α), typically set at 0.05, which determines the threshold for rejecting the null hypothesis. Decision criteria are based on the type of test (e.g., one-tailed or two-tailed) and critical values from a distribution table (e.g., z distribution).
3. *Check assumptions and compute the test statistic.* Assumptions such as normality and random sampling must be validated before conducting the test. The test statistic (e.g., z) is calculated using sample data, allowing researchers to determine whether the observed sample mean is significantly different from the population mean.
4. *Draw a conclusion.* Researchers compute the p value to determine the likelihood of obtaining the observed result if the null hypothesis is true. If the p value is below α, the null hypothesis is rejected. This chapter emphasized reporting findings accurately, including the proper use of APA format.

We also discussed the advantages and risks of one-tailed tests, which focus on effects in a single direction, versus two-tailed tests, which consider both extremes. Researchers must decide on the type of test before data analysis to avoid bias and ensure the validity of results.

As we move forward, the steps for hypothesis testing introduced in this chapter will serve as the foundation for nearly all statistical methods we will learn. Whether *t* tests, analysis of variance (ANOVA), or more advanced techniques, the process will follow the same basic structure. This consistency will help make your analyses clear and organized, building on what you've already learned as we explore more tools for answering research questions in the upcoming chapters.

KEY TAKEAWAYS FOR EDUCATIONAL RESEARCHERS FROM CHAPTER 7

- Hypothesis testing helps distinguish between real patterns in data and those that occur due to random chance, ensuring that conclusions drawn are reliable and evidence-based. This is critical for assessing the effectiveness of programs, policies, and teaching methods in education practices. For example, a school administrator evaluating a new reading intervention program uses hypothesis testing to determine whether observed improvements in literacy scores are meaningful. This ensures that evidence-based decision-making and resources are allocated to strategies that genuinely impact student outcomes.
- The null hypothesis represents the default assumption (e.g., no difference or effect), while the alternative hypothesis reflects the researcher's prediction or question. Properly formulating these hypotheses is crucial for meaningful analysis. For example, an education researcher might test whether students taught using a hands-on STEM curriculum outperform students in a traditional lecture-based curriculum. The null hypothesis would state that there is no difference in performance between the two groups, while the alternative hypothesis would predict that the hands-on curriculum leads to higher scores.
- The sampling distribution of means, centered on the population mean assuming the null hypothesis is true, provides a reference for evaluating whether an observed sample statistic (i.e., sample mean) deviates significantly due to systematic effects or random chance. For example, as explored in this chapter, a teacher comparing third graders' math scores to the district-wide average uses the sampling distribution to calculate the standard error. By applying the one-sample z test, the teacher determines whether the observed difference between her district's average score and the state average is statistically significant or if it could likely occur due to random variation in sample selection. This process helps decide whether the district's scores reflect true performance gaps or are consistent with expected variation.
- Setting an alpha level (e.g., $\alpha = 0.05$) establishes a clear rule for deciding whether to reject the null hypothesis before analyzing the data. This threshold determines how much evidence is required to conclude that the observed results are unlikely to occur by chance. Using an alpha level ensures that decisions about the results are made consistently and with a predefined level of certainty. In education research, setting an alpha level helps researchers maintain fairness and reliability in their conclusions. For example, a researcher studying gender differences in science scores decides on an alpha level of $\alpha = 0.05$ *before running the test*. This means they will only conclude that there is a significant difference if the likelihood of the result occurring by random chance is less than 5%. By following this rule, they ensure that their findings are based on solid evidence, reducing the risk of incorrectly identifying a difference that doesn't truly exist.

- Adequate sample size and thoughtful research design are essential for achieving sufficient statistical power, which is the ability of a test to detect a true effect when one exists. A well-powered study reduces the likelihood of Type II errors (failing to reject a false null hypothesis) and ensures that meaningful effects in the data are identified. Without enough participants, even real differences or effects may go undetected, leading to incorrect conclusions. Sufficient sample size is critical for running hypothesis tests effectively. For example, in a study evaluating the impact of a new math curriculum on student test scores, a small sample might not reveal real improvements simply because there isn't enough data to detect the effect. Additionally, the required sample size often depends on the statistical test being used and the quality of the sample (e.g., randomness and representativeness). However, if the sample quality is poor (e.g., nonrandom or biased), even a large sample may not yield valid conclusions. For studies relying on the central limit theorem, a sample size of at least 30 is generally recommended, as it ensures that the sampling distribution of the mean is approximately normal even if the underlying population distribution is not. Ensuring a sufficient and representative sample size allows researchers to draw reliable, generalizable conclusions while minimizing the risk of both Type I and Type II errors. Careful planning is crucial for designing robust and impactful education research studies.

KEY DEFINITIONS FROM CHAPTER 7

The **alpha level (α)** in statistics is the threshold probability set by researchers to determine whether a result is statistically significant.

Alternative hypothesis (H_1 or H_a) is a statement about what we suspect is true instead of the null hypothesis.

Critical values serve as thresholds or cutoff points that define the rejection region in hypothesis testing.

A **directional test**, also known as a one-tailed test and a one-sided test, is a hypothesis test used when the research question or hypothesis specifies the direction of the effect or difference being tested. A directional test assesses whether the sample statistic is significantly greater than or less than the population parameter, but not both.

A **hypothesis** is an empirically testable proposition about some fact, behavior, relationship, or the like, usually based on theory, that states an expected outcome resulting from specific conditions or assumptions (American Psychological Association, 2018).

Hypothesis testing is a statistical procedure used to make decisions or draw conclusions about a population based on sample data.

A **nondirectional test**, also known as a two-tailed test and a two-sided test, is a hypothesis test used when the research question or hypothesis does not specify the direction of the effect or difference being tested. A nondirectional test assesses whether the sample statistic is significantly different from (greater than or less than) the population parameter.

Null hypothesis (H_0) is the statement that is being tested and most commonly means some version of "nothing to see here" (e.g., no difference, no effect, no relationship).

The ***p* value** tells us how probable it is to have obtained the sample mean we obtained or something even more extreme (meaning, even farther away from μ) if the null hypothesis is true.

The **one-sample *z* test** is a method used to determine whether a specific mean from a single sample is significantly different from a known or hypothesized population mean, assuming that the population standard deviation is known.

Statistical power is the probability that a test will correctly reject a false null hypothesis, essentially measuring a test's ability to detect an effect when one truly exists.

A **statistical test** is a mathematical tool used to analyze data and evaluate a hypothesis.

A **Type I error** (false positive) occurs when we reject the null hypothesis even though the null hypothesis is really true.

A **Type II error** (false negative) occurs when we fail to reject the null hypothesis even though the alternative hypothesis is actually true.

CHECK YOUR UNDERSTANDING

1. What is the purpose of setting an alpha level (α) in hypothesis testing?
   a. To determine the sample size required for the study
   b. To set the threshold probability for rejecting the null hypothesis
   c. To calculate the confidence interval for the population mean
   d. To ensure the sample data are normally distributed

2. A researcher hypothesizes that students from a particular school will outperform the state average on a given test. What type of test should be conducted?
   a. Nondirectional test (two-tailed/two-sided)
   b. Directional test (one-tailed/one-sided)

3. A critical value defines the cutoff point that separates the rejection region from the nonrejection region in a hypothesis test.
   a. True
   b. False

4. What does a *p* value indicate in hypothesis testing?
   a. The probability that the null hypothesis is true
   b. The likelihood of obtaining the observed result or something more extreme if the null hypothesis is true
   c. The size of the effect in the population
   d. The risk of committing a Type II error

5. What is the primary purpose of hypothesis testing in research?
   a. To organize data into meaningful groups
   b. To differentiate real patterns from random chance in data
   c. To calculate the average of a sample population
   d. To ensure data is normally distributed

8

EVALUATING GROUP AVERAGES AGAINST A BENCHMARK: ONE-SAMPLE *T* TEST AND GOODNESS-OF-FIT ANALYSIS

LEARNING OBJECTIVES

By the end of this chapter, you will be able to:

- Describe the differences between *z* tests and *t* tests, including when to use each based on knowledge of population parameters and sample characteristics;
- Calculate and interpret the test statistic, *p* value, and effect size for one-sample *t* tests and goodness-of-fit analyses;
- Clearly explain the distinction between statistical significance and practical significance and evaluate both to provide a comprehensive understanding of research findings; and
- Use confidence intervals to quantify uncertainty in sample mean estimates and evaluate the plausibility of a null hypothesis.

We just learned about *z* tests in Chapter 7, which you can conduct if you want to compare a sample mean to a population with a known mean and standard deviation. Often, however, we do not know the population standard deviation (σ), and we must estimate it with the sample statistic, in this case sample standard deviation (s). When this happens, we can no longer rely on the standard normal distribution, also called the *z* distribution, as it assumes precise knowledge of σ. Instead we use a *t* distribution, which accounts for the added uncertainty of estimating σ. In these cases we use a *t* distribution, which closely resembles its *z* distribution counterpart but with two key differences.

First, the tail ends of the *t* distribution take longer to approach the X-axis compared to the *z* distribution. This is because we are estimating the population standard deviation, and that means we have more uncertainty. If you look at Figure 8.1, imagine drawing a vertical line out toward one of the ends. Can you see how there would be

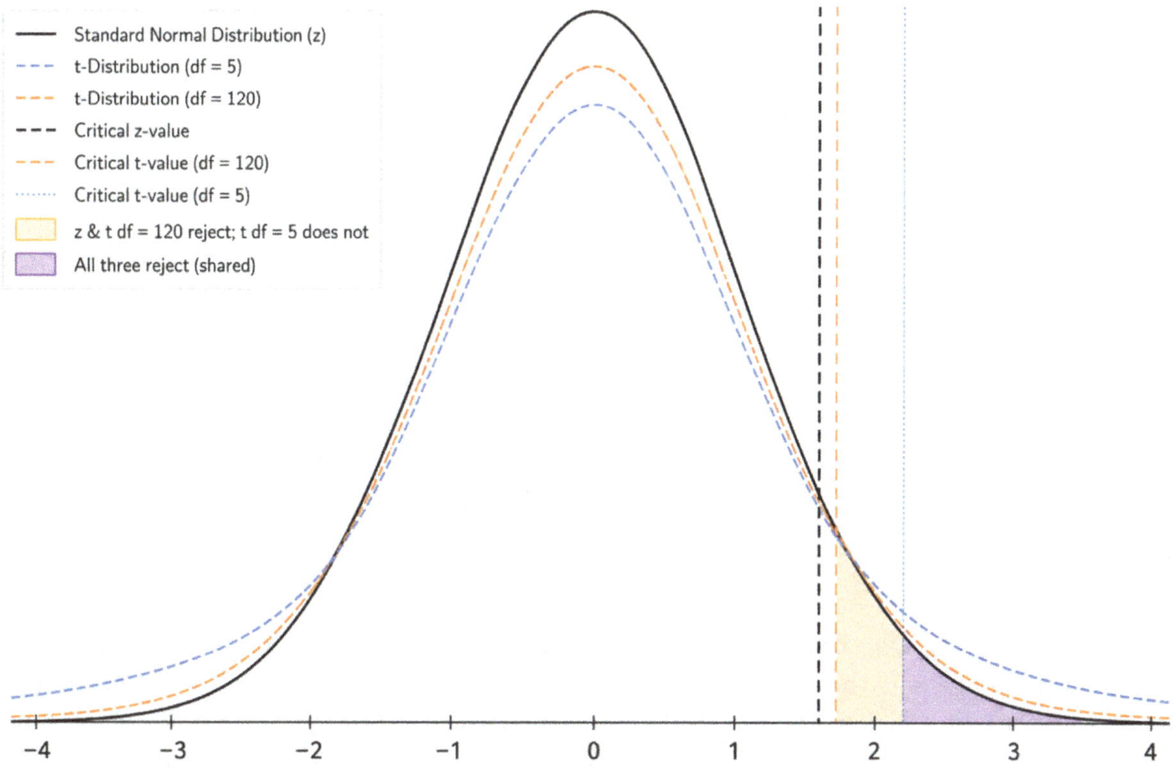

FIGURE 8.1 Comparison of Standard Normal and *t* Distributions

more area under the curve for the *t* distribution with just five degrees of freedom? Shaded areas under the curve represent 5% of the total area. This means that the sample mean must be farther away from the population mean before we can confidently conclude that the observed difference reflects something other than random chance.

Second, there is more than one *t* distribution, and the shape of the distribution depends on your sample size. Recall that a larger sample size leads to a smaller standard error, which in turn increases your statistical power to detect a significant effect. You can see how different *t* distributions look in Figure 8.1. The larger the sample size, the closer the *t* distribution becomes to a *z* distribution. Once $n > 120$, the two distributions are almost identical. Having such a large sample makes up for the fact that we are estimating the population standard deviation.

Another way to look at this is that the *t* distribution is related to the degrees of freedom. You must know the degrees of freedom to utilize a *t* table. Since sample standard deviation is used to compute the estimate of standard error, the degrees of freedom for a *t* test are calculated by using $n - 1$.

CONDUCTING A ONE-SAMPLE *t* TEST

Now, let's walk through an example of conducting a one-sample *t* test. Let's think back to the example from Chapter 7 where we compared a sample mean to a population mean with the *z* test. We have a similar but slightly different scenario this time. We have a sample of just 35 test scores from a single third grade class. The teacher is concerned that the students are underperforming in English Language Arts and hypothesizes that the class will score below the state average of 440. When the results were returned to the school, the class's average score was 425.26. Is this average significantly lower than the expected value of 440? Let's conduct a hypothesis test to find out.

STEP 1: DETERMINE THE NULL AND ALTERNATIVE HYPOTHESES.

Just like last time, the null hypothesis is that there is no difference between the teacher's class and the overall state population. So, the null hypothesis is:

$$H_0: \mu = 440$$

This time, however, we are going to use a *directional*, or *one-tailed, test* because the teacher is specifically concerned that the students are scoring lower than the general population. So, the alternative hypothesis is:

$$H_1: \mu < 440$$

There are 35 students in this class, which might be why they are not doing so well, so $n = 35$. The sampling distribution in this case would be composed of the means of all possible samples of size $n = 35$ taken from the state population of third graders (Figure 8.2). The distribution would be centered at $\mu = 440$ (the population mean, which we know), but we must calculate the standard error using the sample standard deviation. The sample standard deviation (s) happens to be 44.994.

$$\frac{s}{\sqrt{n}} = \frac{44.994}{\sqrt{35}} = 7.605$$

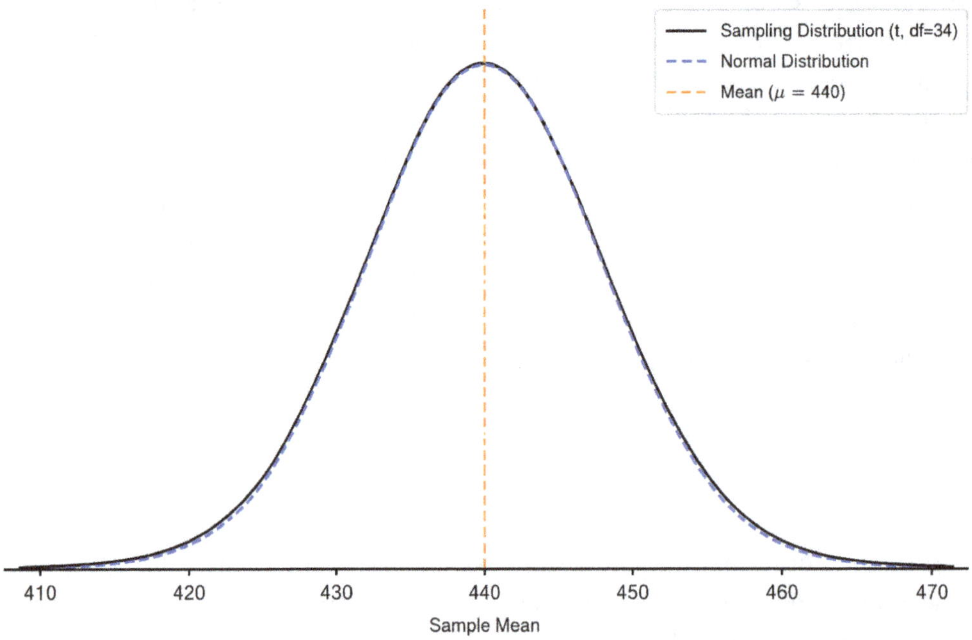

FIGURE 8.2 Comparison of Sampling Distribution (t, df=34) and Normal Distribution

STEP 2: SET THE CRITERIA FOR A DECISION

We will set the alpha level (α) to 0.05. So, we will reject the null hypothesis if the p value is less than 0.05. Since we are conducting a one-sided directional test, the entire rejection region is in the left tail of the distribution. To find the critical value, we first need to consider our degrees of freedom. For a t test, $df = n - 1$, so we have 34 degrees of freedom. In Table A2 in the appendix, there is no line for $df = 34$, so round down to the nearest degrees of freedom (30) as a conservative estimate. If you use an online calculator, which allows you to enter precise degrees of freedom, you will find that the critical value for 34 degrees of freedom at a .05 alpha level is -1.69. It is negative because we expect the sample mean to be lower than the population mean. Figure 8.3 provides a visual representation.

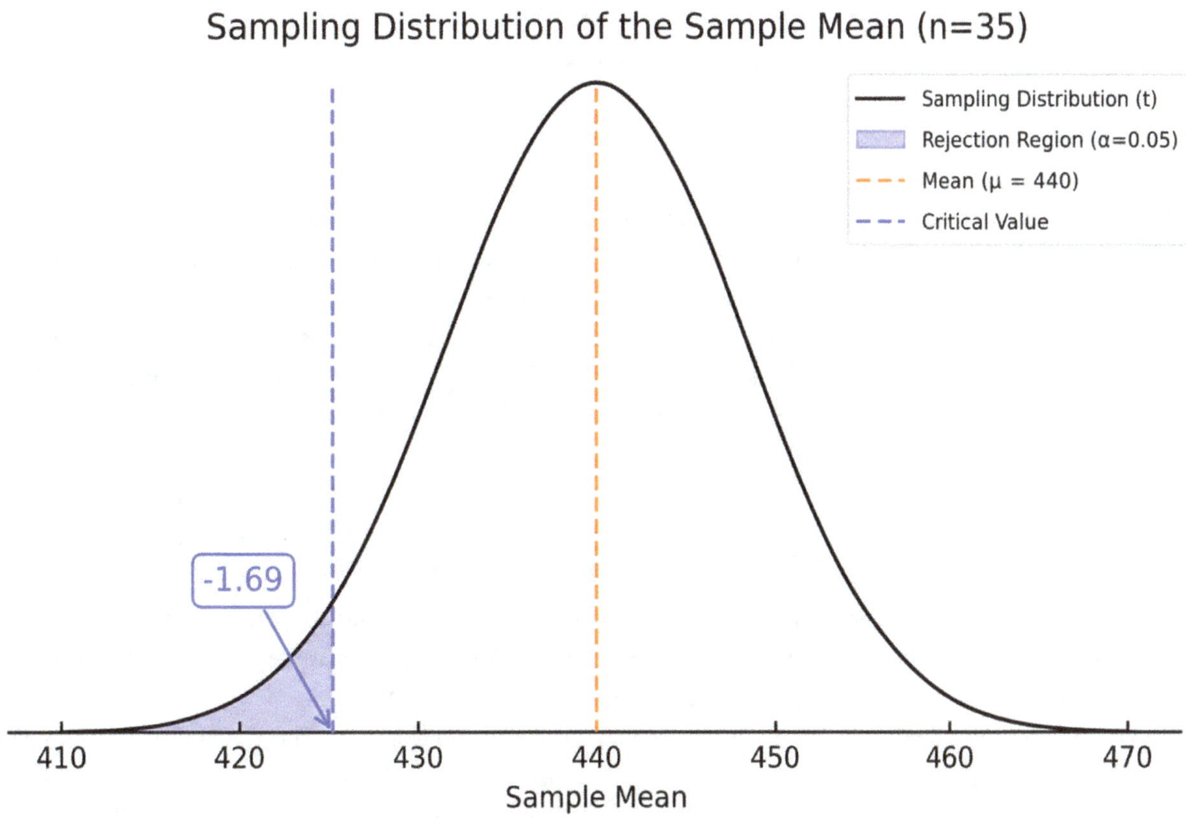

FIGURE 8.3 Critical t Value and Rejection Region for a One-Sided t Test ($\alpha = 0.05$, $df = 34$)

STEP 3: CHECK ASSUMPTIONS AND COMPUTE THE TEST STATISTIC

There are three assumptions underlying this test:

1. *Random sampling.* The scores come from a random, representative sample.
2. *Independence.* The scores are independent of one another.
3. *Normality.* The population score distribution is approximately normal.

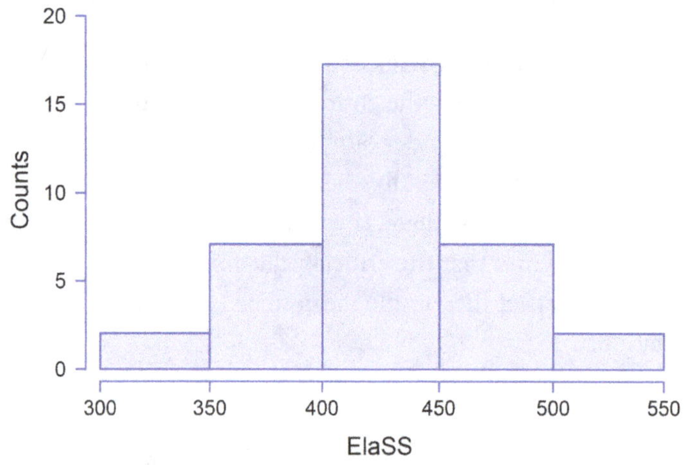

FIGURE 8.4 Histogram of Sampled Student ELA Scores (n=35)

It is okay that we didn't choose a random sample representing all third graders in the state, because the inferences we will draw are only about this one class. We can't see the entire population distribution, but the histogram for the scores looks almost perfectly normal (Figure 8.4).

If the distribution was way off, we could still proceed but must be very cautious about our interpretation of the results. Since our sample size is greater than 30, we would be okay even without such a normal distribution due to the central limit theorem. So, we are all set to calculate our test statistic.

For the one-sample t test, we use the formula:

$$t = \frac{\bar{X} - \mu}{\frac{s}{\sqrt{n}}}.$$

Notice that the formula is very similar to what we used to calculate the z statistic. The only difference is that we now estimate the standard error using the sample standard deviation and sample size in the denominator because we don't know the value of the population standard deviation (σ).

The numerator represents the difference between the observed and test values specified in the null hypothesis. The symbol μ represents the population parameter specified in the hypothesis, often referred to as the *null value*, as it reflects the assumed true population mean under the null hypothesis. The other values for the test statistic come from our sample data:

$$t = \frac{425.26 - 440}{\frac{44.994}{\sqrt{35}}} = \frac{-14.74}{7.605} = -1.938.$$

STEP 4A: FIND THE p VALUE

We must now determine whether our t statistic is significant, which would mean that it is very unlikely that we obtained this value due to chance alone. In Step 2, we said that the critical value for this test was -1.69. Picture

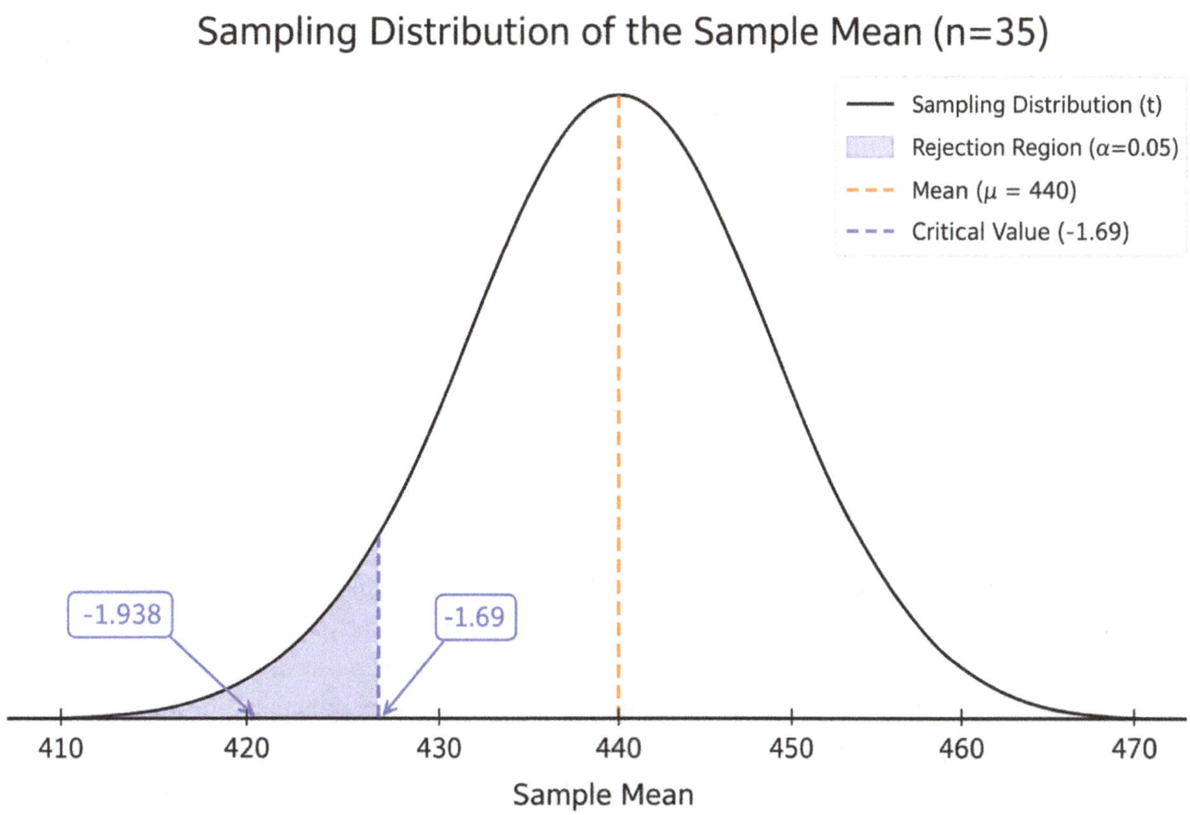

FIGURE 8.5 Obtained *t* Statistic Compared to Critical Value in a Left-Tailed Test

a *t* distribution and draw a vertical line about 1.7 standard deviations below the mean. Anything to the left of that is in the rejection region. Is our *t* statistic of -1.938 in the rejection region? Yes, Figure 8.5 shows that it is farther away from the center than our critical value. This means that our *p* value is less than 0.05.

STEP 4B: DRAW A CONCLUSION AND REPORT YOUR CONCLUSION

Given that our *p* value is less than 0.05, the appropriate decision is to *reject the null hypothesis*. In other words, we have found evidence in favor of the alternative hypothesis. Students in this class scored significantly lower than the overall state population on the test. Let's assume that we use statistical software to land on a *p* value of 0.03 for this analysis.

REPORTING RESULTS IN APA FORMAT

The American Psychological Association (APA), 7th edition (American Psychological Association, 2019), established standardized guidelines for reporting the results of a *t* test. In this case, you would report the results of our test like this: Scores for students in this third grade class (M = 425.26) were significantly lower than scores of third graders statewide, $t(34) = -1.938, p = 0.03$.

The most important part of the reporting is the section that starts with the observed t statistic. You put the degrees of freedom in parentheses immediately following the t; all statistics (e.g., M, t, p) are in italics, and you report the p value after the exact t statistic. Remember that even if the statistical software says $p = 0.000$, you should never report that. Instead, you should report $p < 0.001$. You can also just report that $p < 0.05$ (your alpha level) if you do not have access to statistical software.

CAUTIONS WITH ONE-SIDED TESTS

One caution with directional one-tailed tests is that you need to decide on one-tailed versus two-tailed tests up front. It is really not okay to decide after you see the results that you are going to conduct a one-tailed test. You should only be conducting a one-tailed test if you have good reason to expect the sample mean to be lower or higher than rather than just different from the population mean. And while a one-tailed test has greater power to detect a significant result, there is risk involved. Imagine that we expected the sample mean to be significantly lower, but it turned out to be significantly higher than the population mean. If we had been conducting a two-tailed test, that would be a significant difference from the population mean. But since our alternative hypothesis was that $\mu < 440$, we would fail to reject the null (nonsignificant result). Most of the time you will be using a two-tailed test for the possibility of a difference in either direction. Only use a one-tailed test if you have a very good reason and if you decide ahead of time to do so.

Another caution is that **Type III error** is introduced when using directional tests. **Type III error** occurs when a correct conclusion is drawn for the wrong reason or when the wrong question is asked, and a statistically valid result is obtained for that incorrect question. For example, a school district wants to determine whether lessons learned from a new professional development program will reduce the time teachers spend on grading. After collecting data and running a statistical test, the null hypothesis (that the program has no effect on grading time) is rejected, and a significant difference was found. District leadership interpreted this result as evidence that the program was reducing grading time, so they decided to expand the program. However, a few weeks later teachers were demanding that the district take a closer look at the data because they claimed to *be spending more time* grading. Sure enough, district leadership found that they made a mistake: while the statistical test correctly identified a significant difference in grading time (rejecting the null hypothesis), they *misinterpreted the direction* of the effect. The program *increased* grading time instead of reducing it. This misinterpretation of the direction of the significant effect is a Type III error: correctly rejecting the null hypothesis but misunderstanding what the result actually implies.

UNDERSTANDING EFFECT SIZE

Hypothesis testing is important for knowing if there is a significant difference between a sample and the population, but it will not tell you how big the difference is. A hypothesis test will almost always return a significant result when you have a large sample size because a test statistic is a function of standard error, which decreases as sample size increases.

Given that the mean difference and standard deviation remain the same as sample size increases, the test statistic always increases. For example, if you wanted to conduct a one-sample t test with a sample mean of

22 and an *SD* of 2 compared to a population mean of 23.5, you are sure to get a statistically significant result when $n = 100{,}000$ even if the p value is > 0.05 when $n = 10$. It is somewhat paradoxical that with a sufficiently large sample size, the statistical power becomes strong enough to flag even small differences as significant even if those differences are practically trivial or meaningless. Therefore, it is important to have another way to represent the meaningfulness, or practical significance, of a statistically significant result. That is why we calculate and report effect size.

Of course, one way of understanding the size of the difference is to simply compare the sample mean and population mean on the original measurement scale. Someone who understands a particular testing instrument could make their own judgment as to whether a difference of about 15 points was meaningful or cause for concern. But it helps to have a common language for reporting effect size, and the way we do that is to standardize the effect.

Effect size is a numeric measure that indicates the strength or meaningfulness of the relationship or difference between variables in a study. Unlike p values, which only tell you whether an effect exists, effect size tells you how large or meaningful that effect is. Effect size provides a standardized way to understand the practical significance of research findings, which is particularly useful when comparing results across studies.

Estimated Cohen's d is a common measure of effect size for t tests. We hope you remember back in Chapter 5 when we learned about z scores, which are also known as standardized scores because they are measured in standard deviation units. When we use estimated Cohen's d as a measure of effect size, we are doing the same thing. **Cohen's *d*** quantifies the magnitude of the difference between two means relative to the variability in the data and provides a standardized measure of practical significance, helping to interpret the importance of the result by allowing us to express the effect size in terms of standard deviation units. This approach provides a common scale, making it easier to compare results across different tests or measures rather than relying on raw score differences.

CALCULATING ESTIMATED COHEN'S *D*

One quick note to get us started: We are being careful to refer to the effect size we use for t tests as *estimated* Cohen's d to distinguish it from Cohen's d used in z tests. This distinction arises because effect sizes for t tests are calculated using sample data, whereas z tests rely on known population parameters. But out in the world of educational research, you will normally just hear it called Cohen's d regardless of the type of analysis used.

Here is the formula for estimated Cohen's d for a one-sample t test:

$$d = \frac{\bar{X} - \mu}{SD}$$

In our example above, the sample mean $(\bar{X}) = 425.26$ and $SD = 44.994$. The population mean $(\mu) = 440$. Using the formula:

$$d = \frac{425.26 - 440}{44.994} = -0.33$$

our effect size is -0.33, which according to Table 8.1 (from Privitera, 2018) is a medium effect size. In our sample research question, this means that the students in the third-grade class had scores that were about one-third of

TABLE 8.1 ***Cohen's Effect Size Conventions***

DESCRIPTION OF EFFECT	EFFECT SIZE (D)
Small	$d < 0.2$
Medium	$0.2 < d < 0.8$
Large	$d > 0.8$

Source: Adapted from Privitera (2018).

a standard deviation below the state average. A positive effect size indicates that the sample mean was larger than the population mean, and a negative effect size indicates that it was smaller. Since effect size can be negative, you need to consider the absolute value when making a judgement about the strength of the effect.

Before we move on, we want to remind you of a few other similar formulas because we had a tendency to get them confused in the past. First, here is a reminder of the formula for calculating z scores:

$$z = \frac{X - \mu}{\sigma}$$

Do you see the differences between that and the formula used to calculate Cohen's d? First, with a z score, we only have one X (i.e., a particular value) rather than a sample mean, or $\bar{X}$. But even more importantly, we use the population standard deviation (σ) rather than the sample SD or s to calculate a z score. But overall, the concept is exactly the same: both z scores and Cohen's d involve converting a raw difference from the mean into standard deviation units. The key difference lies in the specific inputs used for calculations.

Another place we have seen learners, including ourselves, get tripped up is confusing the formula for Cohen's d with the formula for the t statistic itself. Here they are, side by side:

$$d = \frac{\bar{X} - \mu}{SD} \qquad t = \frac{\bar{X} - \mu}{\frac{s}{\sqrt{n}}}$$

Even though the first formula, Cohen's d, uses SD to represent the sample standard deviation and the second formula, t statistic, uses s, they represent the same value. So, the only difference is that d has nothing to do with sample size, but sample size (n) is hugely important in calculating standard error, which is the denominator for the t statistic. For Cohen's d, the mean difference is divided by the standard deviation, which represents the variability in the actual scores within the sample. For the t statistic, the mean difference is divided by the standard error, which represents the standard deviation of the theoretical sampling distribution. By definition, it is going to be much smaller than the original sample standard deviation.

In this example, the same mean difference (-14.743) served as the numerator for both d and t. When we divided it by standard error for the t statistic, we got -1.938. This told us how far our sample mean was from the center of the sampling distribution. When we divided it by the sample standard deviation for Cohen's d, we got -0.33, which told us how far our sample mean was from the population mean, the center of the population distribution.

If nothing else, we hope this helps you see why we need both effect size and hypothesis tests. Effect size tells us how big a difference is, while the hypothesis test tells us whether we can be confident that the difference is a result of something other than chance. Effect size does not depend on sample size, but you are more likely to get a significant result from your hypothesis test as you increase the sample size.

It is common to report effect size after the results of your hypothesis test. For example, scores for students in this third grade class ($M = 425.26$) were significantly lower than scores of third graders statewide, $t(34) = -1.938$, $p = 0.03$, $d = -0.33$. You might also choose to report the significant result followed by a statement that Cohen's d (-0.33) indicated a medium effect size. Keep in mind that if you have a small sample size, you might be able to calculate a large effect size even without a significant result. This can be interesting to learn and might suggest that you should try to conduct your analysis again with a larger sample size. But *we don't report effect size after a nonsignificant result*, since it does not make sense to describe the size of an effect you just concluded does not exist.

CONFIDENCE INTERVALS

How do confidence intervals fit in? **Confidence intervals (CIs)** provide an additional layer of transparency by offering a range of values likely to contain the true population parameter.

The way we think about it, the reason we use **CIs** is to be up front about our level of uncertainty. We take a sample and report the mean, but how confident are we that the mean of our sample is equal to the true population mean? Or we have two samples and we find the difference between the two means (stay tuned for Chapter 9), but how confident are we that we know the true difference? If we report a point estimate (e.g., a mean) along with an interval estimate, we are being completely transparent about how good our estimate is likely to be.

Another high-level concept to be aware of before we dive into the details is the balance between precision and accuracy. Say we want to estimate the average height of all people in the entire world who are 18 and older. We could be highly confident in our accuracy if we estimated that the mean height of all adults was somewhere between 2 feet (60.96 cm) and 8 feet (243.84 cm). Don't you feel pretty comfortable accepting that the true population mean is somewhere within that interval? And yet, do you feel like you have learned any useful information? Probably not, because our estimate has almost zero precision. We could go in another direction and estimate that the average height was exactly 65.8 inches (167.13 cm). That estimate is much more precise, but who knows how accurate it is?

What we often do is report both point and interval estimates. We commonly see this with survey results. We start with a sample mean and then use the standard error to calculate a CI. The more confident we are (higher level of accuracy), the wider the interval (less precision). We can decide on any confidence level we want and calculate the exact interval, but it is most common to use 90%, 95%, or 99%.

INTERPRETING CONFIDENCE INTERVALS

Now that you have the general idea about the purpose of CIs, we are going to get a little pedantic and refer you to the words of Dr. Paul Savory (2008) in "How Do You Interpret a Confidence Interval?" Say you want to know the mean of some variable in the population. You take a sample, find the mean, and then create a 95% CI around it. It is incorrect to say that "there is a 95% probability that μ falls within this interval." This is incorrect because it implies that μ is random and could take on different values. In fact, μ is a constant—we just don't know what it is—and it either falls within our interval or doesn't. The correct interpretation is "we are confident that if μ were known, this interval would contain it." This may seem like a trivial difference, but it makes an important point that the probability or confidence level we report is referring to the interval, not the population parameter itself.

We highly recommend visiting a CI applet (https://digitalfirst.bfwpub.com/stats_applet/stats_applet_4_ci.html) for a helpful visualization of the concept. The way the applet works is that you select the confidence level and sample size, and then it will randomly pull 1 or 25 samples from the population distribution. If you request 25 samples, you will see them all lined up. Each sample will have a mean and a corresponding CI around it.

In the example in Figure 8.6, we used a 90% CI with a sample size of 10 and asked the applet to take 25 samples. Almost all resulting samples had a CI that contained the true population mean, but two did not. Note that 23/25 is 92% and not 90%, but when we try again, sometimes we get 24/25 or 22/25. Even our 90% CI is just an estimate. The applet provides a great visualization of the difference between the original population distribution and the sampling distribution. We recommend playing around in the applet to see how increasing the confidence level and/or the sample size will typically result in a greater percentage of possible sample intervals containing the true population mean.

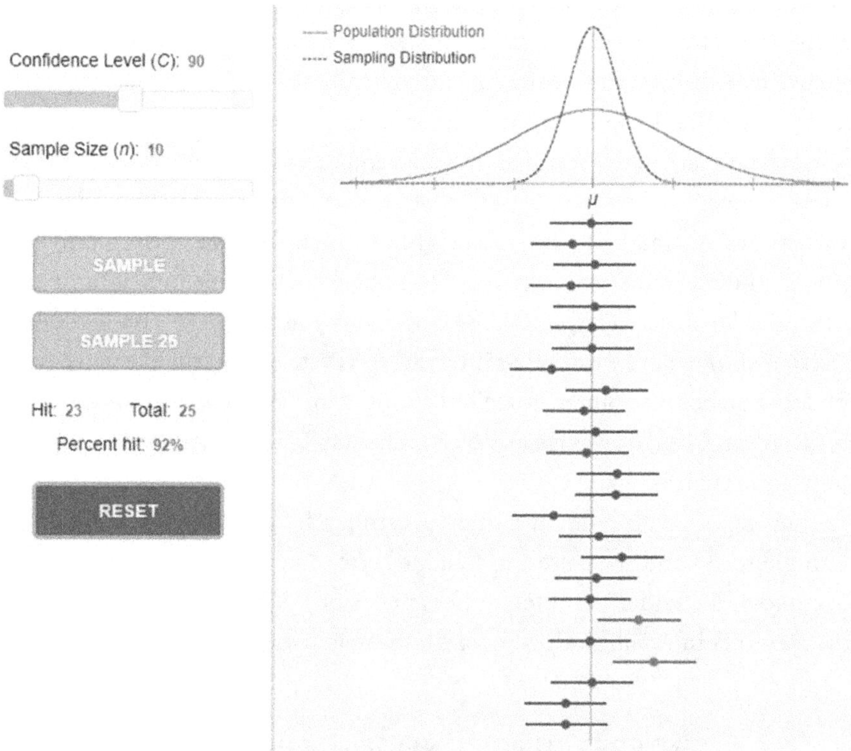

FIGURE 8.6 Screenshot of Applet: Simulation of 90% Confidence Intervals from 25 Samples ($n = 10$)

CALCULATING CONFIDENCE INTERVALS

We are going to be calculating a CI around a sample mean, but keep in mind that you can estimate a CI around other statistics as well, such as a mean difference or an odds ratio. Here is how we estimate an unknown population mean using a CI when population standard deviation is unknown:

$$CI\ for\ Population\ Mean = Sample\ Mean \pm Level\ of\ Confidence \times Standard\ Error$$

In mathematical terms:

$$\mathrm{CI}(\mu) = \bar{X} \pm t_{critical} \times \frac{s}{\sqrt{n}}$$

So, there are three steps to estimate a CI for a population mean.

1. Compute the sample mean and standard error.
2. Choose the level of confidence and find the critical values of the *t* distribution for that level.
3. Compute the estimation formula to find the upper and lower confidence limits.

The width of our interval depends on several factors: how confident we want to be, the size of our sample, and the variability (or standard deviation) of scores within our sample.

Let's work through an example. Suppose you selected a sample of 100 participants and computed sample statistics. You found a sample mean of 4 and an *SD* of 5. Now, you want to estimate a CI around that mean.

STEP 1: COMPUTE THE SAMPLE MEAN AND STANDARD ERROR

From the information above, we already know that $\bar{X} = 4$, $SD = 5$, and $n = 100$, so we can calculate the standard error as follows:

$$\frac{s}{\sqrt{n}} = \frac{5}{\sqrt{100}} = 0.5$$

STEP 2: CHOOSE THE LEVEL OF CONFIDENCE AND FIND CRITICAL VALUES

How do we know what level of confidence to choose? This is where you must balance precision and accuracy. In social science research we typically use a 95% level of confidence, which as you saw is consistent with our typical practice of choosing an alpha level of 0.05 for hypothesis testing.

We look to Table A2 in the appendix to find the critical value. Find the *t* value for a 95% level of significance for a two-tailed test with $df = 99$. There is no row for 99 degrees of freedom in the table, so we will use $df = 100$ for an approximation, and we get the *t* value of 1.98.

STEP 3: COMPUTE THE ESTIMATION FORMULA TO FIND UPPER AND LOWER CONFIDENCE INTERVAL LIMITS

Using these numbers in the formula from earlier in the chapter, we get:

$$4 \pm 1.98 \times 0.5 = (3.01, 4.99)$$

We can conclude that we are 95% confident that the interval of 3.01 to 4.99 contains the true population mean.

USING A CONFIDENCE INTERVAL TO EVALUATE A NULL HYPOTHESIS

We are going to walk through one more scenario using our original one-sample *t* test example. This example will demonstrate how estimation through CIs is connected to hypothesis testing, showing that both methods can provide related insights into evaluating a null hypothesis.

STEP 1: COMPUTE THE SAMPLE MEAN AND THE STANDARD ERROR

We already calculated the relevant values, but to get everything back in one place, our sample mean was $\bar{X}$ = 425.26, our standard error was 7.605, and we wanted to know if our sample mean could be considered equal to the population mean of 440.

STEP 2: CHOOSE THE LEVEL OF CONFIDENCE AND FIND CRITICAL VALUES

When we conducted the *t* test, we decided to use a directional test. This time for creating a CI, we will use the critical value associated with a two-tailed test. So, we need to find the *t* value for a 95% level of significance for a two-tailed test with $df = 34$. Table A2 doesn't have a row for $df = 34$, so we will use $df = 30$ for a conservative approximation, and we get the *t* value of 2.042. Note that the exact critical *t* value for $df = 34$ at a 95% confidence level is $t = 2.032$.

STEP 3: COMPUTE THE ESTIMATION FORMULA TO FIND UPPER AND LOWER CONFIDENCE INTERVAL LIMITS

Using the numbers from earlier in the chapter, we get:

$$425.26 \pm 2.042 \times 7.605 = (409.73, 440.79)$$

If the value of the population parameter we started with ($\mu = 440$) is contained within our 95% CI, then we cannot reject the null hypothesis. As you can see, 440 is just barely contained within our 95% CI. If we had used a two-tailed test instead of a one-tailed test earlier, we would have failed to reject the null hypothesis. This reinforces the importance of identifying one-tailed or two-tailed tests before running your statistical analysis.

CHI-SQUARE GOODNESS-OF-FIT TEST

Most of the methods we will learn throughout this book are parametric, meaning we are using a sample to test hypotheses about parameters (e.g., means and variances) in a population. But not all research questions involve population parameters, so parametric methods don't always meet our needs. Most of the methods we learn also come with certain assumptions, but real-world data won't always meet those assumptions. **Nonparametric statistical tests** are methods used in statistics that do not require the data to follow a specific distribution, such as the normal distribution. These tests are particularly useful when:

- The data violate assumptions required for parametric tests (e.g., normality, homogeneity of variance);
- The data are ordinal or ranked instead of being measured on an interval or ratio scale; and
- The data contain outliers or are severely skewed.

The first nonparametric technique we will learn is the **chi-square goodness-of-fit test.** If you have data on a single variable and you want to compare it to known or expected values, you will use the chi-square goodness-of-fit test. This test will allow you to evaluate how well your observed data align with these expected values. While the expected values may sometimes represent an even distribution across groups, it is not a requirement. The key is that you already know the expected values.

Use this test when your goal is to determine whether a discrepancy between a set of proportions (frequencies) that you are expecting to see and a set that you have observed is not simply due to a sampling error or random chance. Let's look at the goodness-of-fit test more closely with an example.

Suppose you want to investigate whether a state-level STEM enrichment program provides equitable access based on gender. Across the state, 474 students were selected to participate in the program. You would expect 237 males and 237 females, assuming a 50/50 split if there were perfect gender balance in the selection process. However, the actual frequencies are displayed in the table below.

TABLE 8.2 *Gender Distribution of Participants*

	FREQUENCY	PERCENT
Female	216	45.6
Male	258	54.4
Total	474	100

The distribution is clearly not 50/50, but is it different enough that the outcome is likely not a result of random chance? To answer this question, we apply a chi-square goodness-of-fit test. This is another way to test a hypothesis, so we follow the same steps we have already learned.

STEP 1: STATE THE RESEARCH QUESTION AND DETERMINE THE NULL AND ALTERNATIVE HYPOTHESES

Typically, we can make two types of predictions for goodness-of-fit tests. We often expect all categories to be equal. We would expect a 50/50 gender split if participants are limited to only selecting between female and male. Similarly, we would expect 25% of students to choose each menu item if limited to four quality lunch options. But other times, we have reason to expect distributions to not be evenly split across options. For example, imagine we wanted to investigate the proportion of students identified as gifted by race/ethnicity. If we had participants from five different groups, we would not necessarily expect 20% of the gifted students to come from each group. Instead, we would expect that the distribution of gifted students match the percentage of students from each group in the student body overall.

In this example, our null hypothesis is that there is no difference in proportions by gender, or, mathematically, the proportions of males and females selected are equal to the expected proportions (i.e., 50% each).

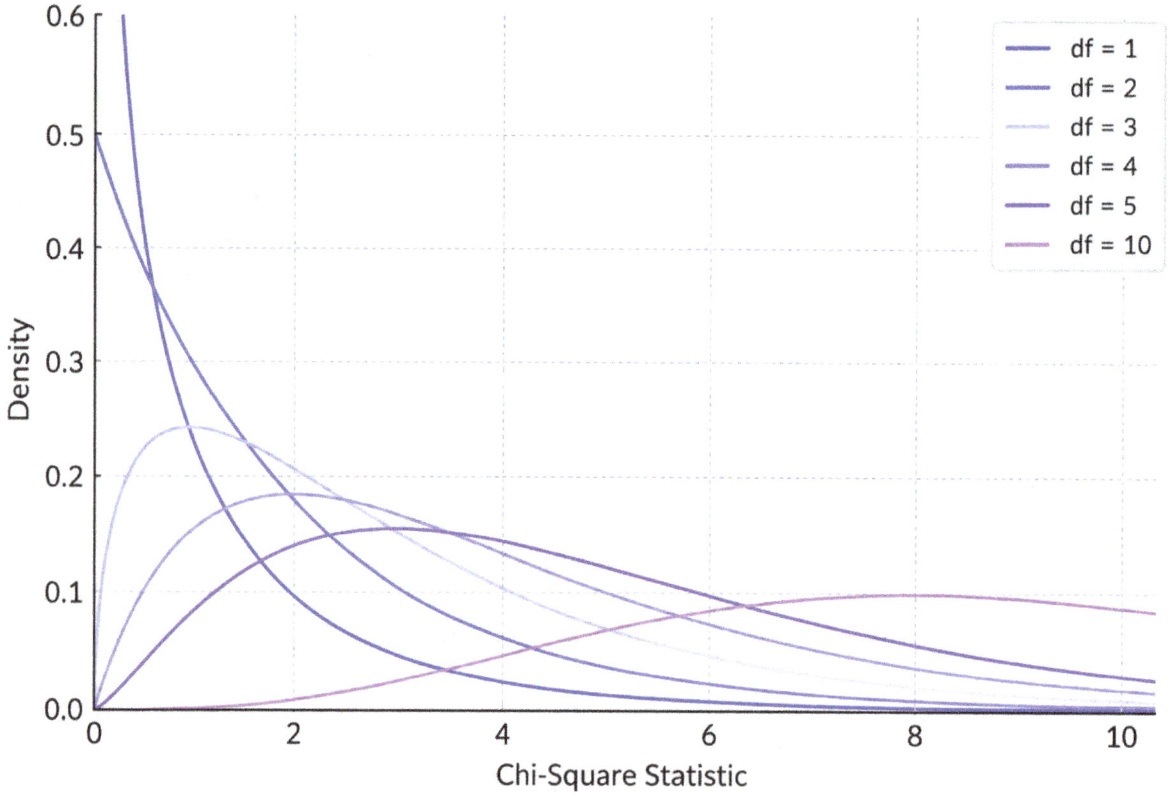

FIGURE 8.7 Chi-Square Distributions with Different Degrees of Freedom

The alternative hypothesis is that there is a significant difference by gender, or the proportions of males and females selected are NOT equal to expected proportions.

STEP 2: SET THE CRITERIA FOR A DECISION

We will once again use our typical alpha level of 0.05. If the p value is smaller than 0.05, we will reject the null hypothesis. The next part of this step is finding the critical value, and that brings us to a new distribution: the **chi-square distribution.** Like the t distribution, the chi-square distribution is a whole family of distributions that varies depending on the degrees of freedom (see Figure 8.7 for examples). The chi-square distribution is a probability distribution that arises when a sum of the squares of k independent standard normal random variables is calculated. If $Z_1, Z_2, \ldots, Z_k$ are independent standard normal random variables, the sum: $\chi^2 = Z_1^2 + Z_2^2 + \ldots + Z_k^2$ follows a chi-square distribution with k degrees of freedom (df).

For the chi-square goodness-of-fit test, the degrees of freedom are equal to $k - 1$, where k = the number of categories or levels within your variable. Our variable is gender, and participants were confined to just two categories ($k = 2$), so we have just one degree of freedom. Since we are using $\alpha = 0.05$, we can flip Table A6 in the appendix and find that the critical value is 3.84.

One thing you might notice about Table A6 is that unlike most of the tables we have used, the critical value increases as degrees of freedom increase. The reason for this will make sense when you see the formula for the test statistic. Another important point is that the chi-square test is *always one-tailed*, even though our alternative hypothesis does not imply any specific direction for the difference. Instead, we test for the total deviation from the expected values. We reject the null hypothesis if our observed chi-square statistic exceeds the critical value.

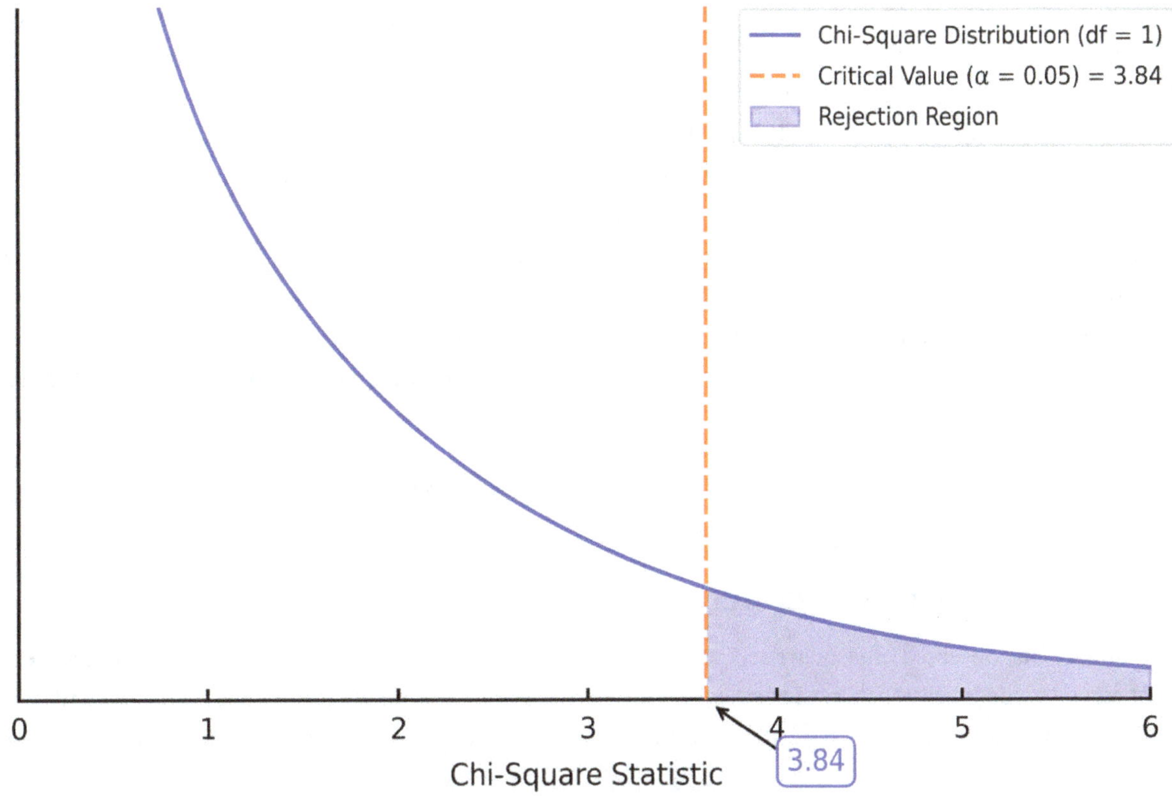

FIGURE 8.8 Chi-Square Distribution with *df* = 1

STEP 3: VERIFY THAT DATA MEET NECESSARY ASSUMPTIONS FOR THE STATISTICAL METHOD AND CALCULATE THE TEST STATISTIC

There are two data conditions that are required for the use of a chi-square test:

1. Observations are *independent*, meaning that the same individual or case cannot be categorized into more than one group.
2. The expected frequency count (n) for each cell or category should be up to 5 or more. If your expected values are too small, this will cause problems with the calculation of the test statistic. The most common solutions to this problem are increasing your sample size to ensure sufficient observations in each cell or merge categories to achieve larger expected frequencies. You might also choose to conduct a Fisher's exact test, which we will not detail in this book.

When we talk about chi-square tests, we use the word "cell." Think of this as if you are creating a table with rows and columns. A cell is just one square in the table. It belongs to one row and one column. In this example, our table is one column (count of students) by two rows (female/male). So, we have two cells altogether.

The formula for the goodness-of-fit test is

$$\chi^2 = \sum \frac{(O_i - E_i)^2}{E_i}$$

In this formula, O_i = the observed count of ith cell (i.e., how many individuals we observed actually fall within the cell), and E_i = the expected count of ith cell (the number of individuals that we think *should* be in the cell). For each cell, we would find a difference between observed and expected values, square the difference (remember, if we are adding up differences and want the answer to be something other than zero, we must square the differences first), and divide it by the expected value. We would then take these squared values for every comparison between the expected and observed pair and add them together.

Can you see why it is problematic if expected counts are very small? If you look at the formula for the chi-square test statistic, the expected value goes in the denominator. If we have a very low expected value, we can end up with an artificially inflated test statistic, which increases the risk of Type I error.

In our example, we already know the expected frequencies (a 50/50 split of 474 employees would be 237 in each group). We also have the observed frequencies from Table 8.2 earlier in the chapter, so we add those values to the equation:

$$\chi^2 = \frac{(216 - 237)^2}{237} + \frac{(258 - 237)^2}{237} = 3.72$$

STEP 4: DRAW A CONCLUSION FOR YOUR RESEARCH QUESTION

We compare our observed chi-square test statistic to the critical value we determined earlier. Remember, if our observed value is greater than the critical value, we reject the null. In this case $3.72 < 3.84$, so we fail to reject the null. While there is not an exact 50/50 split in employees along gender lines, the difference is not large

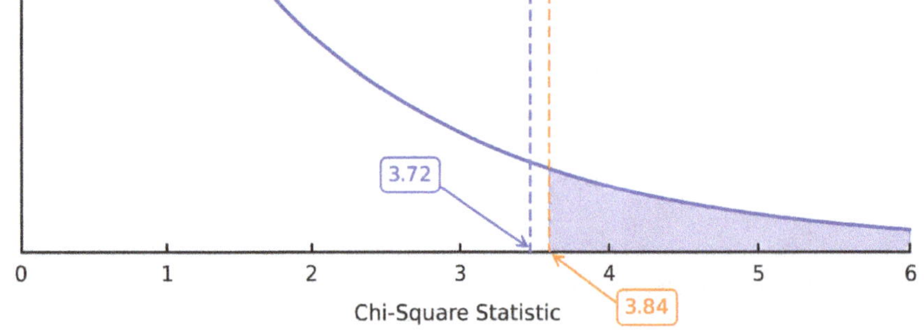

FIGURE 8.9 Observed χ^2 Statistic Compared to Critical Value (df = 1, α = 0.05)

enough to be considered significant. The exact *p* value associated with χ^2 (1) = 3.72 is p = 0.0537. Figure 8.9 shows where our observed *p* value values on the curve.

> Here, we present an example of a chi-square goodness-of-fit result written in APA style. A chi-square goodness-of-fit test was conducted to determine whether the state-level STEM enrichment program provides equitable access based on gender, with expected proportions for male and female participants. All assumptions for the analysis were met. The results indicated no statistically significant difference in the proportions of males and females selected for the program (χ^2 (1) = 3.72, p = 0.054). These findings suggest that the enrichment program provides equitable access by gender.

CONCLUSION

Hypothesis testing bridges theoretical research questions and practical data analysis, offering a structured framework for making evidence-based decisions. By understanding statistical significance, effect sizes, and confidence intervals, researchers can draw meaningful conclusions while accounting for uncertainty and variability. For educational researchers, it is particularly important to assess both statistical and practical significance when interpreting empirical results.

While *p* values indicate whether a result is statistically significant, they do not reveal the magnitude or practical relevance of the effect. Measures of effect size, such as Cohen's *d*, provide valuable context by quantifying the size of an effect, helping researchers determine its importance in real-world settings. For example, a small *p* value paired with a negligible effect size may lack practical significance, limiting its applicability. Therefore, when statistically significant findings are observed, it is essential to consider their implications within the research context, ensuring that the results are both statistically sound and practically meaningful. This dual evaluation enhances the relevance and impact of research findings in education and beyond.

KEY TAKEAWAYS FOR EDUCATIONAL RESEARCHERS FROM CHAPTER 8

- Two statistical methods, one-sample *t* tests and chi-square goodness-of-fit tests, are for comparing the observed statistic to a benchmark. Use a one-sample *t* test when comparing a sample mean to a known population mean with unknown standard deviation. Use a chi-square goodness-of-fit test when comparing observed proportions or frequencies in a single categorical variable to known or expected distributions.
- Statistical significance (e.g., *p* values) indicates whether an observed effect is unlikely due to chance, while effect sizes (e.g., Cohen's *d*) quantify the magnitude and practical relevance of the effect. By reporting both, education researchers can ensure that their findings have actionable value, informing policy, curriculum design, and classroom interventions.
- The correct interpretation of a 95% CI for a population mean is as follows: *We are 95% confident that the interval contains the true population mean.* If we were to take many random samples of fixed sample size (e.g., $n = 50$) from the population and compute a CI for each sample, approximately 95% of those intervals would contain the true population mean. Confidence intervals help researchers communicate the degree of certainty in their findings, ensuring that conclusions account for variability and are not overly dependent on a single estimate. They

are particularly beneficial when researchers want to evaluate the reliability of a point estimate or compare results across studies.

- Larger sample sizes increase statistical power but may detect trivial differences as significant. Carefully designing studies to balance power with meaningfulness enables education researchers to use resources effectively and avoid misinterpreting results as practically relevant when they are not.

KEY DEFINITIONS FOR CHAPTER 8

The **chi-square distribution** is a probability distribution that arises when a sum of the squares of k independent standard normal random variables is calculated. If $Z_1^2, Z_2^2, \ldots Z_k^2$ are independent standard normal random variables, the sum: $\chi^2 = Z_1^2 + Z_2^2 + \ldots + Z_k^2$ follows a chi-square distribution with k degrees of freedom (*df*).

The **chi-square goodness-of-fit test** is a nonparametric method used to determine whether observed data frequencies match expected frequencies for a single categorical variable. This test is useful when comparing observed proportions to known or hypothesized distributions, such as evaluating gender equality in a workforce. By calculating the chi-square statistic and comparing it to a critical value, researchers can assess whether deviations from expected frequencies are due to random chance or are statistically significant.

Cohen's *d* quantifies the magnitude of the difference between two means relative to the variability in the data and provides a standardized measure of practical significance, helping to interpret the importance of the result by allowing us to express the effect size in terms of standard deviation units.

Confidence intervals (CIs) provide an additional layer of transparency by offering a range of values likely to contain the true population parameter. This chapter highlighted the interplay between CIs and hypothesis testing, explaining that if a null value (e.g., a population mean) falls within the CI, the null hypothesis cannot be rejected.

Effect size is a numeric measure that indicates the strength or meaningfulness of the relationship or difference between variables in a study.

Nonparametric statistical tests are methods used in statistics that do not require the data to follow a specific distribution, such as the normal distribution.

Type III error occurs when researchers correctly calculate statistical results but misinterpret or apply them to the wrong question or hypothesis.

CHECK YOUR UNDERSTANDING

1. When should a one-sample *t* test be used instead of a *z* test?
   a. When the population mean is known
   b. When the population standard deviation is unknown
   c. When the same size is less than 30
   d. When comparing two sample means

2. What are the degrees of freedom (*df*) in a one-sample *t* test?
   a. $df = n + 1$
   b. $df = n \times 2$
   c. $df = n - 1$
   d. $df = \frac{n}{2}$

3. Which of the following is not an assumption of the one-sample *t* test?
   a. Random sampling
   b. Independence of observations
   c. Population variance of 0
   d. Dataset is approximately normally distributed

4. What is the purpose of calculating effect size, such as Cohen's *d*?
   a. To determine statistical significance
   b. To quantify the magnitude of a difference
   c. To check if assumptions are met
   d. To calculate degrees of freedom

5. What is the risk of using a one-tailed test when the actual effect is in the opposite direction?
   a. Type I error
   b. Type II error
   c. Type III error
   d. Increased power

9

ANALYZING DIFFERENCES BETWEEN TWO GROUPS

LEARNING OBJECTIVES

By the end of this chapter, you will be able to:

- Identify and differentiate between different types of tests to compare two groups, understanding their appropriate applications based on the nature of the data and comparisons being made;
- Understand differences between parametric and nonparametric tests;
- Explain the logic behind conducting independent- and related-samples *t* tests on continuous variables, including verifying assumptions, calculating test statistics, and interpreting results within the context of hypothesis testing;
- Apply the Mann-Whitney *U* test and Wilcoxon's matched-pairs signed-ranks test, demonstrating an understanding of their logic, appropriate use cases, and result interpretation; and
- Evaluate the practical significance of statistical results using measures such as Cohen's *d* and eta squared, providing context for the findings of *t* test analyses.

In Chapter 8 we explored the one-sample *t* test, a method used to compare a single group's mean to a known population mean. This gave us the foundation for understanding how *t* tests work and their role in hypothesis testing. Now, we are moving on to explore additional types of *t* tests to compare two groups. Before diving into the details of the independent-samples *t* test, which is commonly used in educational research, let's take a brief look at three types of *t* tests. These include the one-sample *t* test (covered in Chapter 8), two independent-samples *t* test, and two related-samples *t* test, each tailored to different types of comparisons. Understanding the distinctions between these tests will help clarify when and how to use them effectively.

TYPES OF *t* TESTS

All three types of *t* tests involve comparing the values of a variable from one group of individuals to something else—and that *something else* tells you what kind of *t* test you should use. To explain the three tests, imagine that you have a class of third grade students, and you have a list of each student's reading achievement scores. Table 9.1 explains how you might analyze those scores with each type of *t* test, depending on what you're comparing them to.

TABLE 9.1 ***t Test Characteristics***

WHAT TO COMPARE	*t* TEST USED	EXAMPLE
One single number	One-sample *t* test	You want to see if this group's scores are, on average, significantly different from the state average for reading achievement scores.
A different group's scores on the same variable	Two-independent-samples *t* test	You want to compare this group's reading achievement scores to those from another third-grade class.
The same group's scores on a different variable or at a different time	Two related-samples *t* test	You want to compare this group's reading scores to their math achievement scores, to see if their achievement levels are different across the two subjects. OR You want to compare this group's reading scores after third grade to the same group's reading scores after fourth grade to see if they change over time.

Now, we will provide a very brief overview of each type.

ONE-SAMPLE *t* TEST

This should be very familiar, since we just learned it in Chapter 8. You use this *t* test if you know the population mean (μ) but *do not know* the population standard deviation (σ). You are trying to find out if your sample mean ($\bar{X}$) is significantly different from a known population mean. To run this analysis, we will use only one continuous variable or column of data—each individual in the group has one value or score for the variable of interest.

TWO-INDEPENDENT-SAMPLES *t* TEST

Commonly in practice, you do not know a population mean to use as a comparison point. You are instead comparing sets of numbers obtained from two separate groups in order to draw inferences about their populations. Maybe you have two groups of students, one of which received some kind of intervention, and you want to compare end-of-year achievement scores between them. You might want to know if female college students have higher grade point averages than male students. This time, your dataset will need two columns to run the test: one to indicate which group the individual is in and one with their value for the dependent variable.

RELATED-SAMPLES *t* TEST

You are still comparing two sets of numbers, but this time both sets belong to the same people or two groups that are somewhat related. For example, you might be comparing achievement scores at the beginning and end of a year to see if a class improved. You might also compare outcomes under two different conditions, such as how many times students were absent when class met in person compared to the number of absences for the same students when the school shifted to remote learning. Sometimes instead of having the same group of people, you have a matched group (e.g., siblings or a set of individuals who have been matched based on their scores on other variables). Either way, your dataset will look different yet again. As with the one-sample *t* test, you only have a single list of individuals or a single set of matched pairs. However, this time they each have two columns of information (e.g., pretest and posttest scores or English scores and math scores).

We hope this helps you avoid getting confused about the three different types of *t* tests. The details for how to conduct each analysis are similar, but there are important differences. The names are actually pretty good clues:

- *One*-sample: You only have one set of data (e.g., a set of reading scores from a sample of third graders), so you need a population value (e.g., the national average of reading score) for comparison.
- Two *independent*-samples: You have two distinct groups that are not related or connected in any way—one might even say they are two independent groups. Membership in a group is mutually exclusive, meaning that no individual or case can belong to both groups simultaneously.
- Two *related*-samples (also known as a paired-sample or repeated-measures test): Here you have two sets of numbers that are connected—either (1) the same people at different times or different conditions or (2) people who are paired together.

And now we will dive into the details of the independent-samples *t* test. This is probably one of the most used methods to compare two groups in educational research.

OVERVIEW OF THE TWO-INDEPENDENT-SAMPLES *t* TEST

As we did when learning about the one-sample *t* test, let's again walk through an example from educational research. Similar to the previous example, the teacher is now concerned if her class average of English Language Arts (ELA) scores ($\bar{X} = 425.26$, $SD = 44.99$) is significantly different from another class of third graders whose teacher joined the pilot implementation of a new reading intervention. When the result was returned, the class average of the fellow teacher was 440 with a standard deviation (*SD*) of 35. Figure 9.1 displays the class distributions. Are these two averages significantly different? Let's find out.

Similar to the previous one-sample *t* test example, we will complete four steps of hypothesis testing.

STEP 1: DETERMINE THE NULL AND ALTERNATIVE HYPOTHESES

With a one-sample test, the null hypothesis states that there is no difference between the sample mean and the population mean. We are now comparing the means of two independent samples to determine statistically if their population means are identical or different. So, the null hypothesis about population means is that there is no difference between the population means of the two groups. The way we indicate this is that one population mean (μ_1 for Group 1) minus the other (μ_2 for Group 2) is zero.

$$H_0: \mu_1 - \mu_2 = 0$$

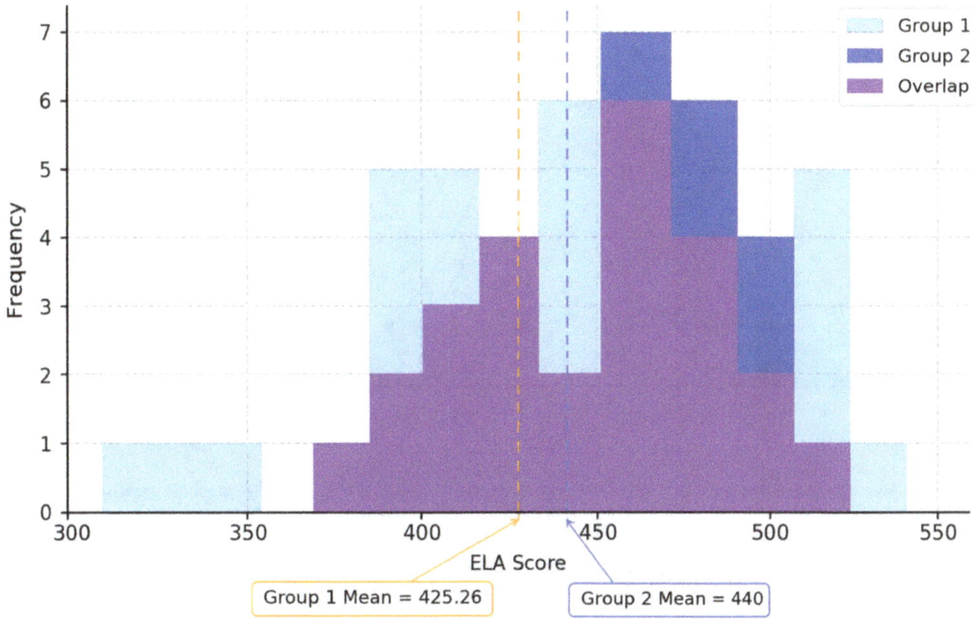

FIGURE 9.1 Comparing Two Group Averages on ELA Scores

Therefore, the alternative hypothesis is that there is a difference between the two groups:

$$H_A: \mu_1 - \mu_2 \neq 0$$

Note that this is the alternative hypothesis for a two-sided or two-tailed test. We are not assuming that one mean is larger than the other, just that they are different. If you had reason to believe that one group mean would be larger than the other, your alternative hypothesis would be:

$$H_A: \mu_1 > \mu_2$$

or

$$H_A: \mu_1 < \mu_2$$

STEP 2: SET THE CRITERIA FOR A DECISION

The next step in testing the null hypothesis is to set the criteria for a decision: How much of a difference is significant enough that we can conclude that it is not only due to random chance? We decide that $\alpha = 0.05$ as usual, but the degrees of freedom are slightly different for an independent-samples t test compared to a one-sample t test. This time, we must subtract one from each group, so $df = (n_1 - 1) + (n_2 - 1) = n_1 + n_2 - 2 = n - 2$, where n_1 and n_2 refer to the sample size of Group 1 and Group 2, respectively, and n refers to the total number of subjects in both groups combined.

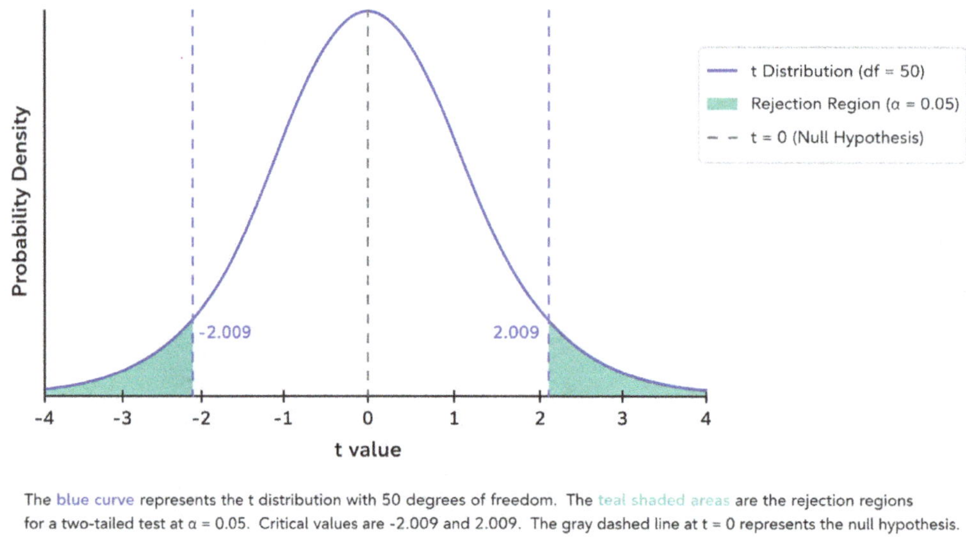

FIGURE 9.2 *t* Distribution with Critical Values ±2.009 for $\alpha = 0.05$ and $df = 50$

Flip to Table A2 in the appendix and look for the intersection between your alpha level and your degrees of freedom to find the critical value. For our example, both groups have a sample size of 35, so the $df = 70 - 2 = 68$. Since we don't have $df = 68$, round down to 50 *df* to be conservative. The critical *t* value associated with 50 *df* at $\alpha = 0.05$ for a two-tailed test is ±2.009 (Figure 9.2).

STEP 3: CHECK ASSUMPTIONS AND COMPUTE THE TEST STATISTIC

The third step has a few parts. First, we verify that our data meet the assumptions required for this method of analysis, and then we calculate the test statistic. The assumptions for an independent-samples *t* test are:

- Normality: The dependent variable is approximately normally distributed (as before, this requirement is less important if $n > 30$).
- Random sampling: You are using a random sample of independent observations.
- Independence: The scores are independent of one another.
- **Equal variances**: The variances of the dependent variable are equal across two groups or populations being compared.

Equal variance is new, but it is important to keep in mind for this method. Remember in Chapter 3 when we said that means are important, and so is standard deviation (or variance, which is the same basic concept)? Here is one place where that comes into play. Imagine you are comparing scores of two groups on the same test. In one group, all scores are tightly clustered near the mean. In the other group, they are completely spread out with a much wider range than compared to the first group. It is harder to make a solid conclusion about the differences between two group means when the distributions are completely different. Figure 9.3 shows examples of equal and unequal variances.

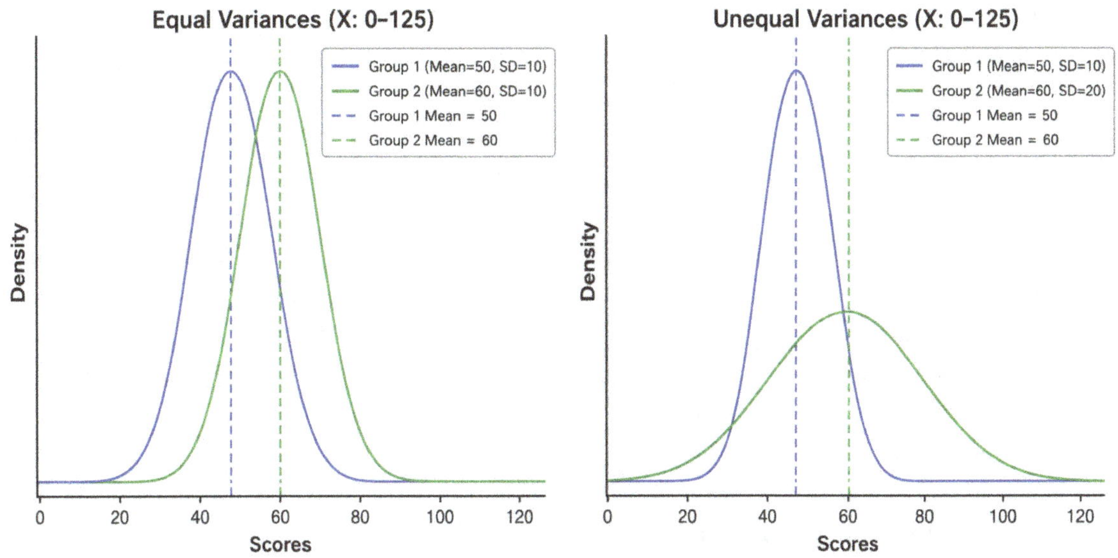

FIGURE 9.3 Examples of Equal and Unequal Variances

When analyzing data from two groups, it is essential to compute and evaluate the size of variance for each group. This helps in understanding the spread of the data and determining if the variances are similar or different. Drawing histograms for each group is particularly useful for visually inspecting the data spread and identifying any noticeable differences between two distributions.

LEVENE'S TEST FOR ASSESSING EQUAL VARIANCES

Levene's test is a statistical procedure used to assess whether the variances of a dependent variable are equal across different groups. This test is typically performed before conducting analyses that assume equal variances, such as *t* tests and ANOVA, to determine if this assumption holds. We won't show you how to calculate Levene's test by hand because statistical software packages will do this for you.

However, the important thing to remember is that if the result of Levene's test is statistically significant ($p < .05$), your variances *are not equal*. If your variances are not equal, you should consider using a different test or, at minimum, report the statistical results adjusted for unequal variances between two groups, which is provided by most software packages. If your Levene's test results are not statistically significant ($p > .05$), your group variances are considered equal.

STEP 4A: FIND THE *P* VALUE

Next, we calculate the test statistic. We are going to show you the formula first, but please don't panic. It is not as bad as it seems.

$$t = \frac{(\bar{X}_1 - \bar{X}_2) - (\mu_1 - \mu_2)}{\sqrt{S_p^2\left(\frac{1}{n_1} + \frac{1}{n_2}\right)}}$$

Let's take that a little at a time, starting with the numerator. The first part is the difference between group means in your sample. Then we are comparing that to the expected difference between group means in the population (i.e., the mean differences specified in the null hypothesis). Look back at the null hypothesis for this type of test. What is the expected difference between group means in the population? Zero! For practical purposes, the top half of the fraction is just the difference between group means.

Do you remember what was in the denominator of the t statistic for a one-sample t test? It was the standard error, or the sample standard deviation divided by the square root of the sample size. This denominator is more complicated because we have two separate groups. Under the null hypothesis, we assume that both sample groups came from the same population, but most likely the standard deviations of these two groups are not exactly the same. So, which sample variance do we use as an estimate for the population variance? The optimal solution is to make full use of the variance estimates from both samples to obtain a more accurate estimate of the population variance.

There is another new value in the denominator, s_p^2, which represents the pooled sample variance. The formula for this looks very complicated, but it is just a way of weighting the two sample variances by the sample size. If one group is bigger than the other, their sample variance is weighted more heavily by this formula. If you notice that this looks a little different from what you might find elsewhere, that is because we are being more specific about how you get to the degrees of freedom, with $n - 1$.

$$S_p^2 = \frac{(n_1 - 1)s_1^2 + (n_2 - 1)s_2^2}{(n_1 - n_2) - 2}$$

What does all that mean? You are multiplying the sample variance for group one by the degrees of freedom for Group 1 and adding that to the same thing for Group 2. Then you are dividing the outcome by the degrees of freedom for the whole test. In our example, the sample size of each group is 35. Thus:

$$S_p^2 = \frac{(34)s_1^2 + (34)s_2^2}{68}$$

Now you have a measure of sample variance that you can use for the whole combined sample of both groups. If both groups have the same sample size, like our example, there is a much simpler version of the formula you can use:

$$S_p^2 = \frac{s_1^2 + s_2^2}{2}$$

In our example, $s_1^2 = (44.99)^2$, and $s_2^2 = (35)^2$. Let's plug in these values and compute the s_p^2:

$$S_p^2 = \frac{s_1^2 + s_2^2}{2} = \frac{(44.99)^2 + (35)^2}{2} = 1624.55$$

Either way, once you have the value for s_p^2, you bring that back to the t statistic formula and do what we always do: standard deviation divided by the square root of the sample size. It just looks more complicated when we have two separate sample sizes and because this formula starts with sample variance and then takes the square root rather than starting with standard deviation, which is already the square root of the variance.

$$t = \frac{(\bar{X}_1 - \bar{X}_2) - (\mu_1 - \mu_2)}{\sqrt{S_p^2\left(\frac{1}{n_1} + \frac{1}{n_2}\right)}} = \frac{(425.26 - 440) - 0}{\sqrt{1624.55\left(\frac{1}{35} + \frac{1}{35}\right)}} = \frac{-14.74}{9.634} = -1.53$$

If you are completely lost, try not to get too caught up in the math and focus on the concept. Figure 9.4 will also help you visualize this example. We are finding the difference between the two group means and dividing it by the standard error for the difference. It is just more complicated to calculate this version of standard error than it has been for other tests. The good news is that statistical software programs will do the math for you.

The last step in testing a hypothesis is more straightforward: compare the value of the t statistic you just calculated to the critical value you identified in Step 2. If the t statistic is larger, meaning farther out from the center of the distribution in either direction, then a significant difference exists between the groups.

The observed t value is = -1.53 which is closer to 0 than the lower critical t value of -2.009. Therefore, we fail to reject the null hypothesis. The actual p value for t = -1.53 in a two-tail test is 0.131.

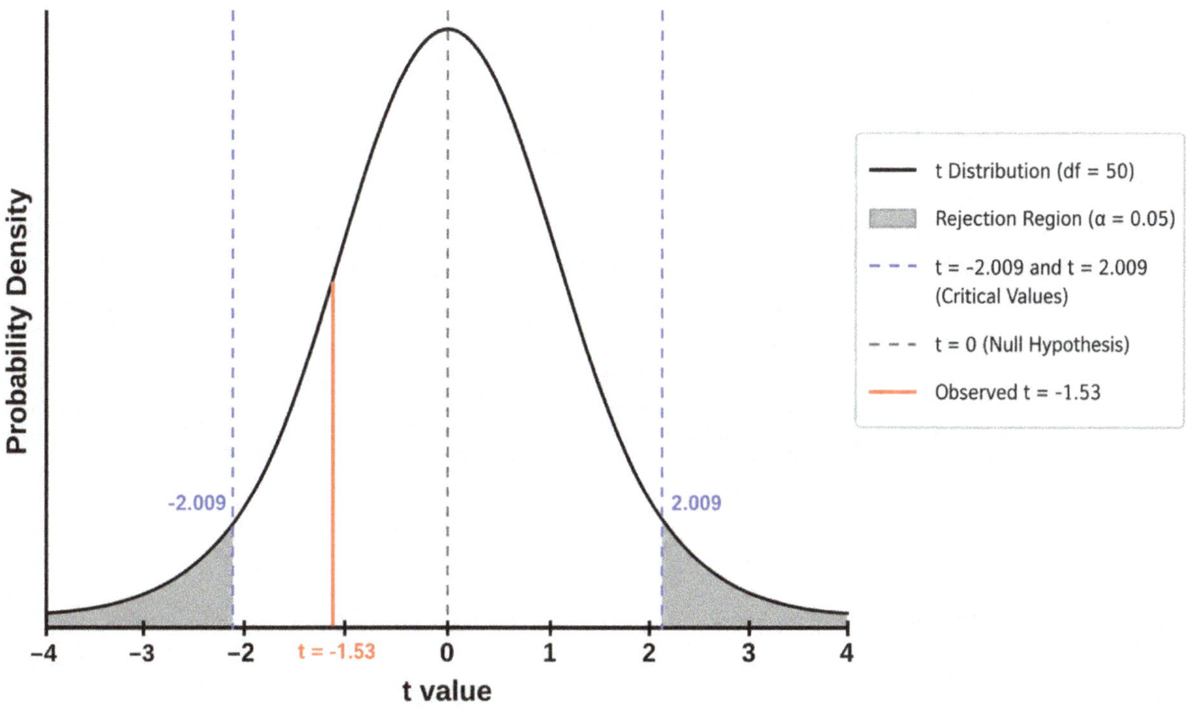

FIGURE 9.4 Observed t Statistic Compared to Critical Values in a Two-Tailed Independent-Samples t Test (α = 0.05, df = 50)

STEP 4B: DRAW A CONCLUSION AND REPORT YOUR CONCLUSION.

Since we failed to reject the null hypothesis, which states that there is no difference in population means of the two groups, we conclude that the ELA classroom averages taught by two different teachers are not significantly different. The observed mean difference between two groups is simply by chance due to sampling. Note that we do not calculate the effect size to assess practical significance, as no statistically significant difference was found.

How might we report the result in APA style? An independent-samples *t* test was conducted to compare ELA classroom averages between students taught by two different teachers. The results indicated that there was no statistically significant difference in classroom averages between the two groups, $t(68) = -1.53$, $p = .131$, suggesting that the mean ELA classroom scores for students in these two classrooms are comparable.

EFFECT SIZE FOR THE INDEPENDENT-SAMPLES *t* TEST

If you find a significant difference, you should calculate the effect size. We most commonly report effect size for this method using estimated Cohen's *d*. If you remember back to Chapter 8, Cohen's *d* quantifies the magnitude of the difference between two means relative to the variability in the data. Similarly, we can use Cohen's *d* as a standardized measure of practical significance, helping to interpret the importance of the result. However, the formula for Cohen's *d* that you use for an independent-sample *t* test is slightly different from what we have already learned:

$$d = \frac{M_1 - M_2}{\sqrt{S_p^2}}$$

The numerator is just the difference in sample means, and the denominator is the pooled sample standard deviation; you can find the formula earlier in this chapter.

Eta squared (η^2) is another measure of effect size used to quantify the proportion of variance in a dependent variable that is associated with one or more independent variables in an analysis. Eta squared provides an estimate of how much of the variability in the dependent variable is explained by the grouping variable and can be calculated using the following formula:

$$\eta^2 = \frac{t^2}{t^2 + df}$$

Where

- t^2 is the square of the *t* test statistic; and
- *df* is the degrees of freedom for the test.

Either way, you only report effect size if you find a significant result when conducting your hypothesis test.

OVERVIEW OF THE RELATED-SAMPLES *t* TEST

We have finally arrived at the last type of *t* test, and to make things extra interesting, it goes by several different names depending on how you are using it. **Related-samples *t* test** is typically used as a general term for a *t* test that involves comparisons of related or dependent groups. Additionally, it is used when you have the same participants measured on two dependent variables that use the same or comparable scales, such as math tests versus reading tests. For example, a researcher may examine whether there is a significant difference between students' performance on standardized tests in two subjects.

Two specific types of the related-samples *t* test are often used in research, each corresponding to a particular research design:

- **Repeated-measures *t* test** is used when we measure *one group of participants twice* on the same or comparative measure at different points in time. In educational research, this test is often applied in pretest-posttest design, where the goal of the analysis is to assess change in scores before and after an intervention or at the start and end of an academic year. For example, a study might measure students' reading comprehension before and after a literacy intervention to evaluate its effectiveness.
- **Paired-samples *t* test** is used when individuals or cases are matched in pairs based on key characteristics before comparing their scores. Unlike repeated-measures design, the paired individuals are not the same person measured twice; rather, they are two different people who are statistically *equivalent* as pairs except for the dependent variables. For example, you could statistically match students who experience an intervention with a control group of students who are very similar on key variables. Other times family members (siblings, married couples) are paired if it is reasonable to assume that they are similar in important ways. This type of *t* test is less common, as it often requires more advanced statistical methods or poses logistical challenges in finding well-matched pairs.

The related-samples *t* test is actually a special case of the one-sample *t* test. In a related-samples test, you collect data in pairs, and the difference between each pair is calculated. Then, we compute statistics based on these differences. Your beginning assumption, or null hypothesis, is that there will be no difference between the pairs. You could create a new variable in your dataset that was equivalent to the difference between pairs and use a one-sample *t* test to compare that difference variable to the expected value of 0.

Let's assume we are not creating new variables, and we will show you how to directly conduct a related-samples *t* test. However, keep in mind that you would get the same results either way. We know it can be hard to understand why a related-samples *t* test is not the same as an independent-samples *t* test, since we are interested in the differences between two sets of scores in both cases. It really matters statistically that the same or similar people are in both groups in a related-samples study. The short description is that when you have different people in your two groups, it is hard to know for sure if the differences are really due to your grouping factor (e.g., free/reduced lunch status, treatment/control for an intervention) or if you by chance ended up with people who were very different from each other for reasons that had nothing to do with your grouping factor.

With a related-samples *t* test, we don't need to worry about any of that. We know there are not random differences between the people in our two groups because the same people are in both groups. From a statistical perspective, this changes two things: we calculate degrees of freedom differently, and our hypotheses and test statistics are based on the difference between scores. Let's use an example to look at each step of the related-samples *t* test more closely.

Imagine that you developed a new student learning intervention and piloted it with a sample of nine randomly selected students. You have data on their scores *before and after* the intervention, and now you want to determine if student scores significantly changed. Your dataset would look like Table 9.2.

If this was an independent-samples *t* test, you would be thinking about the distribution of *pre* scores and the distribution of *post* scores. However, in a related-samples test, the focus shifts to the distribution of *difference*

TABLE 9.2 *Pre- and Postintervention Scores*

PRE	POST	DIFFERENCE
10	14	4
10	9	-1
11	16	5
13	14	1
14	17	3
15	18	3
16	18	2
16	17	1
17	20	3

scores (i.e., post – pre), which in this example ranges from -1 to 5 and has a sample mean of 2.33 with an *SD* of 1.80. For this example, the research question we will address is whether scores changed significantly between pre and post, which may suggest an effect of the intervention.

STEP 1: DETERMINE THE NULL AND ALTERNATIVE HYPOTHESES

Recall that with an independent-samples t test, there are two separate sample means to consider, so the null hypothesis for a two-tailed test is:

$$H_0: \mu_1 - \mu_2 = 0$$

With a related-samples t test, our null hypothesis for a two-tailed test is:

$$H_0: \mu_D = 0$$

The alternative hypothesis for a two-tailed, nondirectional test is:

$$H_A: \mu_D \neq 0$$

The null hypothesis states that the difference between two dependent variables is zero, and the alternative is that the difference between dependent variables is not zero. Keep in mind that you can also conduct a one-tailed, directional test by making the alternative hypothesis that $\mu_D < 0$ or $\mu_D > 0$.

STEP 2: SET THE CRITERIA FOR A DECISION

The next step in testing the null hypothesis is to set the criteria for a decision: How much of a difference is significant enough that we can conclude it is not a result of chance alone? As usual, we can decide that

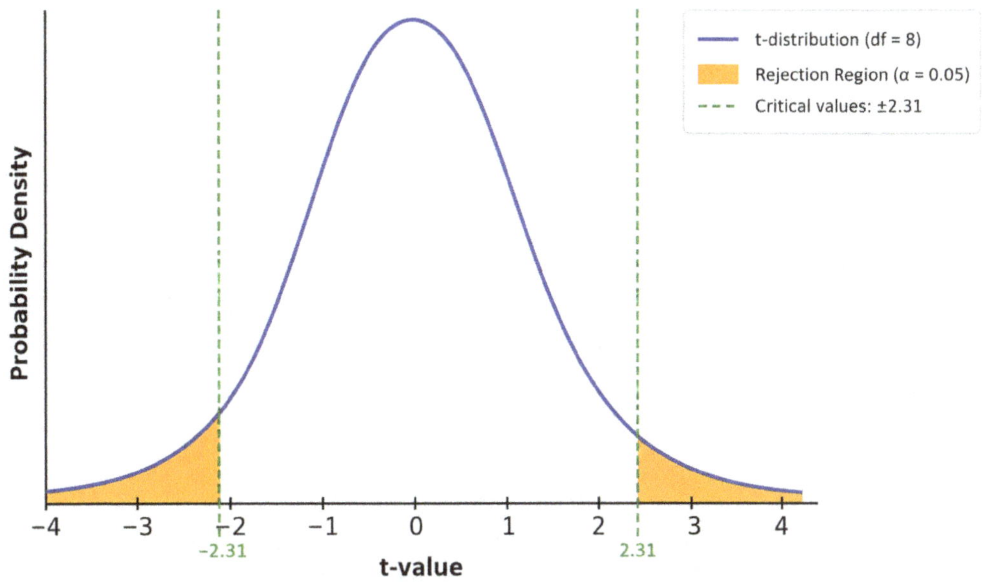

FIGURE 9.5 Rejection Regions for a Related-Samples *t* Test ($df = 8$, $\alpha = .05$)

$\alpha = 0.05$. The degrees of freedom for a related-samples *t* test are simply calculated by taking $n - 1$, where n refers to the number of pairs. Since each participant contributes one pair of scores (before and after), the number of pairs is equal to the number of participants in the study (with no missing values on both tests). You can use a *t* table just as before, looking for the intersection between your alpha level and your degrees of freedom to find the critical value. In our example, because $n = 9$, $df = 9 - 1 = 8$. If you flip to Table A2 in the appendix and look at $df = 8$ and $\alpha = .05$ (two-tailed), you see that the critical value for this test is ± 2.306, which has been rounded to ± 2.31 in Figure 9.5.

STEP 3: CHECK ASSUMPTIONS AND COMPUTE THE TEST STATISTIC

The third step has a few parts. First, we verify that our data meet the assumptions required for this method of analysis, and then we calculate the test statistic. The assumptions for a related-samples *t* test are:

- *Normality*: The distribution of the difference scores is approximately normal. As with other *t* tests, this assumption is less critical when the sample size is greater than 30 due to the central limit theorem.
- *Random sampling*: You are using a random sample of independent observations.
- *Independence*: The scores are independent of one another. Although the two groups in a related-samples *t* test are obviously dependent on each other—since they consist of paired scores from the same individuals or matched pairs—the scores within each group must still be independent of one another. This means that each participant's score should not be influenced by another participant's score within the same group. In other words, the assumption of independence applies to the individual observations within each group, not between groups.

Notice that we don't have the equal variances assumption as we did with the independent-samples t test. We have that assumption with independent-samples t tests because we are concerned that major differences in group distributions mean that the groups are very different, but as we said before, since we are using the same people this time, we are not as worried about that.

STEP 4A: FIND THE p VALUE

Next, we calculate the test statistic. You will notice a lot of Ds in the following formula because we are *focusing on difference scores*, not two separate sample means. Instead of having two distributions of sample means, we have one *distribution of sample differences.*

$$t = \frac{\bar{D} - \mu_D}{\frac{S_D}{\sqrt{n}}}$$

Let's unpack each part of the formula separately to make things easier to digest. The numerator is looking for the difference between the sample mean difference and the expected difference in the population under the assumption that the null hypothesis is true. However, recall that our null hypothesis is that there is no difference in the population, so μ_D is zero. Therefore, for our purposes the numerator is just the sample mean difference. Where does that come from? It is calculated as the average of the individual difference scores (post – pre) across all pairs in the sample, which is 2.33 in our example.

Now we look at the denominator of the test statistic formula. This should look familiar by now because we see a standard deviation divided by the square root of the sample size (i.e., standard error). The first value is S_D, which is something we have not seen before. That represents the standard deviation of the difference distribution, which in this case is 1.80. Conveniently, our sample size is 9, so the square root of n is 3. Using this, we calculate that the standard error of the sampling distribution of difference scores is 0.6. Therefore, we can calculate the test statistic as follows:

$$t = \frac{2.33 - 0}{\frac{1.80}{3}} = \frac{2.33}{0.6} = 3.883$$

The next step in testing our hypothesis is to compare our observed test statistic with the critical value. Our test statistic is definitely farther out into the tails of the distribution than 2.306 (Figure 9.6 on p. 109), so we can reject the null hypothesis. The actual p value associated with the observed t statistic of 3.88 with $df = 8$ for a two-tailed test is 0.0047, which is smaller than 0.05.

STEP 4B: DRAW A CONCLUSION AND REPORT YOUR CONCLUSION

The small p value ($p = 0.0047$) indicates that under the assumption that the null hypothesis is true, the probability of observing a difference as extreme as or more extreme than the one we found in the sample is 0.47%. Because this is highly unlikely, we reject the null hypothesis and conclude that there is a statistically significant difference in scores between pre and post. Because the observed difference is statistically significant, we now want to know how the difference is practically meaningful by calculating the effect size.

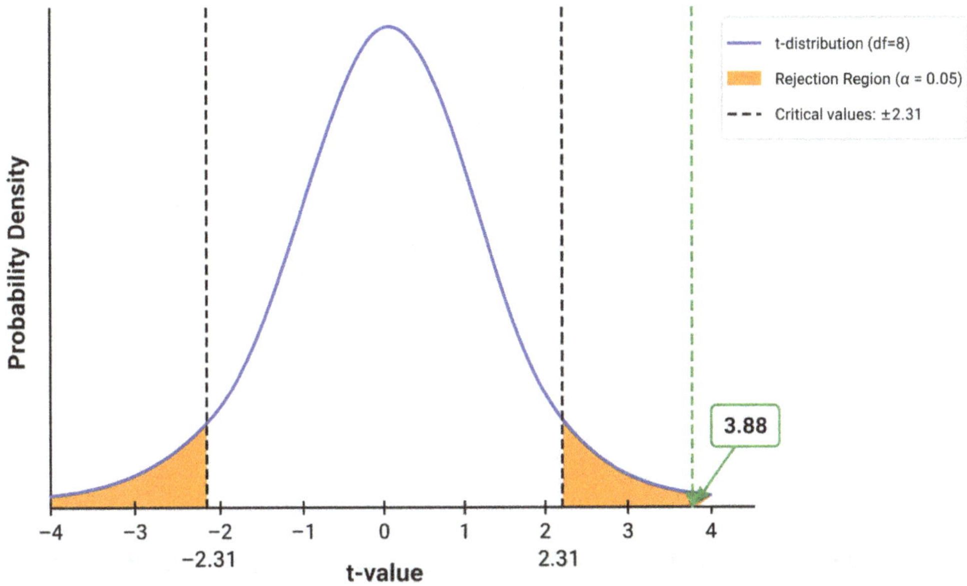

FIGURE 9.6 Observed Test Statistic Compared to Critical Values in a Two-Tailed Related-Sample *t* Test ($\alpha = 0.05$, $df = 8$)

EFFECT SIZE FOR RELATED-SAMPLES *t* TESTS

Once we have found a statistically significant difference, it is useful to examine the practical significance as well. The formula for Cohen's *d* for this type of *t* test is:

$$d = \frac{M_D}{S_D}$$

That is, we take the mean difference and divide it by the standard deviation of the difference. The effect size for the example is calculated as:

$$d = \frac{M_D}{S_D} = \frac{2.33}{1.8} = 1.29$$

This means that on average, scores increased by 1.29 standard deviation units from pre and post, which represents a large effect size. Based on this statistical result and assuming no other confounding factors, we may conclude that the intervention had a meaningful impact on scores!

Here is how you would report this example using APA style. A repeated-measures *t* test was conducted to evaluate the effect of the intervention on scores using a random sample of nine students. Results indicated that, on average, students' scores changed by 2.33 points between pre- and posttests. The difference was statistically significant, $t(8) = 3.88$, $p < .01$. Cohen's $d = 1.29$, which is a large effect size, indicating substantial improvement in scores following the intervention.

There is one extra wrinkle to consider: *What if the expected difference in the population is not zero?* For example, K–12 students take standardized tests in math and reading multiple times each year, and their scores are generally expected to improve over time. Testing companies use population data to provide expected growth benchmarks for each grade level.

Suppose third graders are expected to see a 13-point increase in math scores from the beginning to the end of the school year. If you wanted to determine whether students in your school were experiencing growth that was different from the expected population growth, your null and alternative hypotheses would be as follows:

$$H_0: \mu_D = 13$$

$$H_A: \mu_D \neq 13$$

COMPARING TWO GROUPS WITH ORDINAL DATA: NONPARAMETRIC TESTS

In hypothesis testing, parametric and nonparametric tests differ primarily in their assumptions about the data. In general, parametric tests, such as t tests, are distinguished by the assumption of normality on the data with a continuous dependent variable. Parametric tests typically use the raw data directly in their calculations and require certain conditions, such as homogeneity of variance and either interval or ratio-level measurement of the dependent variable.

In contrast, nonparametric tests are distinguished by the absence of any distributional requirement for the data and are less restrictive in terms of the scales of measurement. These tests are more flexible and can be used with ordinal data or when assumptions for parametric tests are violated. Instead of analyzing raw values, nonparametric methods often involve rank-based transformations of the data before analysis, making them less sensitive to outliers and skewed distributions.

A difference between the two approaches is that parametric tests allow for inferences about population parameters, such as means and standard deviations. However, nonparametric tests do not estimate specific population parameters; they instead assess whether two or more groups share the same distribution. The primary advantage of nonparametric tests is their robustness and fewer assumptions, making them useful alternatives to t tests when data are not normally distributed, have small sample sizes, or contain outliers.

CHOOSING BETWEEN PARAMETRIC AND NONPARAMETRIC TESTS

In general, we prefer to use the test that is the most efficient compared to others and produces the largest statistical power among all alternatives. For example, consider the two independent-samples t test. If the scores are independent and homoscedastic and follow a normal distribution, this test is optimal in that there is no other test with greater statistical power for a given alternative. On the other hand, failure to satisfy the normality assumption means that the t test may not be optimal. In particular, the power of nonparametric competitors often exceeds that of the t test when the population distribution is symmetric but is flatter (or *heavy-tailed*) when the distribution is highly skewed or if the distribution is multimodal and heavy-tailed.

If data assumptions of parametric tests are not adequately satisfied, the actual α of a test (i.e., the probability of having a Type I error) may differ substantially from the user-specified α level (e.g., 0.05), although we never know exactly how much the probability differs. Especially when sample size exceeds 30, there is evidence that many parametric tests, especially those comparing means, maintain control over Type I error rates even when the dependent variable is not normally distributed. If the distribution is skewed and/or multimodal and heavy-tailed, nonparametric tests frequently do a better job of controlling for Type I error rates.

In education research, nonparametric tests are particularly useful when the outcome variable includes responses on a Likert-type rating scale because the distribution is often skewed. You may come across research that uses parametric tests, such as a two independent-samples *t* test, to compare responses on a single Likert scale item between two groups. However, this practice is not generally recommended. Additionally, as we have discussed, outliers can contribute a great deal of error variance in parametric tests, so a nonparametric alternative potentially is more powerful.

Table 9.3 shows parametric tests you are already familiar with and nonparametric alternatives. Below we will describe the Mann-Whitney *U* test and the Wilcoxon signed rank *T* test in more detail.

TABLE 9.3 *Nonparametric Alternatives to Common Parametric Statistical Tests*

PARAMETRIC	NONPARAMETRIC
One-sample *t* test	Sign test
Independent-samples *t* test	Mann-Whitney *U* Test
Related-samples *t* test	Wilcoxon signed rank *T* test
One-way between-subjects ANOVA	Kruskal-Wallis *H* test
One-way repeated measures ANOVA	Friedman test

TWO INDEPENDENT-SAMPLE MANN-WHITNEY *U* TEST

The **Mann-Whitney *U* test** is a statistical procedure used to determine whether the dispersion of ranks is equal across two independent groups. It is used as a nonparametric alternative to the two independent-samples *t* test. The test has three assumptions:

- Both samples are randomly drawn from their respective populations.
- In addition to independence within each sample, there is mutual independence between the two samples (no one belongs to both groups).
- The measurement scale is at least ordinal (no categorical variables).

Let's walk through an example. Suppose that we give nine children a math achievement test and record their scores. They then receive daily practice on arithmetic problems. At the end of each practice session their papers are marked by a teacher's aide, who praises them for the number they got right and states that they might do even better in the next few days. We treat eight other children the same except for the statements made by the teacher's aide. For these children, they point out the problems they missed and tell them that they should be able to do better than that. At the end of two weeks, we retest all children on a parallel form of the math achievement test and calculate their "gain scores" over the two-week period (Table 9.4).

TABLE 9.4 GAIN SCORES BY FEEDBACK TYPE (CRITICISM VS. PRAISE)

Criticism (X)	-3	-2	0	0	2	5	7	9	
Praise (Y)	0	2	3	4	10	12	14	19	21

H_0: The two groups have identical distributions.
H_A: The two groups do not have identical distributions.

These data are measured on a continuous scale, so theoretically we could use a two independent-samples *t* test, but first we should look at the descriptive statistics to see if our data meet the assumptions of that method. You can look at descriptive statistics for each group, but we think that the most helpful way to consider the data is by viewing histograms for each group in Figure 9.7.

Given the very small sample sizes (eight and nine cases for criticism and praise, respectively) and two distributions not looking normal, we should not use the independent-samples *t* test. Since this is a very small dataset, let's calculate the test statistic by hand to illustrate the logic behind the test. In practice, statistical software can perform this analysis.

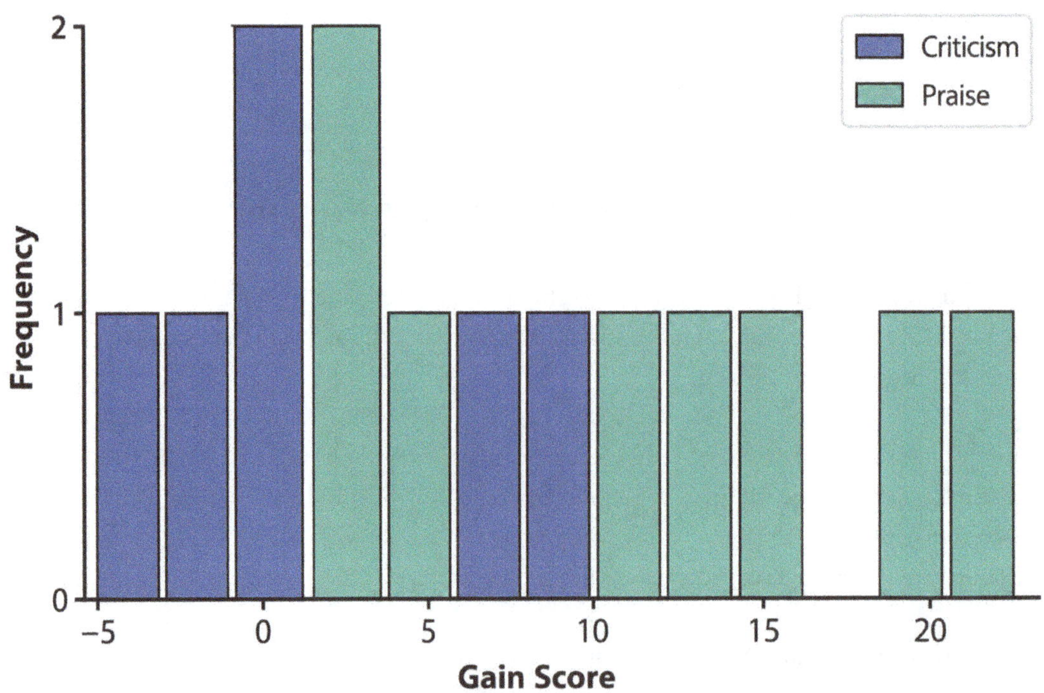

FIGURE 9.7 Distribution of Gain Scores by Feedback Type (Criticism vs. Praise)

Step 1: Label the two groups X and Y. The smaller group should be labeled X; if sample sizes are equal then it doesn't matter which group is labeled X. For this example, X is the criticism condition and Y is the praise condition.

Step 2: Combine all the scores into one distribution of $n_X + n_Y$ cases. Arrange them in order of the dependent variable from lowest to highest and assign ranks, as displayed in Table 9.5.

TABLE 9.5 *Assigned Rank of Gain Scores*

CONDITION	GAIN SCORES	ASSIGNED RANK
X	-3	1
X	-2	2
X	0	4
X	0	4
Y	0	4
X	2	6.5
Y	2	6.5
Y	3	8
Y	4	9
X	5	10
X	7	11
X	9	12
Y	10	13
Y	12	14
Y	14	15
Y	19	16
Y	21	17

When assigning ranks in a dataset, ties occur when two or more scores are identical. In such cases, instead of giving each tied score a unique rank, we calculate the average rank they would have occupied.

For example, in Table 9.5, three students had gain scores of 0. These scores would have been ranked 3, 4, and 5 if assigned individually. Instead, we sum these ranks (3 + 4 + 5 = 12) and divide by 3, giving each student a rank of *4*. The next score is then assigned the rank that follows the tied ranks, which in this case is *6.5* (the average of the next two ranks, 6 and 7).

Step 3: Divide the cases back into their original groups as displayed in Table 9.6.

Step 4: Sum up the ranks for each group.

Sum of ranks for Group X: ΣR_X =1+2+4+4+6.5+10+11+12=50.5

Sum of ranks for Group Y: ΣR_Y =4+6.5+8+9+13+14+15+16+17=102.5

Step 5: Use the summed ranks to calculate the test statistic (i.e., Mann-Whitney *U* Statistic).

$$U_X = n_X n_Y + \frac{n_X(n_X + 1)}{2} - \Sigma R_X = 8 * 9 + \frac{8(9)}{2} - 50.5 = 57.5$$

$$U_Y = n_X n_Y + \frac{n_Y(n_Y + 1)}{2} - \Sigma R_Y = 8 * 9 + \frac{9(10)}{2} - 102.5 = 14.5$$

TABLE 9.6 *Assigned Ranks for Criticism and Praise*

CRITICISM (X)	ASSIGNED RANK	PRAISE (Y)	ASSIGNED RANK
-3	1	0	4
-2	2	2	6.5
0	4	3	8
0	4	4	9
2	6.5	10	13
5	10	12	14
7	11	14	15
9	12	19	16
		21	17

where:

- n_X and n_Y are the sample sizes of groups X and Y, respectively.
- R_X and R_Y are the rank sums for each group.
- U is the smaller of U_X and U_Y.

Step 6: Evaluate the test statistic using Table A7 in the appendix.

We use the value of $U = 14.5$ as the test statistic because it is the smaller of the two U values calculated.

Next, we determine whether we are conducting a one-tailed or two-tailed test. In this case our test is nondirectional (two-tailed), so we divide our alpha level of 0.05 equally between the two tails, resulting in 0.025 in each tail.

We locate the critical value by finding the number of cases in each group along the left side and the top of the table. For group sizes of 8 and 9, the critical value is 15. Since our test statistic ($U = 14.5$) is less than or equal to the critical value (15), we reject the null hypothesis that the two distributions are the same.

Note that unlike many parametric tests where we compare the test statistic to see if it exceeds the critical value, in the Mann-Whitney U test, we reject the null hypothesis when the test statistic is less than or equal to the critical value.

With this result, we conclude that the distributions of the two groups are statistically different, indicating that the scores between the criticism and praise groups differ significantly, $U = 14.5, p = 0.043$.

WILCOXON'S MATCHED-PAIRS SIGNED-RANKS TEST

While the Mann-Whitney U test is used to compare two independent groups, **Wilcoxon's matched-pairs signed-ranks test** is a nonparametric alternative to the related-samples (or repeated-measures) t test.

Let's walk through another example to help you learn more about this method. Suppose we are studying the effect of a new instructional strategy on math performance, specifically on the speed and accuracy of solving arithmetic problems. We form 10 pairs of students, matched according to their pretest score earned on similar math problems, and then randomly assign members of each pair to either the new instructional strategy or the traditional teaching method (control group). After completing a structured learning session, the students take a posttest assessing their arithmetic skills.

The students' posttest scores were recorded in Table 9.7 as *traditional* (control) and *new strategy* (treatment). Although different students are assigned to the two groups, the pairing process creates dependence between their scores, as students were first matched by pretest ability before assignment.

As we are interested in any difference (positive or negative) due to the instructional strategy, we use a two-tailed test. The null and alternative hypotheses for this investigation are as follows:

H_0: There is no difference in student performance between the two instructional groups.

H_A: There is a difference in student performance between the two instructional groups.

Rather than create new tables for each of the steps required to calculate the test statistic by hand, we have added a row to Table 9.7 showing which columns belong to which step in the process.

Step 1: Recode the paired scores in two columns.

Step 2: Obtain the difference between members of each pair (X-Y).

Step 3: Disregard the sign of the difference obtained and then assign ranks to the absolute magnitude of the difference. This means that a difference of -8 would be a higher rank than a difference of 5. If any difference is zero, disregard it in subsequent calculations. For tied ranks, assign the midrank as we did in the prior test.

Step 4: Reassign the appropriate sign to the rank. This is a weird step, but don't take it literally; for example, -2 does not mean the rank of -2 (negative 2nd). It simply provides two types of information about the score at once: the rank of the score and the direction of the difference.

Step 5: Compute the summed rank for both positive and negative differences. The smaller absolute value summation becomes our test statistic (6.5).

TABLE 9.7 *Summary of Steps for Calculating the Wilcoxon Signed-Rank Test Statistic*

	STEPS							
	1		2	3		4	5	
PAIR	TRADITIONAL (X)	NEW STRATEGY (Y)	DIFFERENCE (X-Y)	ABSOLUTE DIFFERENCE	RANK OF ABSOLUTE DIFFERENCE	SIGNED RANK OF X-Y	POSITIVE DIFFERENCE	NEGATIVE DIFFERENCE
1	24	28	-4	4	2	-2		-2
2	39	29	10	10	6.5	6.5	6.5	
3	29	34	-5	5	3.5	-3.5		-3.5
4	28	21	7	7	5	5	5	
5	25	28	-3	3	1	-1		-1
6	32	15	17	17	10	10	10	
7	31	17	14	14	8	8	8	
8	33	28	5	5	3.5	3.5	3.5	
9	31	16	15	15	9	9	9	
10	22	12	10	10	6.5	6.5	6.5	
Sum				**55**			**48.5**	**-6.5**

Note: Each row corresponds to a paired observation. Columns indicate values at each step of the calculation process.

We then use Table A8 to find the critical value for evaluating the test statistic. Since $n = 10$ and we are conducting a two-sided test with an alpha level of .05, the critical value is 8. Our test statistic is 6.5, and similar to the Mann-Whitney *U* test, we reject the null hypothesis if the test statistic is less than or equal to the critical value. In this case, since $6.5 \leq 8$, we reject the null hypothesis.

Alternatively, statistical software can be used to obtain the associated *p* value for the test statistic. In this case, the computed *p* value = 0.037, which is less than 0.05, further supporting the rejection of the null hypothesis. Therefore, we conclude that there is a statistically significant difference in scores between the two instructional strategies. This suggests that the new instructional strategy has an effect. However, it is important to note that the test itself only determines whether a difference exists; it does not indicate the direction of the effect (i.e., whether the new strategy increased or decreased scores).

CONCLUSION

In this chapter, we expanded our understanding of hypothesis testing by exploring various methods for comparing two groups. Central topics included the distinctions among the three types of *t* tests—one-sample, independent-samples, and related-samples tests—and the assumptions underpinning these methods, such as normality, independence, and equal variances. Additionally, the chapter emphasized the importance of effect size measures, such as Cohen's *d* and eta squared (η^2), to evaluate the practical significance of findings beyond mere statistical significance.

The chapter also introduced nonparametric alternatives for scenarios where data do not meet parametric test assumptions. The Mann-Whitney *U* test, a substitute for the independent-samples *t* test, evaluates whether two independent groups have equal distributions, making it ideal for ordinal data or when assumptions of normality and equal variance are violated. Similarly, Wilcoxon's matched-pairs signed-ranks test, an alternative to the related-samples *t* test, is used for paired data to test whether the median differences deviate significantly from zero. These methods are particularly useful for addressing skewed or ordinal data and ensuring robust conclusions when parametric assumptions are not satisfied.

By incorporating both parametric and nonparametric techniques, this chapter equips researchers with a versatile tool kit for hypothesis testing, enabling rigorous analyses of group differences under varied data conditions. These statistical foundations pave the way for more advanced methodologies in subsequent chapters, reinforcing the importance of selecting appropriate tests based on data characteristics and research questions.

KEY TAKEAWAYS FOR EDUCATIONAL RESEARCHERS FROM CHAPTER 9

- Choosing the right *t* test matters. The one-sample *t* test compares a group's mean to a known population mean (e.g., comparing students' reading scores to the national average). The independent-samples *t* test compares means from two separate groups (e.g., evaluating whether students in a new math curriculum perform better than those in a traditional program). The related-samples *t* test (paired *t* test) compares the same group at two time points or under different conditions (e.g., pre- and postintervention reading scores).
- Checking assumptions before running a *t* test is crucial. Normality, independence, and equal variances should be checked before running parametric tests using histograms, *Q-Q* plots, and Levene's test. For example, if two

third grade classrooms have vastly different test score variances, an adjusted *t* test or a nonparametric alternative should be used.

- Interpreting statistical versus practical significance is important. A statistically significant result ($p < 0.05$) does not always mean that the difference is meaningful. Effect sizes such as Cohen's *d* help determine whether the difference is practically important (e.g., does a statistically significant 2-point improvement on a 100-point test really matter in practice?).
- Nonparametric tests are useful for skewed or ordinal data. When working with a single Likert-scale item on a scale or survey (e.g., student engagement ratings: strongly agree to strongly disagree), normality assumptions may not hold. The Mann-Whitney *U* test (for independent groups) and the Wilcoxon matched-pairs signed-ranks test (for paired data) may be preferred in such cases. If assumptions for an independent *t* test are violated, the Mann-Whitney *U* test compares whether two groups have different distributions rather than just mean differences (e.g., comparing test scores between students who received praise versus criticism in a classroom setting). When data are dependent (e.g., before vs. after an intervention) but not normally distributed, the Wilcoxon matched-pairs signed-ranks test is an alternative to the related-samples *t* test (e.g., tracking students' math problem-solving speed before and after using a new instructional strategy).
- Statistical software helps, but understanding the underlying logic is key. Software such as IBM SPSS, R, and Python can quickly compute test statistics and *p* values, but researchers should understand the underlying logic of hypothesis testing. For example, before running an independent-samples *t* test to compare student grade point averages in two different school districts, a researcher should check for normality and equal variances before interpreting results. Also, it is important to interpret statistical results within the context of the study.

KEY DEFINITIONS FROM CHAPTER 9

Distribution of sample differences refers to the theoretical distribution that arises when calculating the differences between sample means (in independent-samples *t* tests) or paired scores (in related-samples *t* tests) repeatedly under the same sampling conditions.

Equal variance refers to the assumption that the variance (the measure of variability or spread) in different groups being compared is approximately the same.

Eta squared (η^2) is a measure of effect size used to quantify the proportion of variance in a dependent variable that is associated with one or more independent variables in an analysis.

An **independent-samples *t* test** is a statistical method used to compare the means of two independent groups to determine whether there is a statistically significant difference between them.

Levene's test is a statistical procedure used to assess whether the variances of a dependent variable are equal across different groups. This test is commonly applied before conducting analyses that assume equal variances, such as *t* tests and ANOVA, to determine if this assumption holds.

The **Mann-Whitney *U* test** is a statistical procedure used to determine whether the dispersion of ranks is equal across two independent groups. This test is used as a nonparametric alternative to the two-independent-sample *t* test.

The **paired-samples *t* test** is used to pair individuals on similar key variables.

The **related-samples *t* test** typically means that you have the same participants measured on two dependent variables that use the same scale, such as math versus reading tests, or under a control condition versus under a treatment condition.

The **repeated-measures *t* test** is used to test one group of participants twice on the same measure at different points in time.

Wilcoxon's matched-pairs signed-ranks test is a nonparametric alternative to the related-samples (or repeated-measures) *t* test.

CHECK YOUR UNDERSTANDING

1. What is the main difference between an independent-samples *t* test and a related-samples *t* test?
   a. Independent-samples *t* tests compare two separate groups, while related-samples *t* tests compare two sets of scores from the same group or matched individuals.
   b. Independent-samples *t* tests compare means to a population value, while related-samples *t* tests compare variances.
   c. Independent-samples *t* tests require categorical data, while related-samples *t* tests require continuous data.
   d. There is no difference; they are used interchangeably.

2. Which assumption is unique to independent-samples *t* tests but not required for related-samples *t* tests?
   a. Normality
   b. Random sampling
   c. Equal variances
   d. Independence of scores within groups

3. When would you use a one-sample *t* test instead of an independent- or related-samples *t* test?
   a. When you have two independent groups to compare
   b. When you want to compare a single group to a known population mean
   c. When you compare pre- and posttest scores from the same group
   d. When you have categorical data

4. You are conducting a study to compare the effectiveness of two teaching methods on student performance. You have a small sample size of just 16 students (two groups of 8 students taught by different teachers in different classrooms, and there is no reason to suggest they are related), and the scores in your dataset are not normally distributed. Which test should you use to compare these scores?
   a. Mann-Whitney *U* test
   b. Wilcoxon matched-pairs signed-ranks test
   c. Independent-samples *t* test
   d. One-sample *t* test

5. Which effect size measure is commonly reported for independent-samples *t* tests?
   a. r^2
   b. Phi coefficient
   c. Cohen's *d*
   d. Omega

10

ANALYZING DIFFERENCES AMONG MORE THAN TWO GROUPS

LEARNING OBJECTIVES

By the end of this chapter, you will be able to:

- Differentiate between one-way ANOVA and factorial ANOVA by identifying appropriate scenarios for each test based on the number of independent variables and their levels;
- Describe the underlying assumptions of one-way ANOVA and factorial ANOVA, including normality, homogeneity of variances, and independence;
- Interpret results from one-way ANOVA and factorial ANOVA to determine group differences, examine interaction effects, and assess the impact of covariates on dependent variables, ensuring accurate statistical conclusions; and
- Use nonparametric tests such as the Kruskal-Wallis test for data that violates parametric assumptions.

One-way analysis of variance (ANOVA) is a statistical method used to compare the means of three or more groups to determine if there are statistically significant differences among them. Before we jump in, let's briefly revisit the independent-samples *t* test, because a one-way ANOVA is really just an extension of the *t* test. You could also say that an independent-samples *t* test is a special case of ANOVA comparing only two groups. In a *t* test, we compare means between two groups. In a one-way ANOVA, we compare means among three or more groups. Maybe you don't just want to compare high school students to college students, but instead you want to compare students from several different age groups. Sure, you could use *t* tests to look for differences between two groups at a time, but that is not a good idea. First of all, it is not efficient because comparing five groups would mean running 10 *t* tests. But even more importantly, you would increase the risk of making a Type I error, and we can avoid that risk by using an ANOVA.

How does running multiple *t* tests increase the odds of a Type I error? **Experiment-wise alpha** (or **family-wise alpha**) refers to the overall probability of making at least one Type I error (false positive) across all hypothesis tests in an experiment. This is particularly relevant in situations where multiple comparisons are made, as the risk of incorrectly rejecting a null hypothesis increases with the number of tests. In other words, when we use an alpha level of .05, we are saying there is a 5% probability that an observed significant difference was due to chance alone. If we run 10 tests on the same data and there is a 5% chance of a Type I error each time, those odds really add up. Using a single ANOVA to compare means across multiple groups at once is better than conducting multiple *t* tests because we are controlling for any compounding of the Type I error rate.

How does an ANOVA work? There is a clue in the name: analysis of variance. With ANOVA, we are comparing the variance in a dependent variable that is due to differences between groups to the variance due to differences within each group. There will always be some differences within groups because people are different in ways unrelated to the groups they are in. The key with an ANOVA is to find out how much of the total difference (or variance) in the dependent variable is due to people being in different groups, as opposed to people just being different from each other. We will go through an example to explain, but this video provides a good visualization of the concepts. (Link: https://www.youtube.com/watch?v=ITf4vHhyGpc.)

Let's now go through the steps of hypothesis testing for a one-way ANOVA using a hypothetical example to illustrate the process.

A researcher conducted a study for language acquisition to examine how different study techniques impact students' performance to recall new vocabulary words. A total of 40 college students were randomly assigned to one of four learning strategies:

- *Repetition Group.* Students repeatedly read through a list of 25 vocabulary words.
- *Visualization Group.* Students created a mental image related to each vocabulary word to aid recall.
- *Technology-Aided Learning Group.* Students used an interactive flashcard app that provided spaced repetition and automated quizzes to reinforce learning.
- *Control Group.* Students were instructed to read through the list without applying any specific memorization strategy.

After studying for 15 minutes, students completed a *recall test*, writing down as many words as they could remember. Imagine that the results were as shown in Table 10.1.

TABLE 10.1 ***Recall Test Results Across Learning Strategies***

REPETITION	VISUALIZATION	TECHNOLOGY-AIDED LEARNING	CONTROL
13	17	17	10
12	15	18	12
12	15	16	9
11	17	16	12
9	18	20	4
11	18	21	11
11	14	18	9
14	15	20	8
13	17	19	9
8	18	17	5
$\bar{X}$ = 11.4	$\bar{X}$ = 16.4	$\bar{X}$ = 18.2	$\bar{X}$ = 8.9

In each group the numbers are clustered very close together, and there are clear differences between groups. It is easy to glance at Table 10.1 and decide that the differences are due to group membership. Now, imagine the results were as shown in Table 10.2.

TABLE 10.2 ***Updated Recall Test Results Across Learning Strategies***

REPETITION	VISUALIZATION	TECHNOLOGY-AIDED LEARNING	CONTROL
14	11	19	13
9	16	22	13
16	13	13	10
8	17	19	13
14	13	19	13
19	21	20	10
9	14	14	18
10	15	14	13
12	18	20	8
10	12	19	14
$\bar{X} = 12.1$	$\bar{X} = 15$	$\bar{X} = 17.9$	$\bar{X} = 12.5$

There are still some differences between groups, but the differences within groups are so great that it is hard to tell what is going on. Is the gap between the *Visualization Group* and the *Control Group* due to the learning condition, or did the Visualization Group just randomly have a few more individuals who are unusually good at remembering words? In this case, the within-group variation is so high that it obscures any between-group variation.

Figure 10.1 provides a visual that may help. Imagine we have three groups, X_1, X_2, and X_3. Each group has its own distribution of scores on the dependent variable, with its own mean and variance/standard deviation. If individual differences within the groups are large, then it is hard to make a conclusion about between-group differences by comparing means because the group distributions may overlap.

Since there is quite a bit of overlap across the three distributions (large spread), mean differences may not be statistically significant. So, how do we know if there are significant group differences?

We are going to keep this very general to start with and then go through the calculations with the actual data from the recall study. An ANOVA is really a ratio of two types of variances. The variance between groups is in the numerator, and the variance within groups (otherwise known as the error variance) is in the denominator. As more and more of the total variance is associated with the groups, the ratio gets larger. This ratio is called the F statistic and can be calculated by the following formula:

$$F = \frac{MS_{BG}}{MS_E} = \frac{\textit{variance between groups}}{\textit{variance within groups}}$$

Unlike a t test, which uses a t statistic and follows the t distribution, an ANOVA uses the F statistic and follows the F distribution. A t test compares means, which can be positive or negative. However, variance is either

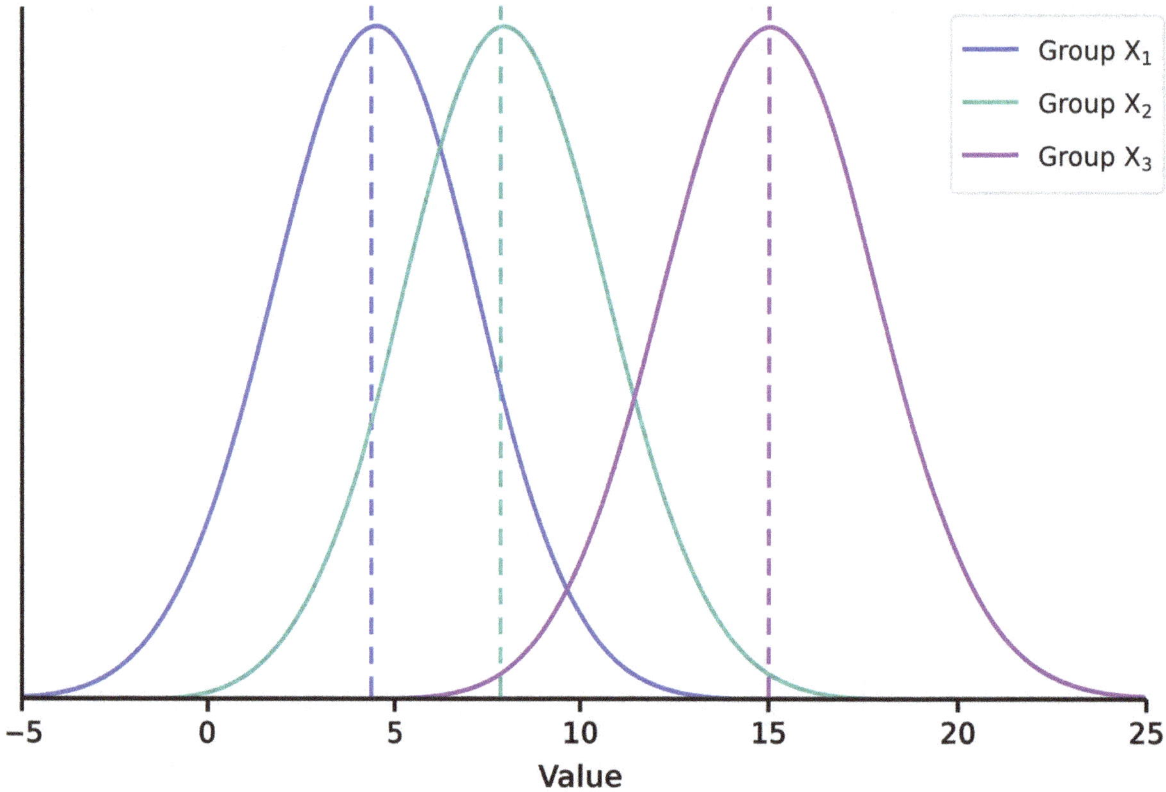

FIGURE 10.1 Within-Group Variability and Its Impact on Between-Group Differences

positive or zero, so the *F* distribution looks quite different than the *t* distribution. And actually, there are multiple *F* distributions. Just like the *t* distribution changes with the degrees of freedom (higher degrees of freedom means the distribution becomes closer and closer to normal), the *F* distribution also changes based on degrees of freedom. This time, however, we have two separate degrees of freedom. Some examples are shown in Figure 10.2.

The degrees of freedom for the numerator (between groups) are the number of groups (k) minus 1. The degrees of freedom for the denominator (within groups, or error variance) are the total sample size (N) minus the number of groups (k). Just like with a *t* test, we use the degrees of freedom to help identify the critical value and then decide if the *F* statistic obtained by our data exceeds that value.

Turn to Table A2 in the appendix, and you will find the critical values for the *F* distribution. Note that you must find the degrees of freedom for the numerator across the top and degrees of freedom for the denominator down the left side. You will see that even when your sample size is very large, which results in high degrees of freedom in the denominator, critical values are still above 1.0. This means that to get a significant *F* statistic, the variance between groups needs to be greater than the variance within groups.

Let's get back to the different sources of variance in the dependent variable. In Figure 10.3, we have individual data points for three groups of four individuals each. There are horizontal lines representing each of the group means as well as the grand mean for the entire dataset. We are going to focus on one case—person one (X_{31}) in Group 3 to explain the different types of variance.

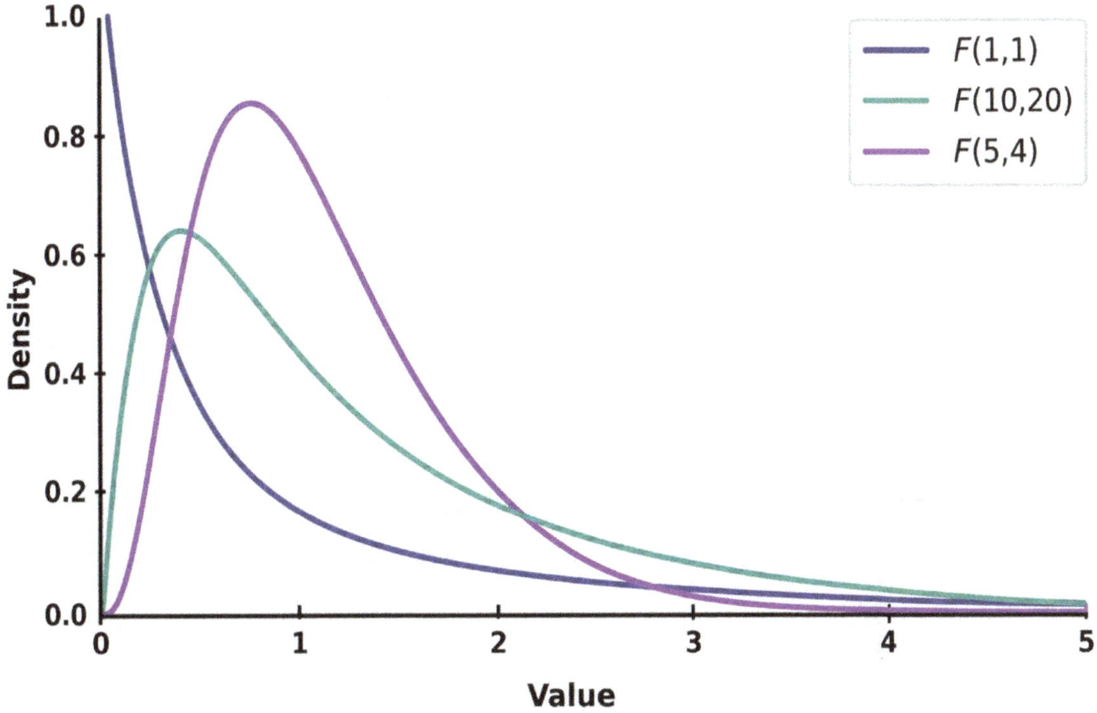

FIGURE 10.2 Examples of *F* Distributions with Varying Degrees of Freedom

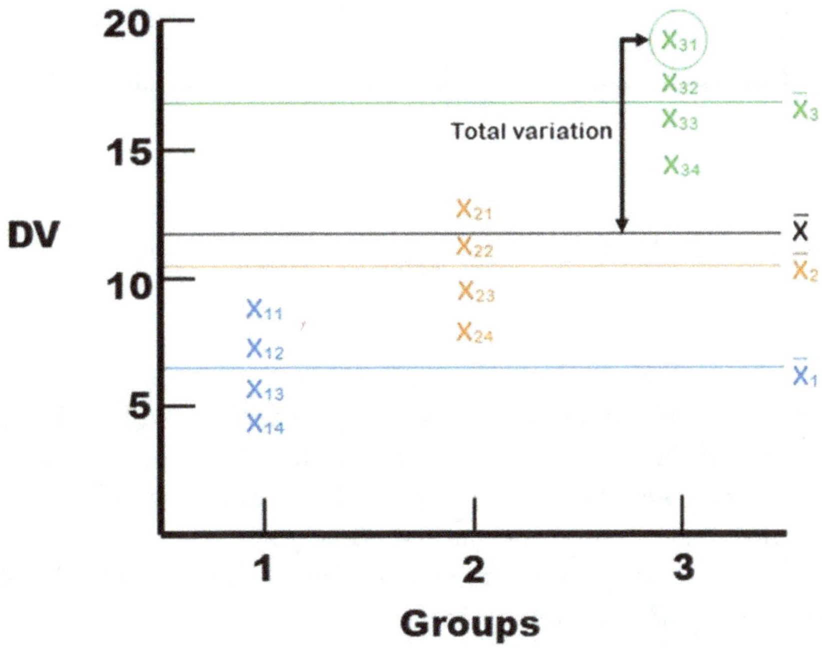

FIGURE 10.3 Total Variation in ANOVA: Individual Score Relative to the Grand Mean

The first thing we see is the distance between individual X_{31}'s score on the dependent variable and the overall average. This is the distance X_{31} contributes to the *total variation*. Imagine squaring that distance and adding it up with the squared distances between the overall mean and every other individual data point, and you have the sum of squares (total).

Next, we can see the distance between individual X_{31}'s score on the dependent variable and the mean for Group 3 in Figure 10.4. This is the distance X_{31} contributes to the *variation within groups*, or the *error variance*. **Error variance** in ANOVA refers to the variation in the dependent variable that cannot be explained by the independent variable(s) and represents the variability within groups or conditions, which arises from random differences among individual observations rather than systematic differences between groups. We square that distance for each observation in each group and add them up to get the sum of squares (error).

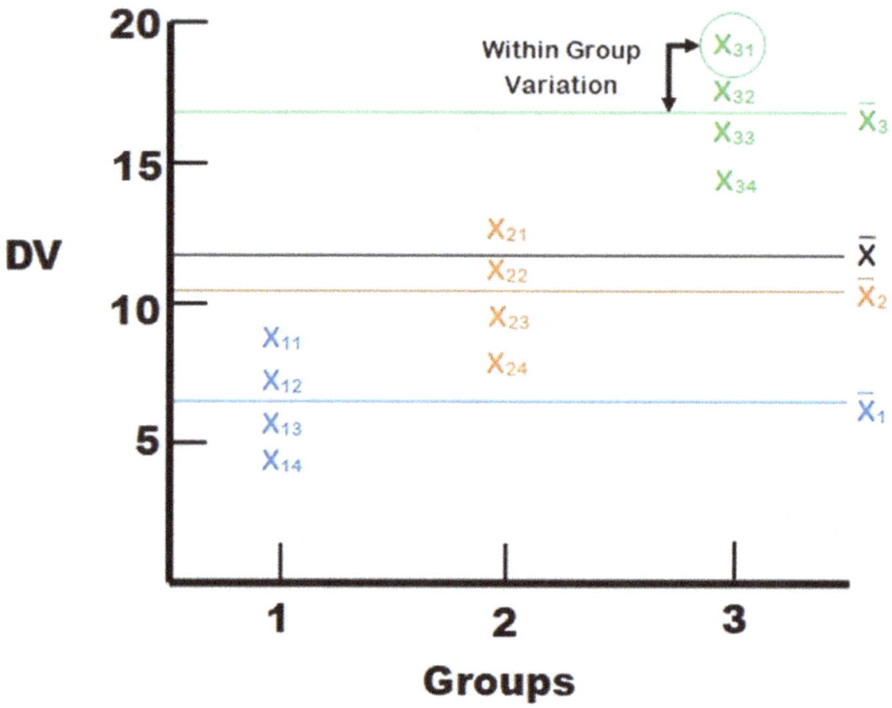

FIGURE 10.4 Within-Group Variation in ANOVA: Individual Score Relative to the Group Mean

Finally, we have the distance between the mean of Group 3 and the overall mean in Figure 10.5. This is the distance that all members of Group 3 contribute to the *between-group variance*. We have three groups, so we have three group differences from the overall mean. To find the treatment effect, we first multiply each group mean difference by the number of observations in the group (in this case, each group mean by 4) and then sum the three group mean differences to get the sum of squares (treatment).

Okay, now we are going to dive into the actual numbers. But before you get overwhelmed by all the *SS*s and *MS*s, remember that a mean square (MS) is just a variance.

Let's work through the example with different learning strategies. Table 10.3 gives the results for each instructional group.

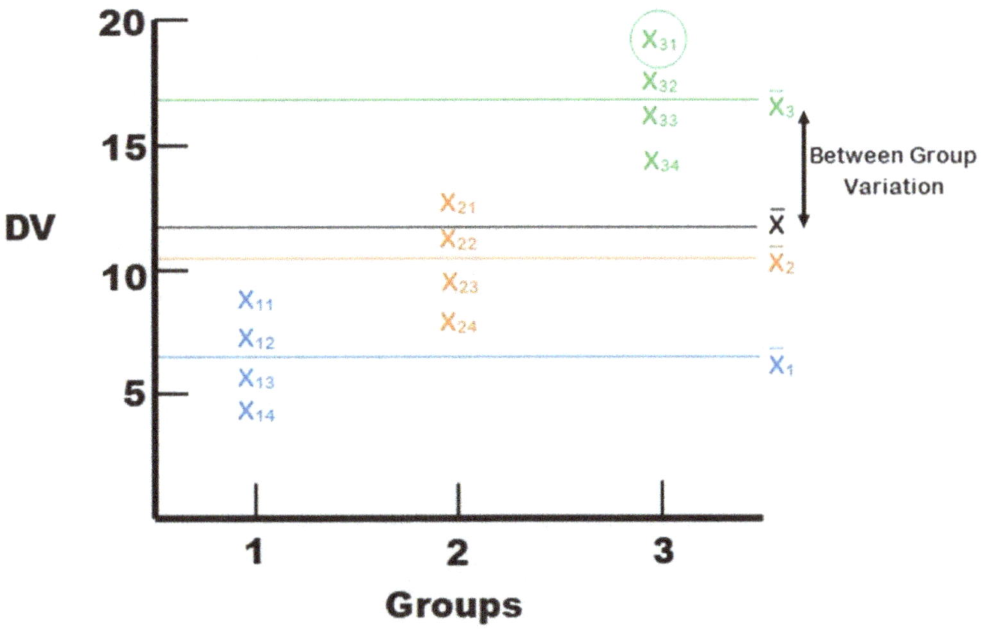

FIGURE 10.5 Between-Group Variation in ANOVA: Group Mean Relative to the Grant Mean

TABLE 10.3 ***Recall Test Results Across Learning Strategies***

REPETITION	VISUALIZATION	TECHNOLOGY-AIDED LEARNING	CONTROL
13	17	17	10
12	15	18	12
12	15	16	9
11	17	16	12
9	18	20	4
11	18	21	11
11	14	18	9
14	15	20	8
13	17	19	9
8	18	17	5
$\bar{X}$ = 11.4	$\bar{X}$ = 16.4	$\bar{X}$ = 18.2	$\bar{X}$ = 8.9

STEP 1: DETERMINE THE NULL AND ALTERNATIVE HYPOTHESES

The first step is to state our hypotheses. The hypotheses for a one-way ANOVA are as follows:
For *k* groups,

H_0: $\mu_1 = \mu_2 = \ldots = \mu_k$ (all group means are equal)
H_a = *not all the* μ_i *are equal (at least one group means differ from the others)*

It is important to note that the alternative hypothesis does NOT imply that all group means are different. Rather, rejecting the null hypothesis indicates that at least one group mean is significantly different from the rest. This could mean that all means are different or that only one differs from the others. In our example k = 4, and thus:

H_0: $\mu_1 = \mu_2 = \mu_3 = \mu_4$

Or

H_0: $\mu_{Repetition} = \mu_{Visual} = \mu_{TechAid} = \mu_{Control}$

H_A = *not all the* μ_i *are equal* (at least one group means differ from the others.)

STEP 2: SET THE CRITERIA FOR A DECISION

Let's set $\alpha = 0.05$ as our criteria for decision as usual. Remember, ANOVA uses the F distribution, which is defined by two separate degrees of freedom for the sampling distribution of the test statistic, $F = \frac{MS_{BG}}{MS_E}$. The degrees of freedom for between-group variation are $df_{BG} = k - 1$. The degrees of freedom for within-group (error) variation are $df_E = N - k$. In this example, $k = 4$ and $N = 40$. Thus, the sampling distribution of F follows an F distribution with $(df_{BG}, df_E) = (3,36)$. Figure 10.6 shows the F distribution for this example. We reject the null hypothesis if our observed F value is greater than the critical F value, which in this case is 2.87.

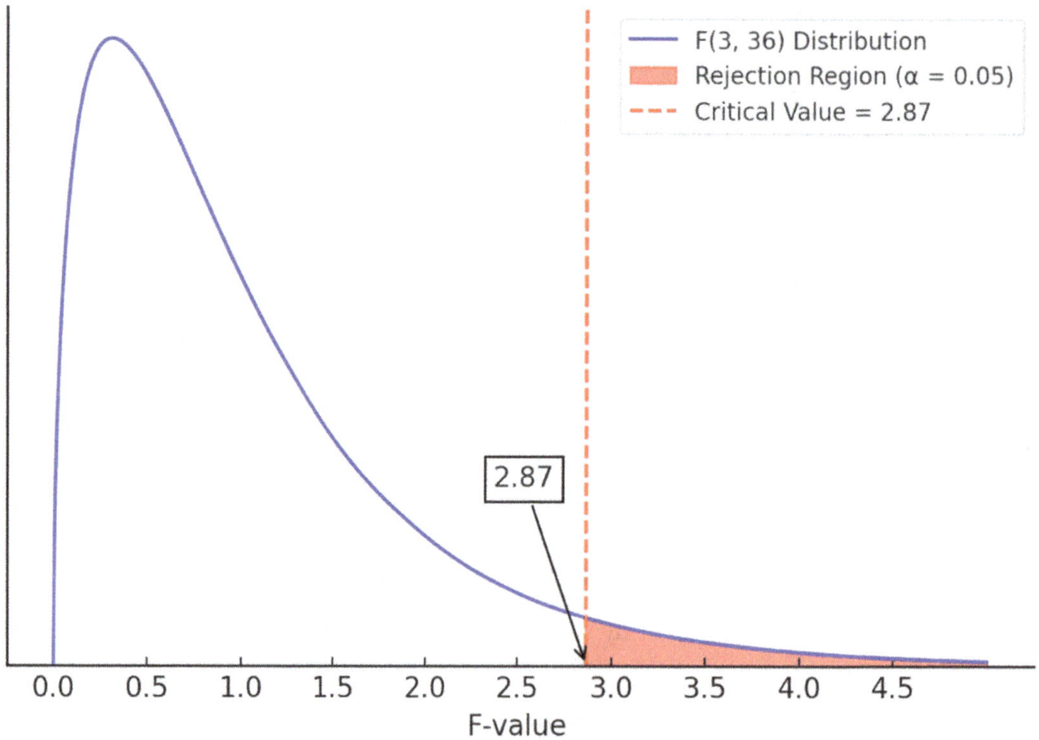

FIGURE 10.6 F Distribution with Critical Value for $\alpha = 0.05$ and $df = (3,36)$

STEP 3: CHECK ASSUMPTIONS AND COMPUTE THE TEST STATISTIC

The next step is to ensure that your data meet the assumptions required for a one-way ANOVA. What are these assumptions?

- *Independence within groups*: The same participants shouldn't be in multiple groups.
- *Normality*: The dependent variable should be approximately normally distributed within each group.
- *Homogeneity of variances*: The variance of the dependent variable should be similar across all groups.
- The dependent variable (the variable of interest) needs to be a *continuous scale* (i.e., the data needs to be at either an interval or ratio measurement).

As we learned, it is important to explore data using both descriptive statistics and visual inspections with figures and graphs. Let's review the descriptive statistics (Table 10.4).

TABLE 10.4 ***Descriptive Statistics for Repetition, Visualization, Technology-Aided Learning, and Control***

LEARNING STRATEGY	$\bar{X}$	*SD*	SKEWNESS	KURTOSIS
Repetition	11.4	1.84	-0.61	-0.02
Visualization	16.4	1.51	-0.37	-1.59
Technology-Aided Learning	18.2	1.75	0.24	-1.23
Control	8.9	2.69	-0.77	-0.08

Although we need to be cautious about interpreting descriptive statistics due to a small sample size in each group, we observed that

- Skewness values range from -0.77 to 0.24, which is within the normality threshold (-1 to 1).
- Kurtosis values range from -1.59 to -0.02, indicating mild to moderate deviations from normality.
- The control group shows the largest deviation in skewness (-0.77) and a relatively higher standard deviation (2.69), suggesting possible nonnormality.

Again, acknowledging small sample size per group, our visual inspection with graphs in Figure 10.7 (histograms and *Q-Q* plots) indicates the following:

- *Repetition*: Slight left skew, mild deviation in *Q-Q* plot.
- *Visualization*: Close to normal, minor deviations.
- *Technology-aided learning*: Slight right skew, generally normal.
- *Control*: Noticeable skewness and deviation in *Q-Q* plot.

Overall, while some distributions—particularly for control—exhibit mild deviations from normality, none display severe nonnormality. Additionally, it is important to note that ANOVA is robust to mild violations of the normality assumption, allowing us to proceed with the analysis.

Now, what about the homogeneity of variances? Recall that we used Levene's test to assess the equality of variances. We can use the same statistical test for equality of variances for more than two groups. Before then,

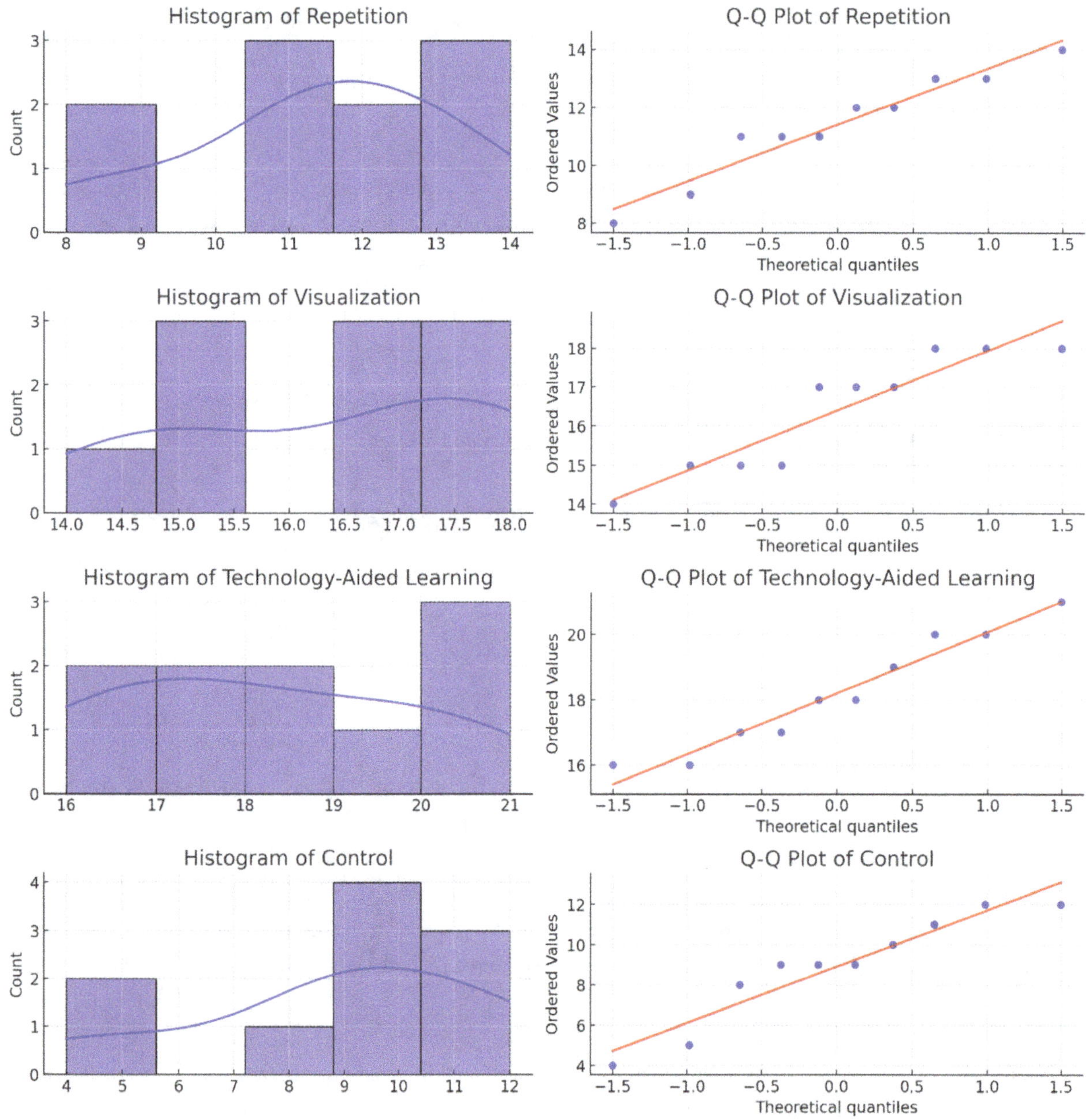

FIGURE 10.7 Histograms and *Q-Q* Plots

let's start with exploring the descriptive statistics one more time by comparing the size of standard deviations across groups. The control group has a slightly higher standard deviation than the others, which suggests possible violation of the homogeneity of variance assumption. To formally determine this, let's run Levene's test.

Remember that Levene's test is to assess whether the variances across groups (in this case four learning strategies) are equal:

H_0: The variances across all groups are equal (i.e., homogeneity of variance assumption holds).

H_A: At least one group has a different variance (homogeneity of variance is violated).

Levene's test can be run with statistical software such as SPSS. For the learning strategies data, the p value for Levene's test is 0.648, which is greater than 0.05. Thus, we fail to reject the null hypothesis, which means that the homogeneity of variance assumption is met. Now, we can proceed with a one-way ANOVA analysis.

To test our null hypothesis, we need to define and estimate the three sources of variation in ANOVA: total variation, error (within-group variation), and treatment (between-group variation). We are not going to walk you through the manual calculations. Instead, we will first present a standard output from statistical software (Table 10.5).

TABLE 10.5 ESTIMATIONS OF VARIANCE

SOURCE	SUM OF SQUARES (SS)	DEGREES OF FREEDOM (df)	MEAN SQUARES (MS)	F
Between groups	558.67	3	186.22	46.78
Within groups (error)	143.3	36	3.98	
Total	701.98	39		

Where are all those values coming from? Start with the first line. Between-groups sum of squares comes from the third type of variation we showed you earlier in the chapter—the difference between group means ($\bar{X}_j$) and the overall mean ($\bar{X}$). The difference for each group was multiplied by the sample size of the group, n_j, and then all summed up across groups.

$$SS_{BG} = \sum n_j (\bar{X}_j - \bar{X})^2 = 10(11.4 - 13.73)^2 + \ldots + 10(8.9 - 13.73)^2 = 558.675$$

Then we have df for the numerator, or df between groups, which is $k - 1$ or 3. Remember, MS = variance, so the mean square value is simply the sum of squares for between-groups (SS_{BG}) divided by the degrees of freedom (df_{BG}), because when we calculate variance we divide SS by df. With this, we now have the numerator for our F statistic.

$$MS_{BG} = \frac{\sum n_j (\bar{X}_j - \bar{X})^2}{k-1} = \frac{558.675}{3} = 186.225$$

Next, we have the line for within-group differences, or error variance. The sum of squares for within-group variance (SS_E) comes from the second type of variation we showed you—the difference between each individual person's score (X_{ij}) and their own group's mean score ($\bar{X}_j$).

$$SS_E = \sum (\bar{X}_{ij} - \bar{X}_j)^2 = 143.3$$

Then we have df for the denominator, or df within groups, which is $N - k$ or 36. The mean square value on this line is SS_E divided by df_E.

$$MS_E = \frac{\sum (\bar{X}_{ij} - \bar{X}_j)^2}{N-k} = \frac{143.3}{36} = 3.981$$

So now you have the numerator for our F statistic.

$$F = \frac{186.225}{3.981} = 46.78$$

STEP 4A: FIND THE *p* VALUE

Based on the decision criteria we set in Step 2, a critical value at α =0.05 is 2.87. Our obtained *F* statistic (F = 46.78) is much larger than this critical value, meaning the associated *p* value is smaller than 0.05. Therefore, we reject the null hypothesis and conclude that at least one group mean is significantly different from the others. The actual *p* value is 1.67×10^{-12}; in such a small *p* value, we report it as $p < .001$.

STEP 4B: DRAW A CONCLUSION AND REPORT YOUR CONCLUSION

As we reject the null hypothesis, we conclude that the average recall performance of college students significantly differs for at least one of the learning strategies used.

> APA-style sample write-up: A one-way ANOVA was conducted to examine the effect of learning strategies on college students' recall performance. The results indicated a statistically significant difference among at least one of the learning strategies, $F(3,36) = 46.78$, $p < .001$.

Before we move on, think about why the homogeneity of variances assumption exists. We are summing up all the differences between individual data points and their respective group means, then dividing that result by the degrees of freedom within groups to get an overall error variance. If some groups have individuals who are widely spread apart from each other while other groups have a much smaller spread, the average variance will not be a good estimate for the error of any of the groups.

A key limitation of ANOVA arises when group sizes are extremely imbalanced, as groups with larger sample sizes will contribute more to variance estimates. This issue becomes more problematic when group sizes are both imbalanced and have unequal variances, leading to inaccurate *F* statistics and an increased risk of Type I or Type II errors. In that scenario, we would suggest taking a random sample from one or more groups with large sample sizes to create similar-sized groups. Of course, if your overall sample size decreases, you lose statistical power, but at least you can improve the reliability of analysis by minimizing bias in variance estimation.

This brings us to the concept of power. In earlier chapters, we mentioned statistical power multiple times. *Statistical power* is the probability of correctly rejecting the null hypothesis when it is false. In other words, it is the likelihood that a study will detect an effect or difference if one truly exists. Look back at the formula for the *F* statistic. What would make the outcome larger and therefore make it more likely to reject the null hypothesis?

1. Make the numerator larger.
2. Make the denominator smaller.

The first is pretty straightforward. If there are greater differences between groups, then the numerator increases. For the second option, we need to look at how we calculate the denominator, which is MS_E. We took the SS_E and divided it by df_E. We now have another fraction, and this time the denominator is $N - k$, which is the formula for within-group degrees of freedom. What happens as this increases?

- When you have a larger number in the denominator of a fraction, the result is going to become smaller.

- Since MS_E is the denominator of the fraction for the F statistic calculation, a smaller MS_E leads to a larger F statistic, making it more likely to detect significant differences between groups.

The conclusion is that pretty much always, a larger sample size gives you more power. Nothing new here, but it helps to see how that actually works out mathematically.

POST HOC TESTS

Think back to the null hypothesis for a one-way ANOVA. If we reject the null hypothesis, all we know is that at least one group mean is significantly different from at least one other group mean. In order to learn which means are different, we need to conduct post hoc tests.

The first step is to look at a means plot for a visual representation of the data, providing an indication of which means may be different. Figure 10.8 shows us that *visualization* and *technology-aided learning* are probably not significantly different from each other, but the other means (*repetition and control*) seem to differ from the first two. We can't tell if they also differ from each other.

Before we run the post hoc tests that will tell us which groups are significantly different from each other, we want to revisit the concept of family-wise error (or experiment-wise error). So far, we have run one analysis on this single dataset, and we used an alpha level of 0.05, indicating that we are willing to accept a 5% chance that we will reject a null hypothesis that is actually true (false positive). If we only perform one test,

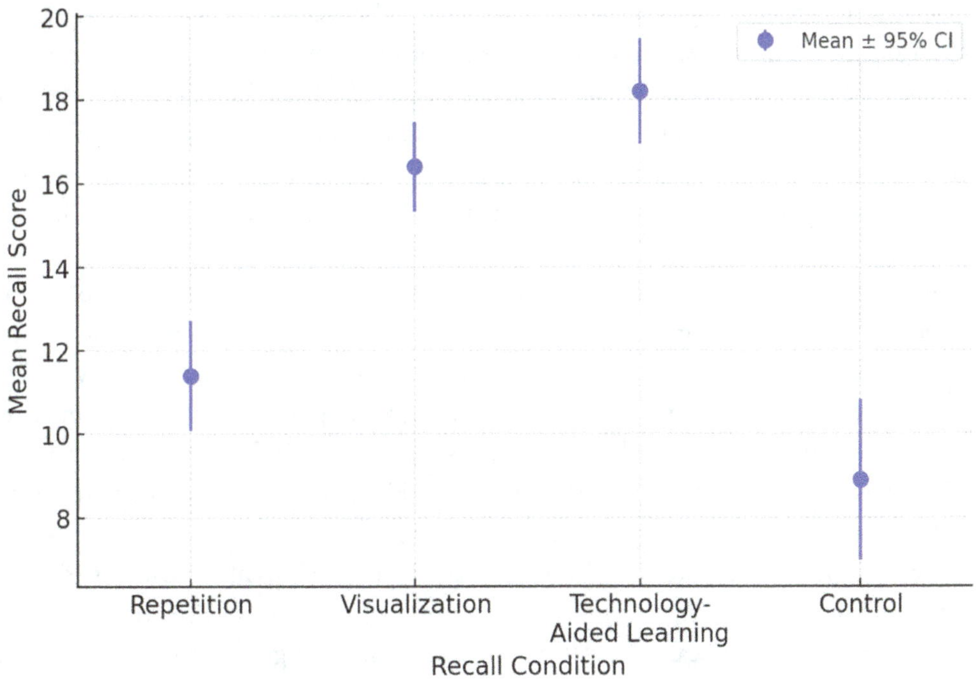

FIGURE 10.8 Mean Recall Scores with 95% CI

then the Type I error rate is equal to our alpha level, but if we conduct more tests on the same data, our chance of getting a false positive increases. Consider all the comparisons we could theoretically make for this recall data:

- Repetition versus visualization
- Repetition versus technology-aided learning
- Repetition versus control
- Visualization versus technology-aided learning
- Visualization versus control
- Technology-aided learning versus control

That is a lot of tests! With four groups, we could make up to six comparisons. The formula for calculating the error rate for an entire set of comparisons is $1 - (1 - \alpha)^c$, in which C = the number of comparisons. For four groups and six comparisons, our experiment-wise error rate is actually 0.26, which is much higher than our theoretical alpha level of 0.05.

There are many different ways to manage this problem. We will highlight three of the most frequently used post hoc tests and provide brief descriptions for several others. **Multiple comparison procedures** make comparisons while controlling for the increased risk of Type I error that occurs due to multiple testing. The multiple comparison procedures use the same general form of the *t* statistic. What distinguishes one multiple comparison procedure from another is the kind of contrasts that can be tested (pairwise vs. complex), how they control the family-wise error rate, and whether they are part of a post hoc or planned analysis.

The **Bonferroni post hoc test** is the most conservative. It simply divides your starting alpha level by the number of tests you are running and then conducts *t* tests between each pair. We have an alpha of .05 and six comparisons in our example, so our effective significance would be 0.0083. This means that under the Bonferroni correction, a comparison would need a *p* value less than 0.0083 to be statistically significant. This is more conservative and reduces the chance of Type I error at the cost of statistical power. This is a pretty tough test, and if you have even more groups, you could end up with a significant ANOVA but no significant pairwise comparisons. If that happens, then you know for sure that your post hoc test was too conservative.

Fisher's least significant difference test is a simple post hoc pairwise comparison procedure used to determine whether the means of two groups differ significantly. This test figures out the smallest significant difference between two means as if a test had been run on those two groups, and then any difference larger than that is also considered a significant result. This test has greater power to detect a significant difference between groups, but the trade-off is that the method does not control for family-wise error rate (the probability of making one or more Type I errors), meaning the likelihood of making a Type I error increases with multiple comparisons.

Tukey's honestly significant difference (HSD) test is a post hoc analysis method used to determine which specific group means are significantly different while controlling the family-wise error rate. Tukey's test is similar to a *t* test but has larger critical values (using the Studentized range statistic), so it is more difficult to reject the null. Table 10.6 provides additional options for multiple comparison tests.

TABLE 10.6 *Other Options for Multiple Comparison Procedures*

TEST NAME	DESCRIPTION
Sidak	Pairwise multiple comparison test based on a *t* statistic. Adjusts the significance level for multiple comparisons and provides tighter bounds than Bonferroni.
Scheffé	Performs simultaneous joint pairwise comparisons for all possible combinations of means. Uses the *F* distribution and can examine all linear combinations of group means, not just pairwise comparisons.
Student-Newman-Keuls (S-N-K)	Makes pairwise comparisons using the studentized range distribution. Compares pairs of means within homogeneous subsets, testing extreme differences first.
Tukey's *b*	Uses the studentized range distribution to make pairwise comparisons. The critical value is the average of the corresponding values for Tukey's HSD and Student-Newman-Keuls tests.
Duncan	Uses a stepwise approach similar to Student-Newman-Keuls but controls the error rate for the entire set of comparisons rather than individual tests. Uses the studentized range statistic.
Waller-Duncan	Multiple comparison test based on a *t* statistic with a Bayesian approach.
Dunnett	Pairwise *t* test comparing a set of treatments against a single control mean. Can perform two-sided tests or one-sided tests to check if means are greater or smaller than the control.

EFFECT SIZE

The last thing we will discuss about one-way ANOVAs is how to measure effect size. The concept is the same as with other types of analyses. Even if there is a statistically significant difference, we want to know if it is practically meaningful. For ANOVAs, effect size is measured in terms of the proportion of variance explained. We have focused on pairwise comparisons, but we are going back to the overall model and estimating to what extent group membership explains changes in the dependent variable for effect size.

To assess the size of the effect in an ANOVA, we can estimate eta squared (η^2), which provides the proportion of variance explained by the treatment:

$$\eta^2 = \frac{SS_{BG}}{SS_{total}} = \frac{558.675}{701.98} = .796$$

This tells us that 79.6% of the variation in the number of words recalled is a function of the learning condition. According to Cohen (1988), this represents a *large effect size*, suggesting that the learning strategy has a strong influence on recall performance. Unfortunately, η^2 is a biased estimator, meaning that it tends to overestimate the true effect size. A less biased estimator for explained variance is omega squared (ω^2), where:

$$\omega^2 = \frac{SS_{BG} - [(k-1)MS_E]}{SS_{total} + MS_E} = \frac{558.68 - (3 \times 3.98)}{701.98 + 3.98} = \frac{546.74}{705.96} = 0.774$$

As you can see, the value for ω^2 is lower, estimating that 77.4% of the variation in the number of words recalled is due to the learning condition. Whenever you need to report effect size, we recommend using ω^2 because it is less biased.

FACTORIAL ANOVA

The first type of ANOVA we learned was *one-way*, meaning there was only one factor. We divided people into groups by condition. Now, we are adding another factor into the mix. A **factorial ANOVA** is a statistical test used to analyze the effects of two or more categorical independent variables (factors) on a continuous dependent variable, including their interaction effects. Factorial ANOVA extends the one-way ANOVA by examining multiple factors simultaneously, allowing for a more comprehensive analysis. A classic example for using a factorial ANOVA is testing outcomes based on an intervention (treatment) and looking at whether the intervention is equally effective across different subsets of the sample.

For example, say you have a new way to coach students on reading and you want to find out if students who experience this treatment earn higher scores on a reading assessment than a control group. There are probably going to be students whose reading ability is high enough to start out that they are not going to show much of an increase—they don't really need the intervention. However, students who started out a little behind on reading could make significant progress based on the intervention. Table 10.7 shows how this would look.

TABLE 10.7 *Two-way Factorial ANOVA Example: Intervention Condition by Reading Ability*

	INTERVENTION CONDUCTION	
Reading Ability	Treatment	Control
High		
Low		

If you did not use a factorial ANOVA and just ran a *t* test comparing reading scores for treatment and control students, you might not see a significant result. But if you divide students into ability groups, you might find that the treatment was very helpful for students in a lower-ability group, as visualized in Figure 10.9. This is called an interaction effect, and it is typically what we are especially interested in when we use factorial ANOVA to analyze our data.

While rare in education research, you can have more than two factors, but for the simplistic nature of this book we are going to stick with only two. This can be referred to as a two-way ANOVA, or an A by B ANOVA, with A and B representing the different number of groups for each factor. In our example we had a treatment group and a control group, and then if we had students grouped in two reading levels, that would result in a 2x2 (2 by 2) ANOVA. There are three basic designs for a two-way ANOVA:

- **2-Between, or Between-Subjects, Design**: A two-way ANOVA where different participants are observed in each cell or group. This means that no individual belongs to more than one cell. For the example in Figure 10.9, students would be in only one of the four groups: treatment/low ability, treatment/high ability, control/low ability, or control/high ability.
- **1-Between, 1-Within, or Mixed Design**: A two-way ANOVA where different participants are observed at each level of the between-subjects factor and are repeatedly observed across the levels of the within-subjects factor. If you have a 2x2 mixed ANOVA, each person will be in two cells. To apply this approach to the example in Figure 10.9, imagine that it wasn't an experiment. Instead, a teacher just used this new method for all students.

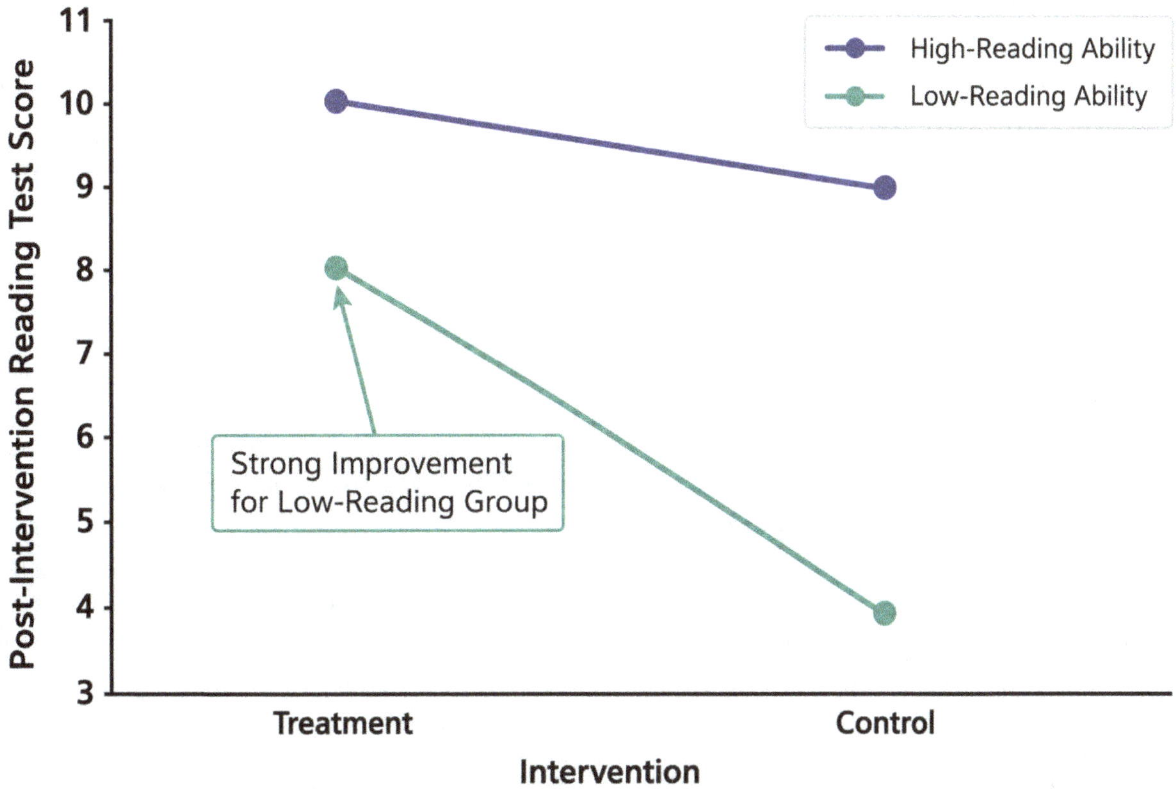

FIGURE 10.9 Interaction Effect: Reading Ability and Intervention

The between-subjects factor would still be the two ability groups, but then students at each ability level would be tested multiple times during the year.

- **2-Within, or Within-Subjects, Design**: A two-way ANOVA where the same participants are observed at each level of two factors. In this case each person is in all the cells; the composition of the cells is the same across the entire grid. If we wanted to apply this design to the example in Figure 10.9, we would have to abandon the idea of grouping students by ability. Instead, imagine that we want to find out if the coaching intervention works better for reading and math. The teacher could test the same group of students on reading and math at the start of the year and then again after a period of time; the idea would be to see if students improved at an equal rate in both subjects.Our example is getting a little weird, so here is a more typical scenario for a 2-within factorial ANOVA: One group of students watches two types of instructional videos on different topics. Students are then asked to recall what they learned from the video right after watching and again one week later. The research question would ask which video was more effective for longer-term recall.

Each factor is usually identified with a letter in a two-way ANOVA, such as A and B, although it doesn't have to be. It is also common to identify the levels of each factor numerically and use the numbers to identify the ANOVA model. For example, if you measure how quickly (in seconds) subjects respond to a stimulus that varies in size (small, medium, large) and color (dark, bright), we might designate Factor A as size and Factor B as color. Using the levels of each factor, you can describe this as a 3x2 ANOVA or 3x2 factorial design.

We often arrange the data from an ANOVA test in a table, and when we do, there is a special notation to indicate each entry. We are going to use another version of the vocabulary recall study as our example.

Imagine that a researcher is investigating the impact of noise level (loud vs. quiet) and learning strategies (repetition, visualization, technology-aided learning, control) on vocabulary recall performance. This study aims to determine whether different learning strategies are more effective for certain types of learners.

The research addresses the following questions:

- Do students who study in loud and quiet environments differ in their vocabulary recall performance?
- Does vocabulary recall vary based on learning strategy (repetition, visualization, technology-aided learning, control)?
- Does the effectiveness of learning strategies depend on the noise level of one's study environment?
    - For example, do students who study in a loud environment benefit more from visualization strategies than those who study in a quiet environment?

To explore these questions, the researcher collected data from 80 college students, with 10 participants assigned to each condition. It is an entirely between-subjects design in that participants were separated into two groups by noise level while studying (loud vs. quiet) as well as the four learning conditions.

- Levels of Factor A, symbolized as "a."
  If there are two levels (e.g., loud vs. quiet), a = 2.

- Levels of Factor B, symbolized as "b."
  If there are four levels (repetition, visualization, technology-aided learning, control), b = 4.

The combination of one level from each factor is called a cell. The number of cells represents the number of groups in a study.

- Total number of cells = ab.

In the vocabulary recall example, there are eight cells.

When we indicate the mean of the specific cell, we use subscripts. As a general form, we use the mean of the cell that is the *i*th category of Factor A and *j*th category of Factor B. For example, the mean of the loud environment group for the repetition condition is represented as $\bar{X}_{11}$. This represents the mean of the first category of noise level while studying and the first category of learning strategy condition.

DESCRIBING VARIABILITY: MAIN EFFECTS AND INTERACTIONS

As we continue learning about ANOVAs, remember that this test is all about the different sources of variance. With a two-way between-subjects ANOVA, we have four sources of variation to consider:

1. We still have *within-group variation*, or error. This is the variation within groups due to measurement error or just people being different from each other even though they are in the same group. As before, you find the sum of squared differences between each individual score in a cell and the mean score for that individual's cell. This is still referred to as SS_E.
2. Our between-groups variation is now split into two factors. First, we have variation due to Factor A, which is known as the *main effect of Factor A*. In the vocabulary recall example, this would be variation in the number of words recalled due to the noise level factor. You compute this just like you would find the between-groups effect in a one-way ANOVA: sum the squared differences between the mean of a specific level of Factor A (including all levels of Factor B) and the grand mean. Instead of calling this SS_{BG}, we call it SS_A.
3. Second, we have variation due to Factor B, or the *main effect of Factor B*. In the vocabulary recall example, this is the effect of learning strategy conditions. Note that the main effects don't take the other factor into consideration. When you are looking for the main effect of learning strategy conditions, you are combining the noise level groups as if there was only one factor. Summing up the squared differences between the mean of a specific level of Factor B and the grand mean gives you SS_B.
4. Finally, we could have variation due to an *interaction effect* between the two factors, which means that the effect of one factor depends on the level of another factor. In our example, you could see a different pattern of mean differences among conditions for the loud group compared to the quiet group. If there is no interaction effect, the pattern in mean differences by learning strategy condition is the same for both noise level groups.

Finding the variation due to the interaction effect takes a couple of steps. First, we find SS_{cells} by summing the squared differences between each cell mean and the grand mean. This is the variability attributed to the combination of levels for Factors A and B. However, this is not the interaction effect, because it also includes main effects that we need to subtract out. Think about it—if we are comparing one cell mean to the grand mean, any difference could be due to the main effect of one or both factors.

What we really need is SS_{AB}, which is the variation solely due to the interaction between Factor A and Factor B and is what we get after subtracting out the variation due to each main effect that we already identified:

$$SS_{cells} = SS_A + SS_B + SS_{AB}$$

Or, slightly rearranged:

$$SS_{AB} = SS_{cells} - (SS_A + SS_B)$$

HYPOTHESIS TESTING IN A TWO-WAY ANOVA

Let's explore our example that investigates the impact of noise level while studying (loud vs. quiet) and learning strategies (repetition, visualization, technology-aided learning, control) on vocabulary recall performance.

Our research questions are the following:

- Do students who study in loud and quiet environments differ in their vocabulary recall performance?
- Does vocabulary recall vary based on learning strategy (repetition, visualization, technology-aided learning, control)?
- Does the effectiveness of learning strategies depend on the noise level of one's study environment?

TABLE 10.8 *Noise Level and Learning Strategies on Recall Performance*

NOISE LEVEL	REPETITION	VISUALIZATION	TECHNOLOGY-AIDED LEARNING	CONTROL
Quiet	16	11	15	10
Quiet	14	9	16	13
Quiet	13	8	13	8
Quiet	13	12	13	13
Quiet	9	14	19	1
Quiet	11	13	21	11
Quiet	12	8	16	9
Quiet	17	10	20	8
Quiet	15	11	18	9
Quiet	8	13	15	3
Mean	**12.8**	**10.9**	**16.6**	**8.5**
Loud	9	16	22	8
Loud	7	16	17	15
Loud	9	18	18	9
Loud	12	12	13	6
Loud	7	12	16	12
Loud	7	16	18	6
Loud	12	14	14	10
Loud	10	18	19	4
Loud	6	15	16	6
Loud	9	13	17	10
Mean	**8.8**	**15**	**17**	**8.6**

The collected data are presented in Table 10.8.

STEP 1: DETERMINE THE NULL AND ALTERNATIVE HYPOTHESES

There are eight possible outcomes that you could obtain when you compute a two-way between-subjects ANOVA:

1. All three hypothesis tests are not significant
2. Significant main effect of Factor A only
3. Significant main effect of Factor B only
4. Significant main effect of Factor A and Factor B
5. Significant A×B interaction only
6. Significant main effect of Factor A and an A×B interaction
7. Significant main effect of Factor B and an A×B interaction
8. All three hypothesis tests are significant

This means that defining the null and alternative hypotheses is somewhat complicated with factorial designs.

HYPOTHESIS FOR MAIN EFFECTS

FACTOR A

H_0: $\mu_1 = \mu_2 = \ldots = \mu_A$ (there is no main effect of Factor A)
H_A: at least one mean of the levels in Factor A is different from other means.

In our example, H_A would be "average number of words recalled differs by noise level group."

FACTOR B

H_0: $\mu_1 = \mu_2 = \ldots = \mu_B$ (there is no main effect of Factor B)
H_A: at least one mean of the levels in Factor B is different from other means.

In our example, H_A would be "average number of words recalled differs by learning strategy."

INTERACTION BETWEEN FACTOR A AND B

It is a bit hard to write an equation as we do for the main effects, but it is okay to say:

H_0: there is no interaction effect between Factor A and B
H_A: there is an interaction effect between Factor A and B

In the recall example, H_A would be "there is an interaction effect between noise level and learning strategy conditions."

STEP 2: SET THE CRITERIA FOR A DECISION

To determine whether the noise level while studying and/or learning strategy significantly affects vocabulary recall, we will use a significance level (α) of 0.05, as usual. This means that we are allowing a 5% chance of incorrectly rejecting the null hypothesis. Since we use two-tailed tests for both main effects (A, B) and the interaction effect, we are considering both directions of possible effects—whether one group performs significantly higher or lower than another.

There are three key effects to evaluate:

- The main effect of noise level while studying,
- The main effect of learning strategy, and
- The interaction effect between these two variables.

Each effect has an associated *F* statistic, which we will compare to a critical *F* value from the *F* distribution table to determine statistical significance. To determine the critical *F* statistics, we first need to know the degrees of freedom for each effect.

TABLE 10.9 *Degrees of Freedom and Critical Values for Each Factor and Interaction Effect*

	DEGREES OF FREEDOM (df_{BG}, df_E)	CRITICAL *F* VALUE FOR TWO-TAILED TEST
Factor A	1, 72	3.97
Factor B	3, 72	2.73
Interaction (A x B)	3, 72	2.73

We will explain later in this chapter how to get these degrees of freedom, but for now we will provide them to move forward. For noise level while studying (which has two levels: loud and quiet), the degrees of freedom are (df_{BG}, df_E) = (1, 72). For the learning strategy factor (which has four levels: repetition, visualization, technology-aided learning, and control), the degrees of freedom are (df_{BG}, df_E) = (3, 72). The interaction between noise level and learning strategy also has (df_{BG}, df_E) = (3, 72).

As we have done before, once we obtain the *F* statistic for each effect from the ANOVA calculations, we will compare it to the corresponding *F* critical value for the given degrees of freedom and significance level. If the calculated *F* value exceeds the *F* critical value, we will reject the null hypothesis, indicating that the effect is statistically significant. On the other hand, if the *F* value is less than the *F* critical value, we will fail to reject the null hypothesis, meaning there is not enough evidence to conclude that the effect is significant. Table 10.9 summarizes the degrees of freedom and the corresponding critical *F* values for conducting *F* tests for each effect.

STEP 3: CHECK ASSUMPTIONS AND COMPUTE THE TEST STATISTIC

ASSUMPTIONS FOR A TWO-WAY BETWEEN-SUBJECTS ANOVA

Because we are talking about a between-subjects ANOVA (no factor is within-subjects), we are back to the same assumptions we started with:

- *Normality*. We assume that the dependent variable is normally distributed in the population or populations being sampled.
- *Random sampling*. We assume that the data measured were obtained from a sample that was selected using a random sampling procedure.
- *Independence*. We assume that the probabilities of each measured outcome in a study are independent or equal.
- *Homogeneity of variance*. We assume that the variance in each group is equal to each other.

We will not go through the detailed steps for testing assumptions, but it is important to use both visual representations of the data and descriptive statistics to assess whether the assumptions of normality and homogeneity of variances are met. Additionally, statistical tests such as the Shapiro-Wilk test for normality and Levene's test for homogeneity of variances can provide further support for interpreting the data.

In our example, Levene's test resulted in $p = 0.772$, meaning we fail to reject the null hypothesis of equal variances. Therefore, we conclude that the variances are approximately equal across all groups, satisfying the assumption of homogeneity of variance.

INTERPRETING TWO-WAY ANOVA RESULTS

Remember that we have more than one hypothesis to test with a two-way ANOVA. We start by looking at the main effects of the two factors. Table 10.10 shows the means for each cell, and the means for each factor.

TABLE 10.10 *Learning Strategy*

FACTOR A: NOISE LEVEL	FACTOR B: LEARNING STRATEGY				FACTOR A
	REPETITION	VISUALIZATION	TECHNOLOGY-AID	CONTROL	
Quiet	$\bar{X}_{11} = 12.8$	$\bar{X}_{12} = 10.9$	$\bar{X}_{13} = 16.6$	$\bar{X}_{14} = 8.5$	$\bar{X}_{1.} = 12.2$
Loud	$\bar{X}_{21} = 8.8$	$\bar{X}_{22} = 15$	$\bar{X}_{23} = 17$	$\bar{X}_{24} = 8.6$	$\bar{X}_{2.} = 12.35$
Factor B	$\bar{X}_{.1} = 10.8$	$\bar{X}_{.2} = 12.95$	$\bar{X}_{.3} = 16.8$	$\bar{X}_{.4} = 8.55$	$\bar{X} = 12.275$

Our main effect tests ignore all the individual cell means because we are essentially running two separate one-way ANOVAs, one for each factor. The test statistic for the main effect of Factor A is:

$$F_A = \frac{\textit{variance of group means for Factor A}}{\textit{variance attributed to error}} = \frac{MS_A}{MS_E}$$

Replace the A with B and you have the test statistic for the main effect of Factor B:

$$F_B = \frac{\textit{variance of group means for Factor B}}{\textit{variance attributed to error}} = \frac{MS_B}{MS_E}$$

Next we test the interaction effect, which determines whether group means at each level of one factor significantly change across the levels of a second factor. Remember from earlier in the chapter that we need to find the variance for the AB interaction, which is the variation due to differences among cells after subtracting out the variation due to each main effect. So, the test statistic for the interaction effect is:

$$F_{AB} = \frac{\textit{variance of AB}}{\textit{variance attributed to error}} = \frac{MS_{AB}}{MS_E}$$

To get all these mean square values, we take the sum of squares and divide it by degrees of freedom. What degrees of freedom do we use? Table 10.11 shows what they would be *if and only if* you have an equal number

TABLE 10.11 *ANOVA Table Formulas*

SOURCE OF VARIATION	SS	df	MS	F
Factor A (Noise Level)	SS_A	$a-1$	$\frac{SS_A}{df_A}$	$F_A = \frac{MS_A}{MS_E}$
Factor B (Learning Strategy)	SS_B	$b-1$	$\frac{SS_B}{df_B}$	$F_B = \frac{MS_B}{MS_E}$
A×B (Noise Level x Strategy)	SS_{AB}	$(a-1)(b-1)$	$\frac{SS_{AB}}{df_{AB}}$	$F_{AB} = \frac{MS_{AB}}{MS_E}$
Error (within groups)	SS_E	$ab(n-1)$	$\frac{SS_E}{df_E}$	
Total	SS_{total}	$abn-1$		

Note: a = the number of categories in Factor A; b = the number of categories in Factor B; n = the number of observations in each cell.

of subjects in each group (n). If you have different numbers of subjects in each group, then your df_{total} is $N-1$, or one less than the total number of subjects, and df_E is the df_{total} minus df for Factor A (df_A), Factor B (df_B), and the interaction (df_{AB}).

STEP 4: FIND A *P* VALUE, DRAW A CONCLUSION, AND REPORT YOUR CONCLUSION

Since we use statistical software to obtain all F statistics in practice, we will not show all the hands calculations of the sum of squares formula. However, in Table 10.12 we present the F table for the vocabulary recall study that summarizes all calculations so we can check our results.

TABLE 10.12 ***2×4 Factorial ANOVA Table Calculations***

	SS	*df*	*MS*	F	*p*
Factor A (noise level)	0.45	1	0.45	0.06	0.81
Factor B (learning strategy)	739.65	3	246.55	31.51	< .001
AxB (noise level x strategy)	164.45	3	54.82	7.01	< .001
Error (within groups)	563.4	72	7.83		
Total	1467.95	79			

The first thing we should check in a factorial ANOVA is the *interaction effect*. The F statistic of 7.01 was obtained by taking the MS_{AB} of 54.82 divided by MS_E, which is 7.83. The F statistic in our example is larger than the critical F value of 2.73, allowing us to reject the null hypothesis. When the interaction is significant, this may change the interpretation of the separate main effects of each independent variable. See Figure 10.10 on the next page.

Next, we look at the main effects of Factor A (noise level) and Factor B (learning strategy), using the MS values for each to calculate the F statistic. In this case, we can reject the null hypothesis for Factor B but not Factor A, and thus we found a significant main effect of learning strategy but not for noise level.

However, the way you interpret these results depends on the type of interaction. In this case, there was a significant interaction effect between noise level and learning strategy on the number of words recalled, $F(3, 72) = 7.01, p < .05$. Therefore, we already know that the effect of strategy conditions on the number of words recalled depends on the noise level, meaning that some strategies work better for students who studied in louder environments, while others work better for students who studied in quieter environments. Also, we know that there is no general advantage for either noise level across learning strategies because we did not find a significant main effect of Factor A. In other words, we cannot say that one type of student overperforms the other consistently across all conductions. At the same time, the significant main effect of Factor B indicates that at least one of the strategies resulted in significantly higher or lower performance than others.

Now, let's revisit the research questions and examine how our statistical results address each one.

- *Do students who study in loud and quiet environments differ in their vocabulary recall performance?*

  The main effect of noise level was NOT statistically significant $F(1, 72) = 0.06, p = .81$. This suggests that learners who studied in loud and quiet learning environments did not differ significantly in their vocabulary recall performance.

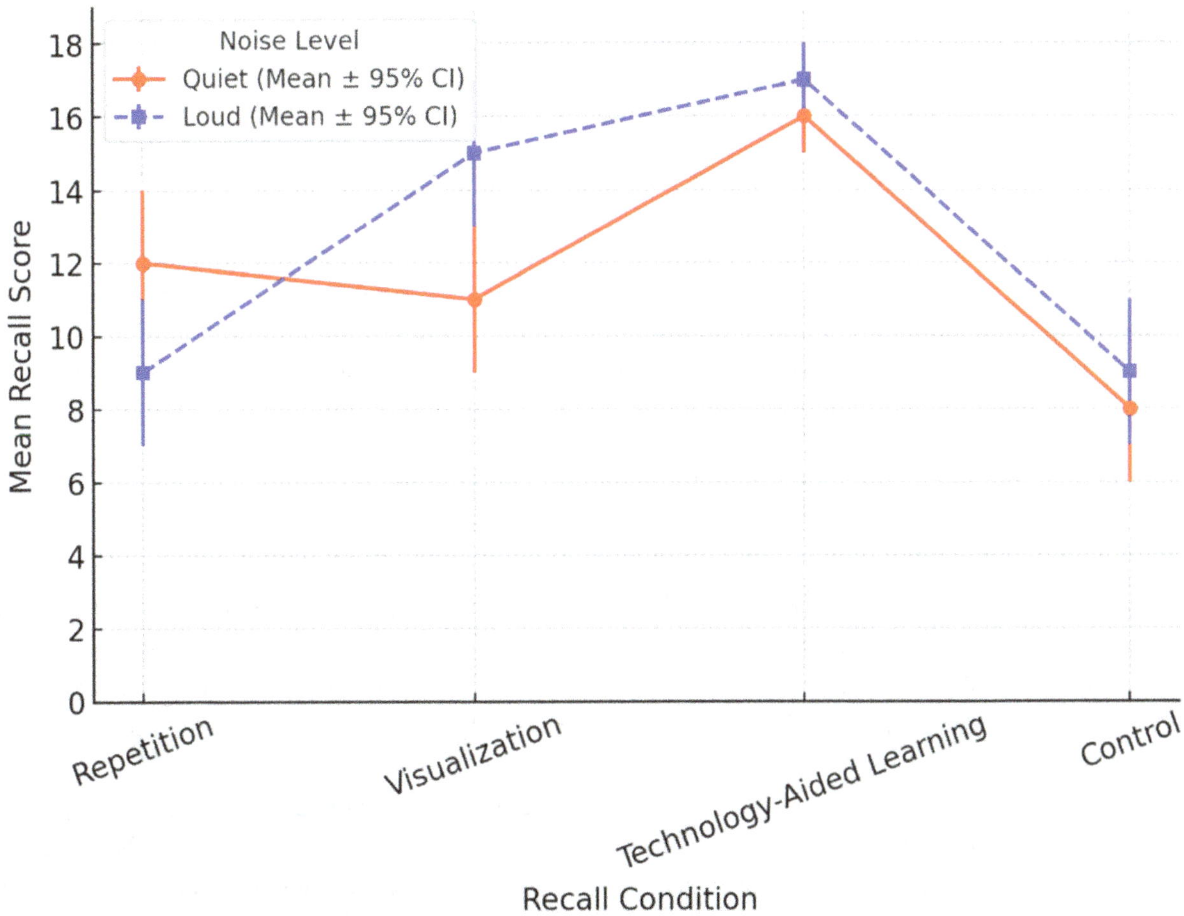

FIGURE 10.10 Interaction between Recall Condition and Noise Level on Mean Recall Score

- *Does vocabulary recall vary based on learning strategy (repetition, visualization, technology-aided learning, control)?*
  The main effect of learning strategy was statistically significant, $F(3, 72) = 31{,}52$, $p < .05$. This indicates that some learning strategies were more effective than others in enhancing vocabulary recall performance. However, a two-way ANOVA test does not tell which strategies are better or worse than others. A post hoc test would be helpful to determine which learning strategies lead to significantly higher or lower performance.

- *Does the effectiveness of learning strategies depend on the noise level of one's study environment?*
  The interaction effect between noise level and learning strategy was statistically significant, $F(3, 72) = 7.01$, $p < .05$. This suggests that the effectiveness of different learning strategies varied depending on whether the learner studied in loud or quiet learning environments.

In summary, some strategies may have benefited students who study in loud learning environments more than those who study in quiet environments and vice versa. The significant interaction effect means that a one-size-fits-all approach to vocabulary learning may not be optimal and that learning strategies should be tailored to individual needs when it comes to noise level while studying.

There are three key points to understand for a two-way ANOVA:

- One group can have significantly higher scores on a dependent variable than the other group across all levels of the second factor, and this doesn't mean that there is an interaction effect. In other words, the two lines can be quite far apart but parallel. The indication of an interaction effect is whether two lines intersect.
- You can have a main effect that is not significant at all, and yet it is very important because of its role in creating an interaction effect.
- If you do have a significant interaction effect, focus on that. Depending on what the means plots look like, you may not end up reporting the main effects.

POST HOC TESTS AND EFFECT SIZE

Post hoc tests are a little different for a factorial ANOVA. In fact, it is common to stop without doing any post hoc tests after finding a significant interaction. If you decide that the main effects are both statistically significant and meaningful, either there is no interaction effect or the main effects are still interpretable even with an interaction effect, and you can move forward with post hoc tests on the main effects just like with one-way ANOVAs.

Instead of running additional tests, just describe the patterns with descriptive statistics. For example, we could describe our vocabulary recall results by saying that both noise level groups were similarly low under the control condition and that both groups had higher scores under the technology-aid conditions, although the quiet group outperformed the loud group under the repetition condition, while the loud group performed better than the quiet group under the visualization condition. Do you notice how long that took to describe? Perhaps the best thing you can do when you find an interaction effect is to display a graph—look back to Figure 10.10.

The effect size calculations for η^2 *and* ω^2 are the same as before. However, there is a third effect size to check because of the interaction term. Here is how you would calculate η^2 and ω^2 for the interaction in our example:

$$\eta^2_{AB} = \frac{SS_{AB}}{SS_{total}} = \frac{164.45}{1467.95} = 0.11$$

$$\omega^2 = \frac{SS_{AB} - [df_{AB} * MS_E)MS_E]}{SS_{total} + MS_E} = \frac{164.45 - (3 \times 7.83)}{1467.95 + 7.83} = \frac{140.96}{1475.78} = 0.096$$

APA-style sample write-up: A two-way ANOVA was conducted to examine the effects of noise level while studying (loud vs. quiet) and learning strategy (repetition, visualization, technology-aided learning, control) on vocabulary recall performance. The assumptions of normality and homogeneity of variance were assessed prior to analysis. Levene's test for equality of variances was nonsignificant, $p = .772$, indicating that the assumption of homogeneity of variance was met. The results of the two-way ANOVA revealed that a significant interaction effect was found between noise level and learning strategy, $F(3,72) = 7.01$, $p < .05$, $\omega^2 = .097$, was relatively small compared to the main effect of learning strategy, $F(3,72) = 31.52$, $p < .05$. These findings suggest that different learning strategies influence vocabulary recall performance, and their effectiveness depends on the noise level of the student's learning environment. However, the interaction effect was smaller than the main effect of strategy, implying that certain strategies were generally more effective regardless of noise level, but some strategies were particularly beneficial for specific learning environment noise levels. Given these results, instructional methods should be tailored to accommodate individual learning environments rather than applying a uniform approach to vocabulary learning.

Whether you use η^2 or ω^2, the effect size of the interaction is small. In this case, the main effect, especially for learning strategy conditions, is much larger than the interaction effect even though the latter is statistically significant.

OTHER TYPES OF FACTORIAL ANOVA

While we have mostly talked about two-way between-subjects ANOVAs, there are other types of factorial ANOVAs you could consider. If you add a third factor in the analysis, the only difference is that you have more effects (main effect A, main effect B, main effect C, interaction A×B, interaction A×C, interaction B×C, and interaction A×B×C). The computation gets more complex, but that's why we use software to run the analysis. The important thing to remember is what the variance components indicate.

KRUSKAL-WALLIS TEST FOR NONPARAMETRIC ANALYSIS

The **Kruskal-Wallis test** is a nonparametric statistical method used to determine whether there are significant differences among three or more independent groups based on a ranked dependent variable. In education research, this test is particularly useful when the assumptions of parametric tests, such as one-way ANOVA, cannot be met. For example, if the data are not normally distributed, contain outliers, or are measured on an ordinal scale, the Kruskal-Wallis test provides a robust alternative. By ranking the data across all groups and comparing the sum of ranks, the text evaluates whether the differences observed among groups are statistically significant.

In educational settings, the Kruskal-Wallis test can be applied to explore differences in student outcomes across instructional methods, grade levels, or demographic groups. For instance, a researcher might use the test to examine whether students from different teaching strategies (e.g., lecture-based, inquiry-based, and flipped classrooms) exhibit varying levels of engagement, as measured by a survey on a Likert scale. Since the test does not rely on assumptions about the underlying distribution of the data, it is particularly suited for analyzing ordinal data such as survey responses, where the distance between scale points may not be equal.

The Kruskal-Wallis test outputs a chi-squared statistic that indicates whether differences among the groups are significant. However, the test does not specify which groups differ from each other. Post hoc tests, such as Dunn's test with a Bonferroni correction, are often employed to identify pairwise differences when a significant result is obtained. The Kruskal-Wallis test thus serves as a valuable tool for education researchers seeking to analyze complex real-world data where traditional parametric assumptions are not tenable, allowing for more inclusive and flexible statistical analyses.

CONCLUSION

The ANOVA techniques and associated statistical methods covered in this chapter underscore the importance of understanding and addressing variability in research designs. Central to these discussions is the one-way ANOVA, which compares the means of three or more groups, highlighting differences in between-group variance and within-group error variance. This distinction is essential for isolating the effects of independent variables and ensuring reliable conclusions, especially when evaluating multiple groups simultaneously without inflating the risk of a Type I error.

Expanding on these foundations, methods such as two-way factorial ANOVA demonstrate the adaptability of variance analysis across various research contexts. Two-way factorial ANOVA allows researchers to examine the interaction effect between two independent variables, providing deeper insights into how multiple factors influence an outcome simultaneously.

Together, the tools and principles detailed in this chapter emphasize the critical role of statistical rigor in identifying meaningful patterns and effects, reinforcing key concepts such as statistical power and the importance of controlling for error through thoughtful experimental design and post hoc testing strategies such as Bonferroni corrections. These techniques provide a comprehensive framework for conducting reliable, interpretable, and impactful research.

KEY TAKEAWAYS FOR EDUCATIONAL RESEARCHERS FROM CHAPTER 10

- ANOVA prevents multiple testing errors. One-way ANOVA is a crucial tool for comparing three or more groups while controlling for Type I errors, which would increase if multiple t tests were conducted. This is particularly relevant when, for example, analyzing student performance across different instructional methods, grade levels, or educational interventions.
- Between-group versus within-group variance matters. ANOVA works by comparing variance between groups (due to instructional differences, policy interventions, or student demographics) with variance within groups (individual differences within the same group). If between-group variance is significantly higher, this suggests that at least one of the groups are statistically different.

- Factorial ANOVA identifies interaction effects. A two-way factorial ANOVA helps researchers examine how multiple independent variables interact to influence an educational outcome. This is particularly useful in evaluating whether interventions work differently for different groups (e.g., high- vs. low-achieving students).
- Checking assumptions is crucial for valid results. Before conducting any type of ANOVA, researchers must check normality, homogeneity of variances, and independence of observations. If these assumptions are violated, adjustments (e.g., using nonparametric alternatives such as the Kruskal-Wallis test) or transformations may be necessary to ensure valid interpretations.
- Post hoc tests determine specific group differences. ANOVA tells us that at least one group differs, but post hoc tests (e.g., Tukey's HSD, Bonferroni) identify which groups differ significantly. For example, in an education study comparing learning strategies, post hoc analysis helps determine whether technology-aided learning is significantly better than visualization or repetition-based methods.
- Effect size is as important as statistical significance. Eta squared (η^2) and omega squared (ω^2) quantify how much variance in the outcome variable is explained by the group membership. Even if a finding is statistically significant, a small effect size may indicate limited practical impact, guiding educators and policymakers in prioritizing interventions that have meaningful effects on student outcomes.

KEY DEFINITIONS FROM CHAPTER 10

The **Bonferroni post hoc test** is the most conservative. It simply divides your starting alpha level by the number of tests you are running and then conducts *t* tests between each pair.

Error variance in ANOVA refers to the variation in the dependent variable that cannot be explained by the independent variable(s). Error variance represents the variability within groups or conditions, which arises from random differences among individual observations rather than systematic differences between groups.

Experiment-wise alpha (or **family-wise alpha**) refers to the overall probability of making at least one Type I error (false positive) across all hypothesis tests in an experiment. This is particularly relevant in situations where multiple comparisons are made, as the risk of incorrectly rejecting a null hypothesis increases with the number of tests.

A **factorial ANOVA** is a statistical test used to analyze the effects of two or more categorical independent variables (factors) on a continuous dependent variable, including their interaction effects. Factorial ANOVA extends the one-way ANOVA by examining multiple factors simultaneously, allowing for a more comprehensive analysis.

Fisher's least significant difference test is a simple post hoc pairwise comparison procedure used to determine whether the means of two groups differ significantly.

Kruskal-Wallis test is a nonparametric statistical method used to determine whether there are significant differences among three or more independent groups based on a ranked dependent variable.

Multiple comparison procedures make comparisons while controlling for the increased risk of a Type I error that occurs due to multiple testing.

One-way analysis of variance (ANOVA) is a statistical method used to compare the means of three or more groups to determine if there are statistically significant differences among them.

Tukey's honestly significant difference (HSD) test is a post hoc analysis method used to determine which specific group means are significantly different from each other. Tukey's HSD controls the family-wise error rate (the probability of making one or more Type I errors) while comparing all possible pairs of group means.

CHECK YOUR UNDERSTANDING

1. What is the primary purpose of a one-way ANOVA?
   a. To compare the means of two groups
   b. To compare the means of three or more groups
   c. To analyze relationships between continuous variables
   d. To assess correlations between independent and dependent variables

2. Which of the following increases the risk of a Type I error when comparing multiple groups?
   a. Failing to apply a correction for multiple comparisons (e.g., Bonferroni correction)
   b. Running multiple *t* tests on the same dataset
   c. Including covariates in the analysis
   d. Increasing sample size

3. Which test is the most conservative post hoc method for controlling Type I error?
   a. Tukey's HSD
   b. Fisher's least significant difference test
   c. Bonferroni test
   d. Kruskal-Wallis test

4. The Kruskal-Wallis test is best suited for which type of data?
   a. Ordinal data or nonnormal distributions
   b. Normally distributed interval data
   c. Binary data
   d. Continuous data

5. What does homogeneity of variance mean in ANOVA?
   a. All groups have the same sample size.
   b. The dependent variable is normally distributed.
   c. The variances of the dependent variable are similar across groups.
   d. There is no interaction effect between variables.

6. What distinguishes a factorial ANOVA from a one-way ANOVA?
   a. It compares means between more than two groups.
   b. It examines the effects of two or more independent variables and their interactions on a dependent variable.

c. It includes covariates to control for variability in the dependent variable.
d. It is used for nonparametric data.

7. Which of the following is an example of a factorial ANOVA design?
   a. Comparing the mean scores of students across three different teaching methods.
   b. Assessing the correlation between test scores and student motivation.
   c. Testing for differences in a single group's test scores across three different time points.
   d. Analyzing the effects of teaching method (lecture vs. inquiry-based) and student age group (young vs. adult) on test scores.

8. What is the effect size measure commonly used in ANOVA to estimate the proportion of variance explained by the independent variable?
   a. Pearson's *r*
   b. Eta squared
   c. Cronbach's alpha
   d. Chi-squared

11

EXAMINING RELATIONSHIPS BETWEEN TWO VARIABLES

LEARNING OBJECTIVES

By the end of this chapter, you will be able to:

- Explain the concept of Pearson's correlation coefficient (r), including its direction and strength;
- Describe and identify the assumptions for correlation analyses;
- Identify scenarios where alternatives to Pearson's r, such as Spearman's rho or point-biserial correlation, are more appropriate;
- Understand when to use a chi-square test for independence and apply it to analyze the relationship between two categorical variables; and
- Conduct a chi-square test for independence and interpret the results to determine whether there is a significant association between the variables.

The t test and ANOVA are statistical methods used to compare means, with a foundational understanding of t tests serving as a prerequisite for learning ANOVA. Similarly, this chapter introduces correlation, a key concept that underpins regression analysis. Rather than examining differences between two or more groups on a single dependent variable, we will focus on determining whether two continuous variables are related within a single group of subjects. The chapter then extends this focus on relationships to categorical data, introducing the chi-square test for association as a method for determining whether two categorical variables are statistically independent of one another.

A fundamental concept to understand is that correlation cannot be calculated between just two individual numbers (e.g., one person's height and weight). Instead, correlation requires two sets of numerical data (e.g., the heights and weights of 50 individuals) to identify patterns. Correlation is fundamentally about detecting patterns in how two variables are related. The chapter then extends this focus on relationships to categorical data, introducing the chi-square test for association as a method for determining whether two categorical variables are statistically independent of one another.

OVERVIEW OF CORRELATIONAL ANALYSIS

Imagine that a math teacher wants to explore whether students who are more motivated to learn mathematics tend to perform better on their math tests. To investigate this, the teacher collects data from 100 students and measures the following:

- Motivation to learn math (X-axis): Students rate their motivation on a scale from 1 to 20.
- Math scores (Y-axis): Final math test scores, ranging from 0 to 100.

A useful starting point when examining potential correlations between two continuous variables is to construct a scatterplot. The scatterplot provides a visual representation of the relationship between the variables, helping to identify patterns or trends. Figure 11.1 provides an example of a scatterplot illustrating the relationship between student motivation to learn mathematics and math scores, based on a random sample of 100 students.

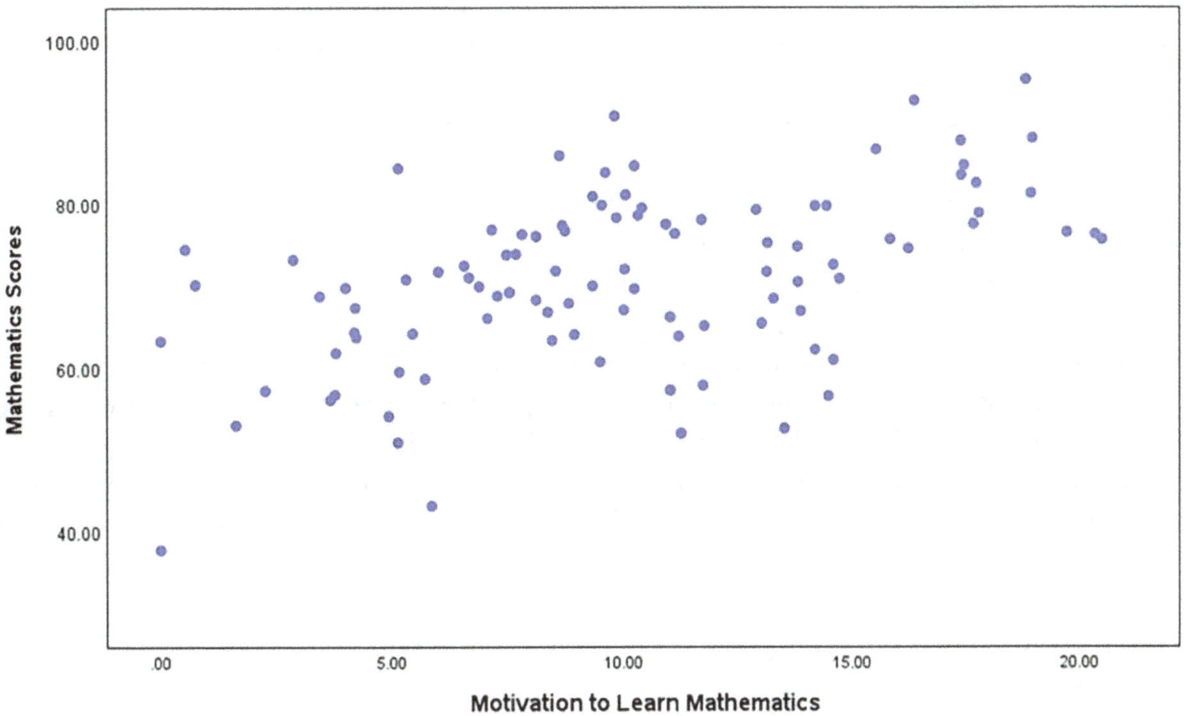

FIGURE 11.1 Scatterplot Showing the Relationship between Math Motivation and Math Performance

We will use this example to explore the fundamental principles of correlation, including the necessary data assumptions and the limitations of correlation analysis. To begin, we will define what correlation measures.

When discussing correlation, we typically refer to **Pearson's correlation coefficient (r)**, which quantifies the linear relationship between two continuous variables. Much like how other sample statistics serve as estimates of population parameters, r is an estimate of the population correlation coefficient (ρ). While statistical software is typically used to compute correlations, understanding the underlying mathematical principles is essential. The following equation provides insight into how correlations are calculated, even if you do not compute them by hand in practice:

$$r = \frac{Cov_{XY}}{\sqrt{S_X^2 S_Y^2}} \text{ or in other words, } r = \frac{\textit{covariance of X and Y}}{\sqrt{\textit{variance of X and Y separately}}}$$

Don't get tripped up by the symbols, but do think about what the equation means. Variance is essentially how much the values of a given variable differ across a sample. If we think about our scatterplot in Figure 11.1, the top half of the equation above reflects the extent to which math scores (Y) and motivation (X) vary together (when one goes up, the other goes up). The denominator reflects how much variance there is in each separate variable: Do math scores vary quite a bit across all students or only a little? What about motivation?

HOW OVERLAP AFFECTS CORRELATION

Imagine two circles, each representing the variance of a variable—one for X (e.g., motivation) and the other for Y (e.g., math scores). Together, these circles contain the total variance in the data. The overlapping area between the circles represents the covariance, which measures how much variance the two variables share.

- *More overlap* → The shared variance (numerator of the correlation formula) gets closer to the total variance (denominator), making the correlation coefficient (r) approach ±1.
- *Less overlap* → The shared variance decreases, and the correlation moves closer to 0.

As the two circles in Figure 11.2 start to overlap more and more, this illustrates the numerator of the correlation formula increasing to match the denominator—bringing the correlation closer to ±1. Conversely, when the circles barely overlap, the shared variance is small, and the correlation remains low.

Pearson's r tells us both the *direction* of the correlation (as values of X increase, do values of Y increase or decrease?) and *the strength* of the correlation (if we drew a best-fitting line through the scatterplot, do

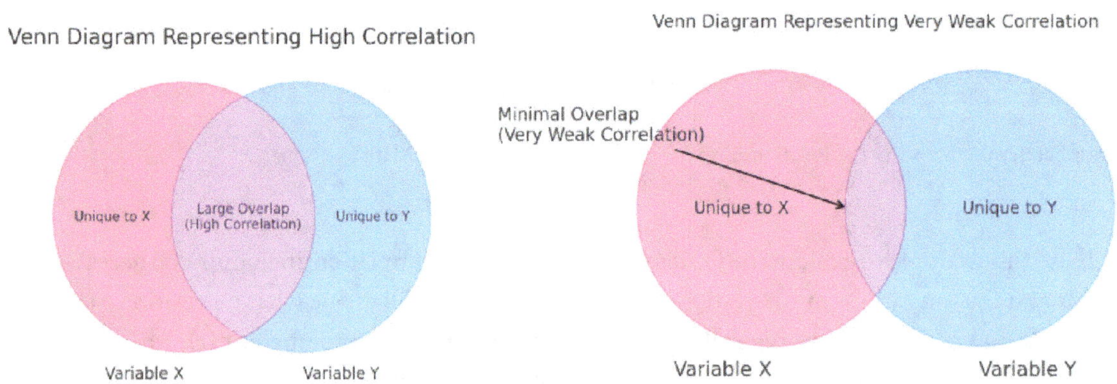

FIGURE 11.2 Venn Diagrams Showing Variance Unique to X and Y

most points fall very close to the line, or are they spread out all over?). As an example, the correlation between math scores and motivation in our dataset is $r = 0.497$ ($p < .001$), which is a moderate, positive correlation. This suggests that as students' motivation to learn mathematics increases, their math scores also tend to increase.

Pearson's r ranges from -1 to +1, with these extreme values representing perfect correlations between two variables. A perfect positive or negative correlation means that knowing the value of one variable allows for the exact prediction of the other. For example, the number of hours spent awake and the number of hours spent asleep in a day are perfectly negatively correlated. In education research, it is extremely rare to observe a perfect correlation, as most relationships between variables involve some degree of variability and external influences. Figures 11.3 and 11.4 show examples of perfect negative and positive correlations, respectively.

Figure 11.5 provides more examples of positive and negative correlations represented as scatterplots, so you can better visualize the relationship between two variables.

The $|r|$ values in the left column of Figure 11.5 are the absolute value of the correlations. Therefore, $r = -0.1$ and $r = 0.1$ are equal in terms of the strength of the relationship. Both would be considered a weak relationship. The interpretation of correlation strength (e.g., weak, moderate, strong) in the far-left column follows conventional guidelines. However, the meaning of a correlation's magnitude should always be considered in the context of your research.

For example, in fields such as education and human behavior, extremely high correlations are rare because multiple factors influence outcomes. A correlation that might be considered moderate in one field could be highly meaningful in another. Therefore, rather than relying solely on standardized thresholds, researchers should interpret correlations based on theoretical expectations, practical significance, and the nature of the variables being studied.

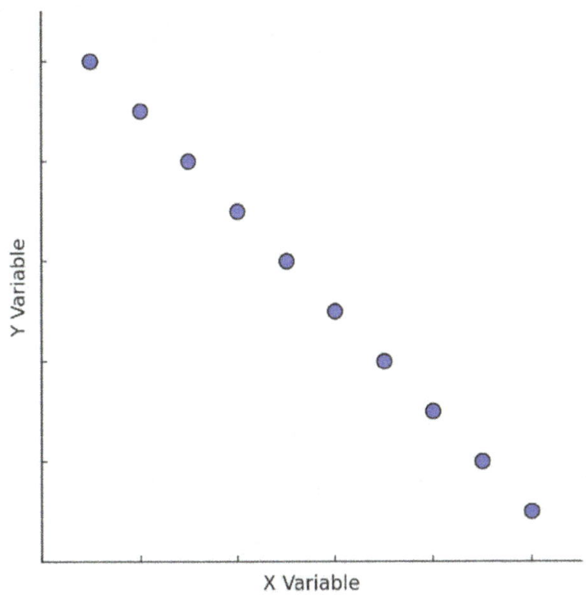

FIGURE 11.3 Scatterplot Showing Perfect Negative Correlation

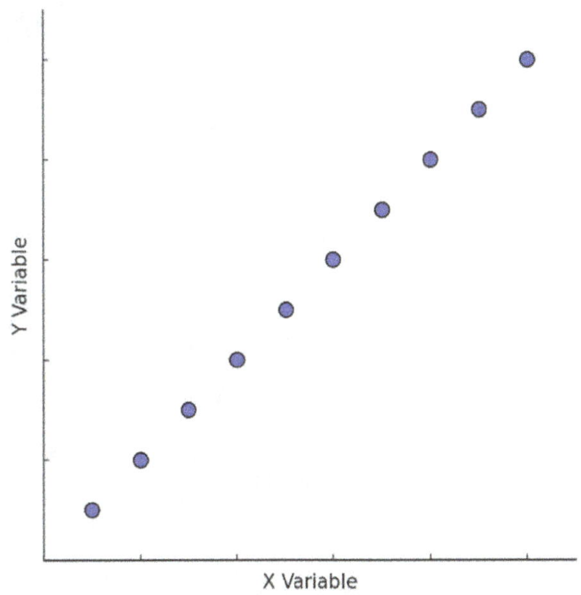

FIGURE 11.4 Scatterplot Showing Perfect Positive Correlation

Correlation Strength	Negative	Positive
|*r*| <.2 **Weak**	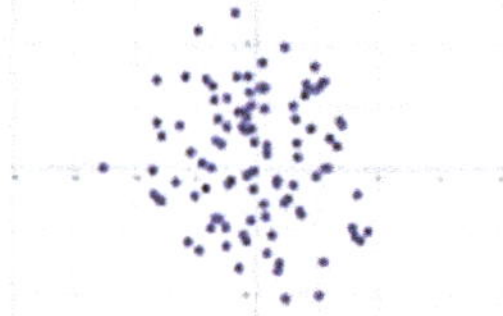 *r* = -0.08	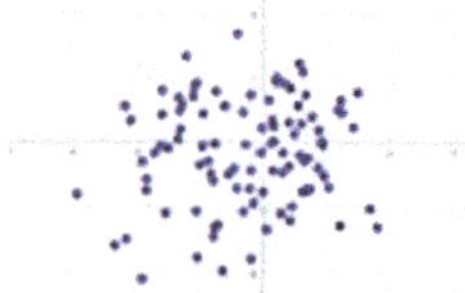 *r* =.10
.2 < |*r*| <.4 **Moderately weak**	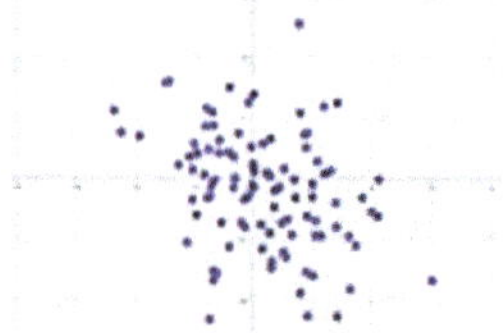 *r* = -0.33	 *r* = 0.35
.4 < |*r*| <.6 **Moderate**	 - *r* = 0.58	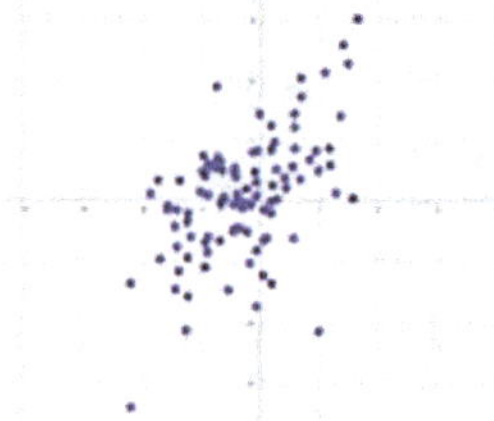 *r* = 0.53
.6 < |*r*| <.8 **Moderately Strong**	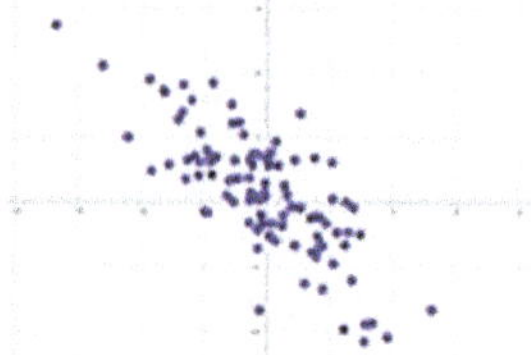 *r* = 0.77	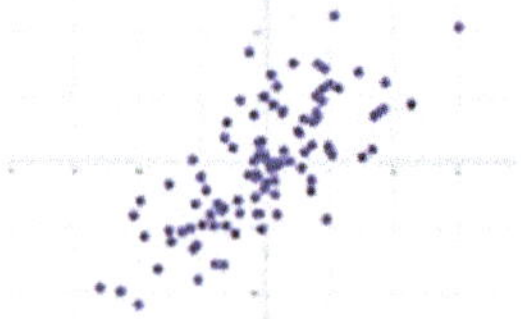 *r* = 0.73
.8 < |*r*| **Strong**	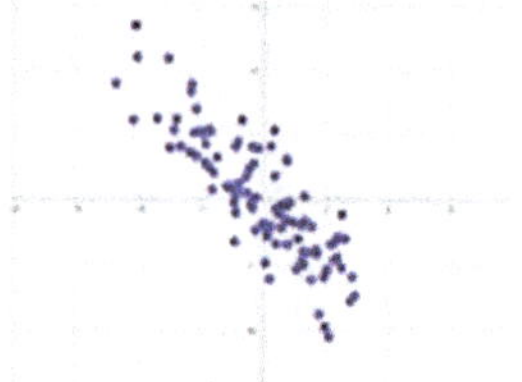 *r* = -0.85	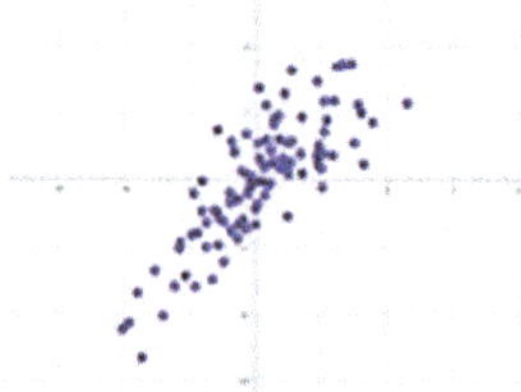 *r* = 0.81

FIGURE 11.5 Sample Scatterplots Illustrating Varying Strengths and Directions of Correlation

DATA ASSUMPTIONS FOR CORRELATIONAL ANALYSIS

We will use our previous example to illustrate the key data assumptions underlying correlation analysis. Figure 11.6 displays the scatterplot once again, depicting the relationship between student motivation and math achievement scores. This time, we have included a best-fitting line to help visualize the overall trend in the data.

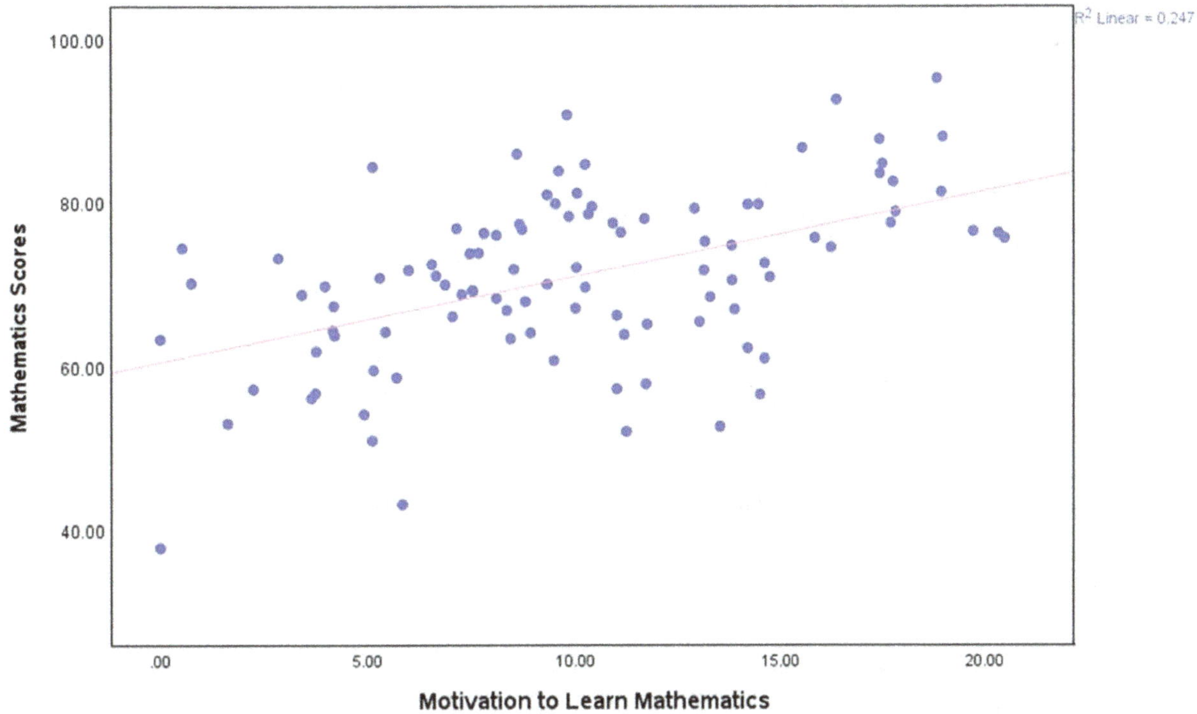

FIGURE 11.6 Scatterplot of Student Motivation and Math Achievement with Best-Fitting Line

Like other parametric methods, correlation comes with a few key assumptions that need to be met for the results to be reliable. These include linearity, homoscedasticity, and bivariate normality. Let's take those one at a time.

Linearity is one of the most important assumptions in correlation analysis and the easiest to check with a scatterplot. For correlation—and later for linear regression, which we will explore in Chapter 12—we must determine whether a straight line is the best way to describe the relationship between the two variables rather than a curved pattern. In Figure 11.7, although the data points do not all fall precisely on the line, a linear relationship is evident. Even without the best-fitting line drawn, one could reasonably estimate its approximate placement within the data.

Homoscedasticity is a new addition to your statistics vocabulary. **Scedasticity** is the extent to which data points are spread out above and below the line. **Homoscedasticity** means that the spread of points above and below the line is the same along the entire line. We think this might be easiest to understand with an example that violates this assumption.

Imagine you plotted the relationship between family income (X-axis) and how much money a family spent on their spring break vacation (Y-axis), as displayed in Figure 11.8. At the lower end of the income scale,

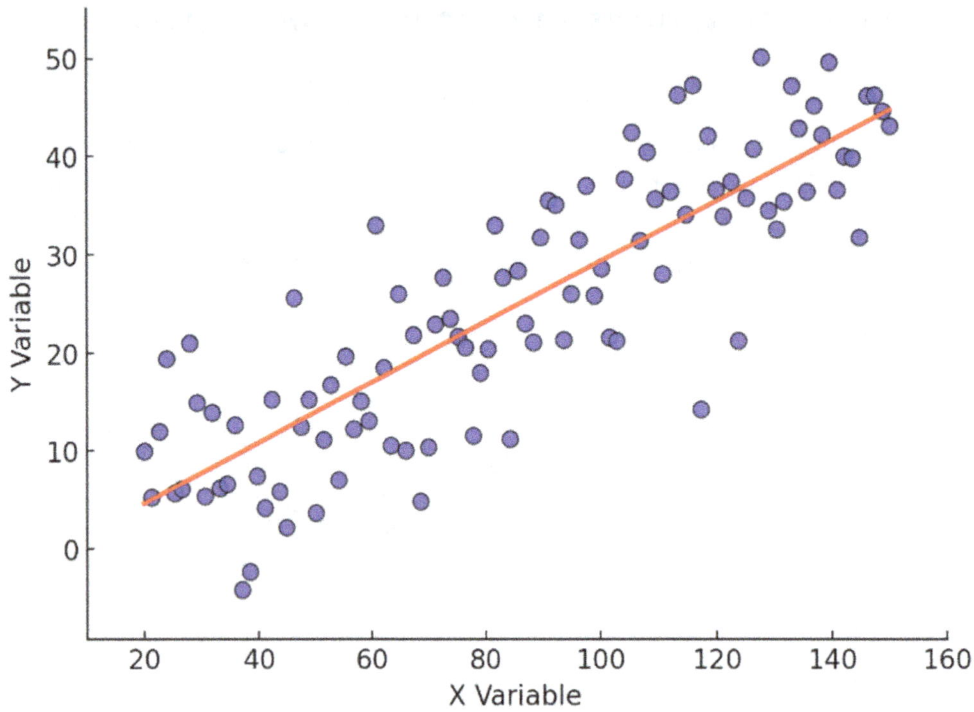

FIGURE 11.7 Sample Scatterplot Showing a Linear Relationship between Two Variables

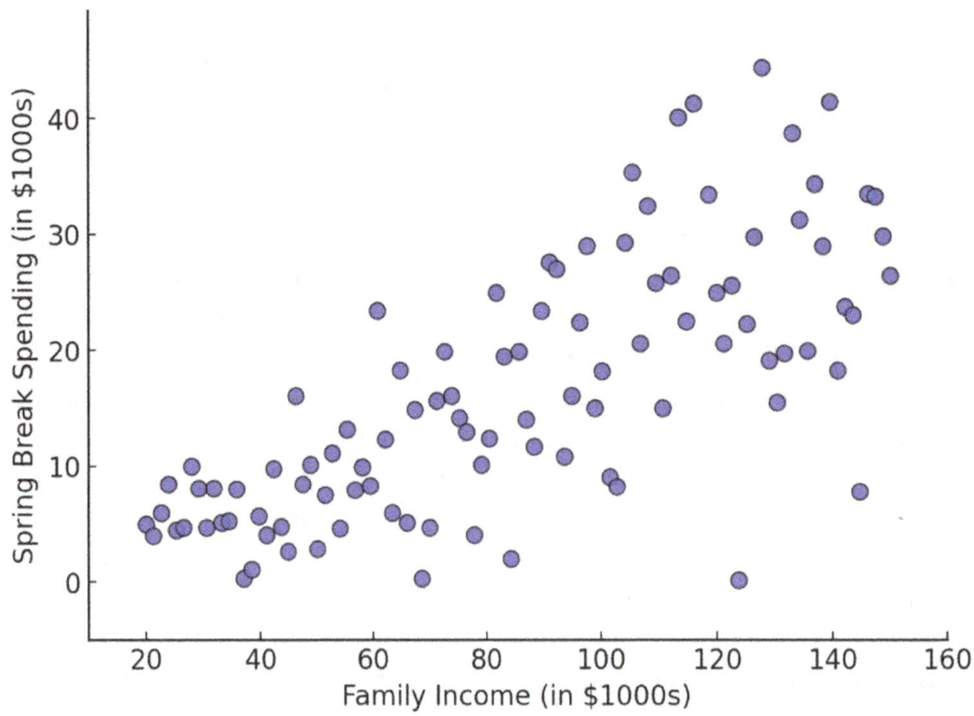

FIGURE 11.8 Scatterplot Showing Heteroscedasticity in Family Income versus Vacation Spending

spending variation is likely minimal, as financial constraints limit discretionary spending. As a result, the data points would be tightly clustered around the best-fitting line. However, at the higher end of the income scale, spending variation increases—some wealthier families may take extravagant vacations, while others may choose to spend little or nothing. This pattern results in **heteroscedasticity**, where the spread of data points is not uniform across values of the independent variable. A heteroscedastic scatterplot often resembles a cone, with the wider end appearing in the upper right portion of the plot in this example.

This is important because when we calculate the value of r, we are assuming that it holds true for all values of each variable. If we don't meet the homoscedasticity assumption, the true correlation is actually lower or higher depending on where we are along the line.

Normality is an assumption you are already familiar with, but it gets a little more complicated with correlation. Start by checking whether each variable is normally distributed. Then we must look at a new concept called **bivariate normality**. Bivariate normality is an extension of the normal distribution to two dimensions where two variables are jointly distributed in a way that their relationship forms a bivariate normal distribution. Figure 11.9 shows yet another version of the scatterplot with the addition of a sideways normal curve representing the frequency of math scores when motivation is 10.

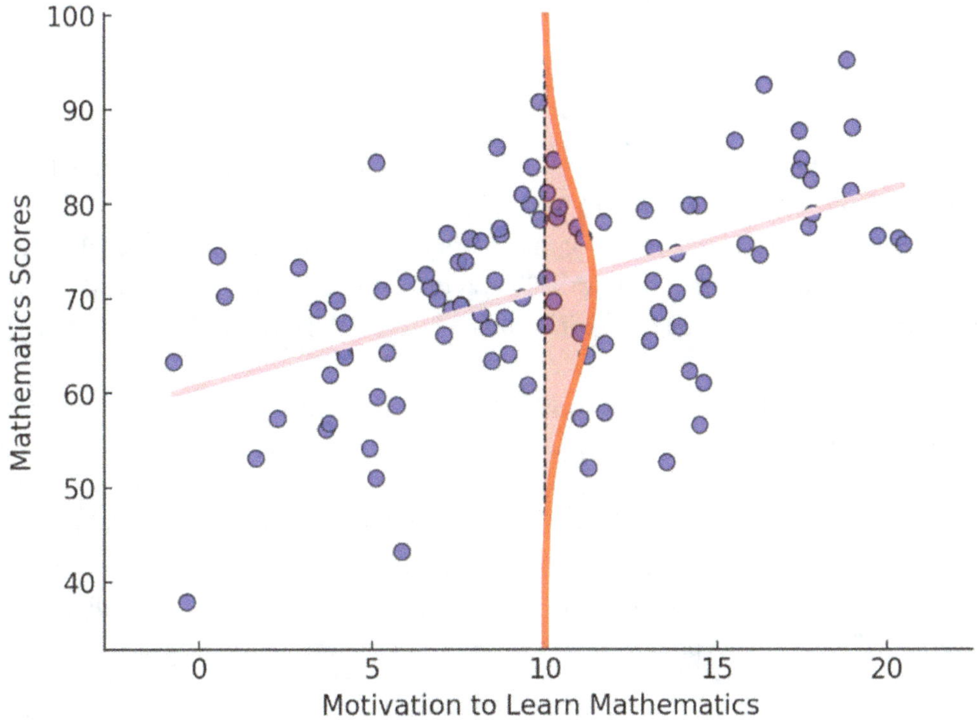

FIGURE 11.9 Scatterplot with Conditional Distribution Curve to Illustrate Bivariate Normality

We hope this addition will help you understand a bivariate normal distribution. First, choose one value of X (motivation)—for this example we chose 10. Then look at all values of Y associated with X = 10. When assessing the distribution of Y values at a given X value, it can be helpful to think of

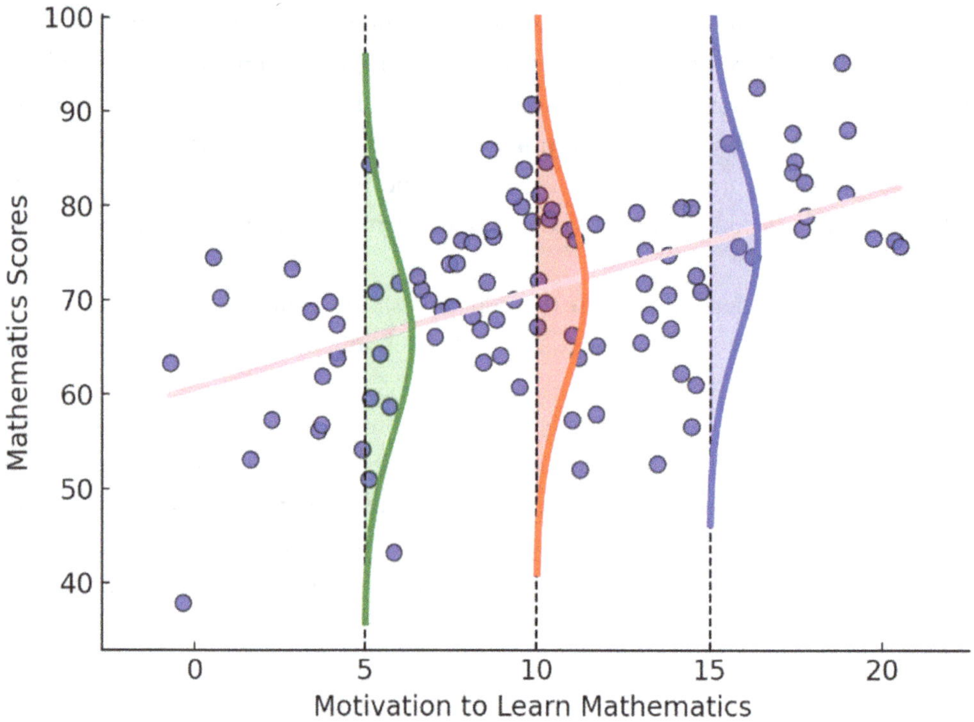

FIGURE 11.10 Visualizing Bivariate Normality in a Scatterplot

looking at the scatterplot from the side. Ideally, the Y values should be normally distributed around a central point for each level of X. In other words, most of the math scores when motivation is approximately 10 should be clustered together in the middle of that distribution. The values that are not close to the middle should be evenly distributed above and below the line, as displayed in Figure 11.10. For X = 10, our example looks good in terms of bivariate normality. That is not the case for all points, but there is no major violation of this assumption in our data.

Another issue to be mindful of with normality is that data can sometimes be uniformly distributed, meaning all values are nearly identical or evenly spread across a range. For example, imagine collecting data on the number of school days per week and the number of hours students spend in school each day. Since most schools follow a five-day schedule with roughly the same daily hours, every data point would be nearly identical. This would result in a scatterplot where all points cluster in the same location, making it impossible to establish a meaningful correlation—even though school days and daily hours are logically related. This highlights how a dataset can appear connected in theory but produce a correlation coefficient of 0 due to lack of variability in the data.

FACTORS AFFECTING THE MAGNITUDE OF CORRELATIONS

While checking data assumptions is essential, there are a few additional considerations before conducting a correlational analysis. Even if data meet the assumptions, other factors can still impact the accuracy and interpretability of correlation coefficients. For example, you need to *check for outliers*. You will also need to make sure you don't have a problem with *range restriction*, and your *sample size needs to be large enough* to ensure that r is a stable estimate of the population parameter ρ.

Range restriction occurs when the variability of a variable (or variables) in a dataset is artificially limited or constrained, such that the range of values is smaller than it would be in the population. This restriction can significantly impact your results, potentially leading to biased estimates and incorrect conclusions. Now, let's examine how each of these conditions can impact the correlation estimate one by one.

Outliers can change both the direction and the strength of a correlation. We have included an example in Figure 11.11. This example adds an outlier of (X, Y) = (8, 16) to the perfect positive correlation we looked at earlier in the chapter. The pink line indicates the new line of best fit caused by the inclusion of this single outlier. In this example, with the outlier, the correlation estimate decreases as the outlier disrupts the perfect linear pattern of the original data. With the shift of the regression line, the best-fit model is now a less accurate representation of the data points.

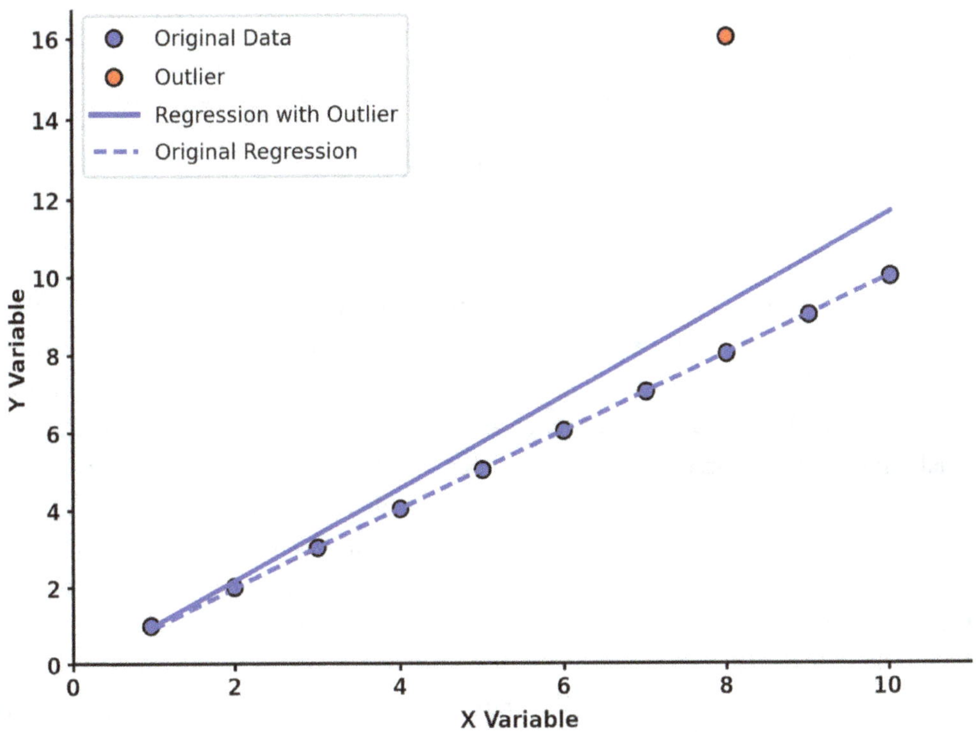

FIGURE 11.11 Impact of Outlier on Correlation Strength and Line of Best Fit

What about the next two graphs?

Figure 11.12 contains an outlier, (X, Y) = (15, 15), as the value is a noticeably larger than the rest of the data points. However, this outlier does not affect the size or direction of the correlation because it aligned perfectly with the original trend. In this case, the outlier follows the same pattern as the rest of the data but represents an extreme value on both variables. Since it maintains the existing relationship, it does not distort the regression line or the correlation estimate.

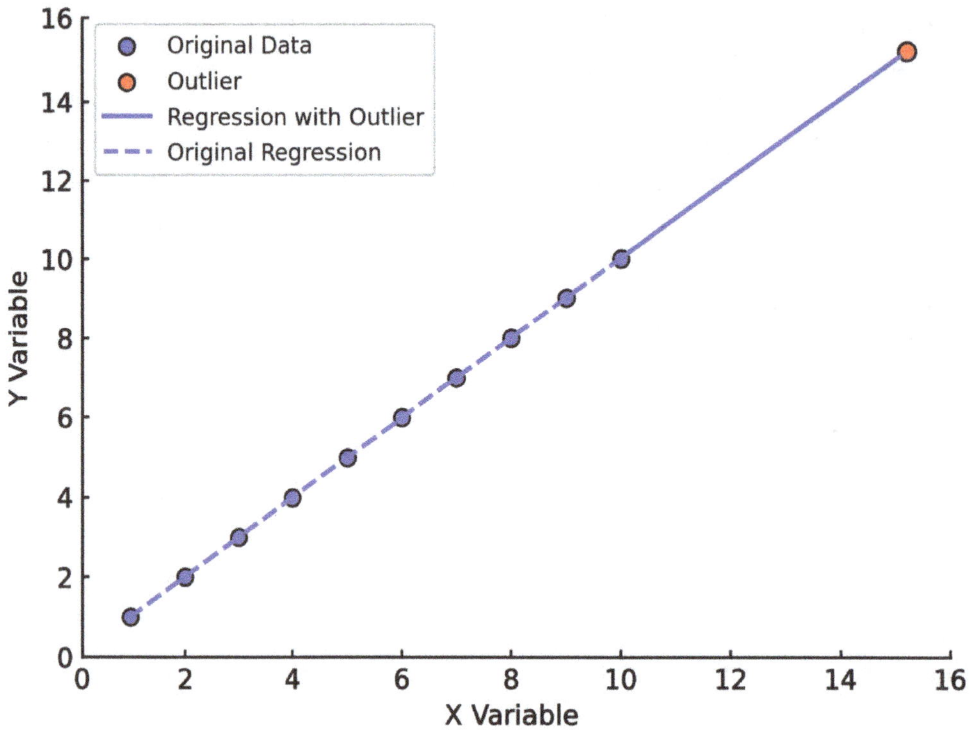

FIGURE 11.12 Noninfluential Outlier That Follows the Original Linear Pattern

Figure 11.13 contains two outliers, but these two are extreme in opposite directions on Y variables. The two outliers at (X, Y) = (9, 17) and (9, 2) do not pull the regression line in any particular direction because they are balanced around the trend. This means that the correlation estimate stays about the same, as these points do not change the overall relationship between the two variables. However, these outliers do increase the spread of the data, making the overall variation (variance) larger. Think of it like measuring students' heights in a classroom. If most students are around the same height but two students are much taller and much shorter, the average height might not change, but the range of heights becomes wider. In statistics, more variance can indicate less precise estimates, which means that the correlation might still be accurate but the confidence in that estimate is lower. This is especially important in smaller datasets, where extreme values can have a bigger impact on the results.

The graphs in Figures 11.11, 11.12, and 11.13 illustrate different types of outliers and their effects on correlation and regression lines. In the first case (Figure 11.11), a single influential outlier can noticeably shift the pink regression line, altering the best-fit model and reducing the strength of the correlation. In contrast, the second and third graphs (Figures 11.12 and 11.13) feature outliers that are not influencing the regression line, keeping the best-fit line close to its original form. However, the outliers may influence the interpretation, the variance, or their estimation precisions. These examples emphasize the importance of visually inspecting data before interpreting correlation results.

A scatterplot provides immediate insight into whether an outlier is truly problematic or simply a natural variation in the data. Without visual inspection, researchers might overlook influential outliers that distort results. With graphical analysis, we can make more informed decisions about how to handle outliers and ensure that correlation estimates are accurate and meaningful.

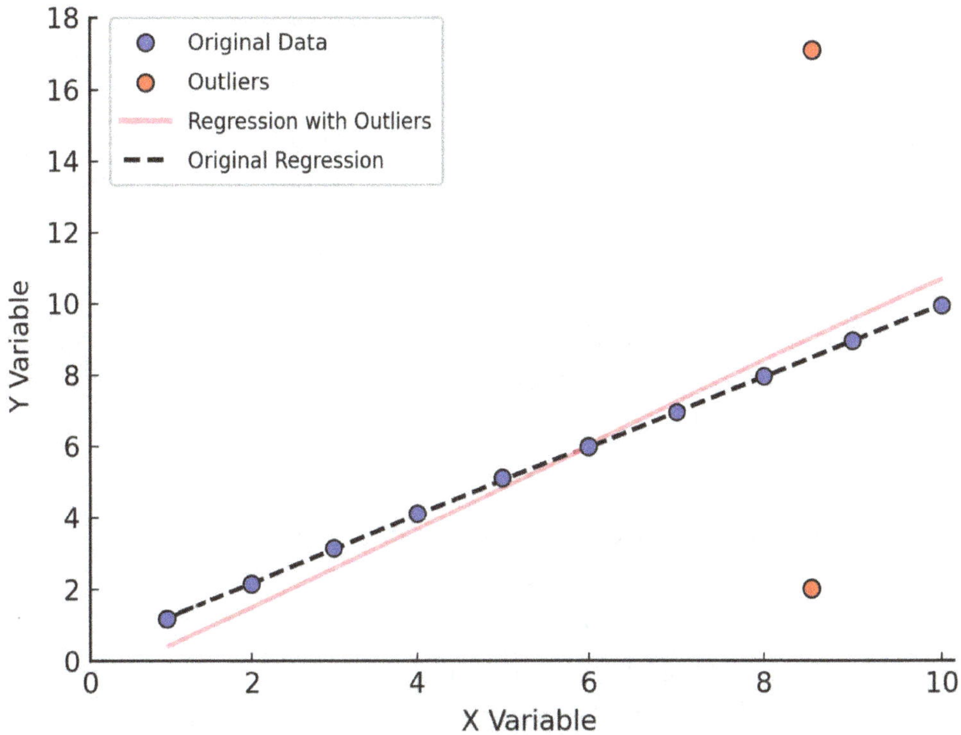

FIGURE 11.13 Balanced Outliers Maintain Correlation but Increase Data Variability

Range restriction is a trickier concept to understand. When interpreting a correlation, it is important to not make assumptions outside of the observed range. This becomes a problem if your data only represent a smaller range of a variable compared to the whole population. For example, there is probably a strong positive correlation between height and reading ability if we are only studying children ages four to eight. However, this does not mean that we can conclude anything about a relationship between height and reading ability among adults. The eight-year-old students are simply taller, on average, than four-year-old students due to normal growth rates among children.

However, this concept has real implications for the kind of analyses you conduct. Imagine if all college students took the Law School Admission Test (LSAT), and we looked for a correlation between final college grade point average (GPA) and LSAT scores. There would not be a perfect correlation, but you can imagine there would be some kind of positive association between the two variables: generally, as GPAs increase, LSAT scores also increase.

We are now going to restrict the range. Instead of looking at GPAs and LSAT scores for all college students, let's only look at the group of students who were admitted to a highly prestigious and selective law school. Of course, a top school is going to require very high GPAs and very high LSAT scores. But for any given admitted student, a slight weakness in one area might be balanced out if another metric is very strong. This could result in students whose LSAT scores are not the highest still being admitted if they have a strong GPA and vice versa. But if you plot those values out, do you know what you will see? You'll see a negative correlation between GPA and LSAT scores. When we restrict the sample to only the highest end of the range for both variables, the correlation doesn't just get weaker, it changes direction.

Here are some other examples of how range restriction can impact correlation in education research:

- The relationship between college entrance scores and GPA among honors college students may appear weak because only high-achieving students are admitted, limiting the variability in both scores.
- The relationship between teacher experience and student achievement may appear weaker in a study conducted within a district that only hires teachers with at least five years of experience. Because the sample excludes novice teachers, the variability in teaching experience is restricted, potentially underestimating the true correlation.

The last factor to consider is sample size. There is no hard-and-fast rule for sample size with a correlation analysis. However, you cannot be confident that your sample is a good estimate of the population with a small sample. This is closely connected to how we test correlations for significance.

STATISTICAL SIGNIFICANCE OF CORRELATIONS

STEP 1: DETERMINE THE NULL AND ALTERNATIVE HYPOTHESES

You can find the correlation between any two sets of continuous variables, but the result may not be statistically or practically significant. We can conduct hypotheses tests with correlational analysis just like with *t* tests. Let's go through the steps, just as we did for *t* tests and ANOVAs, using the math test and motivation example.

The null hypothesis is that there is no correlation in the population, and the alternative hypothesis is that there is some correlation:

$$H_0: \rho = 0$$

$$H_A: \rho \neq 0$$

In our example, the null and alternative hypotheses are:

$H_0: \rho = 0$ (there is NO correlation between math scores and students' motivation to learn math)
$H_A: \rho \neq 0$ (there IS a correlation between math scores and students' motivation to learn math)

STEP 2: SET THE CRITERIA FOR A DECISION

To set the criteria for a decision, let's choose our alpha level at $\alpha = 0.05$ for this two-tailed test and find the critical values. We need to know degrees of freedom, which for tests involving Pearson's r is calculated by $n - 2$, where n is the number of cases used for calculating the correlation. In our example $df = 100 - 2 = 98$.

STEP 3: CHECK ASSUMPTIONS AND COMPUTE THE TEST STATISTIC

We reviewed the scatterplot and observed no violation of assumptions to move forward with our statistical test. The test statistic is simply the value of r that we obtained from sample data. We show the formula of

Pearson's r at the beginning of the chapter, but we will not show how to calculate it here because you will use statistical software in practice. From our analysis, Pearson's correlation between math score and motivation rating score is $r = .497$ with $df = 98$.

STEP 4: FIND THE *P* VALUE, DRAW A CONCLUSION, AND REPORT YOUR CONCLUSION

Turn to Table A4 in the appendix, find the column representing $\alpha = 0.05$ for a two-tailed test, and look down to the line corresponding with the correct degrees of freedom. If the observed r statistic (.497) is higher than what you find in the table, then it is considered statistically significant at $p < 0.05$. As you can see, there is a straightforward relationship between the strength of the correlation (positive or negative) and sample size in terms of what is considered significant. For a smaller sample size, the correlation needs to be stronger to be significant. Since the table does not include the critical r value for $df = 98$, we refer to the closest available values:

- $df = 90 \rightarrow$ critical $r = .205$
- $df = 100 \rightarrow$ critical $r = .195$

Thus, the critical r value for $df = 98$ falls somewhere between .195 and .205. Since our observed correlation ($r = 0.497$) is greater than both values, we can conclude that our p value is smaller than 0.05, meaning that there is a statistically significant correlation between math test scores and motivation ratings.

> APA sample write-up: A Pearson correlation analysis was conducted to examine the relationship between the motivation to learn mathematics and mathematics test scores. The results revealed a statistically significant positive moderate correlation between motivation and math scores, $r(98) = 0.497$, $p < .05$. This suggests that as motivation increases, math scores tend to increase as well.

PRACTICAL SIGNIFICANCE OF CORRELATIONS

That brings us to practical significance, otherwise known as effect size. For Pearson's r, instead of using Cohen's d to measure effect size, we use something called the coefficient of determination, which is mathematically the same as a concept we already touched on with eta squared, proportion of variance. The **coefficient of determination** indicates the proportion of the variance in the dependent variable that is explained by the independent variable. The calculation is very simple:

$$\textit{coefficient of determination} = r^2$$

While r (the correlation coefficient) can be positive or negative, r^2 is always positive. Going back to our original example before removing outliers, we can calculate the coefficient of determination:

$$r^2 = (0.497)^2 = 0.247$$

You interpret this value by saying that 24.7 % of the variance in math scores can be explained (not caused!) by changes in motivation to learn. You will find that r^2 is important when we are evaluating the practical significance of regression equations, which we will learn in Chapter 12.

ALTERNATIVES TO PEARSON'S r

While we will not go into great detail, there are alternatives to Pearson's r that may be helpful to your future research.

SPEARMAN'S RANK-ORDER CORRELATION (r_S)

Spearman rank-order correlation coefficient (r_s), also known as Spearman's rho, measures the relationship between *two ranked (ordinal) variables*. This method is appropriate when the data are measured on an ordinal scale, such as responses on a Likert-type scale (e.g., *strongly agree* to *strongly disagree*). Like Pearson's correlation, r_s ranges from -1 to +1, where values closer to -1 or +1 indicate stronger monotonic relationships.

POINT-BISERIAL CORRELATION (r_{pb})

The **point-biserial correlation coefficient (r_{pb})** is used to measure the relationship between a *continuous variable* (e.g., motivation to learn math, achievement scores) and a *dichotomous variable* (e.g., pass/fail, enrolled/not enrolled, full-time/part-time). In education research, point-biserial correlation is particularly useful for examining whether higher achievement scores are associated with course completion or whether students with greater motivation are more likely to enroll in advanced courses.

FINAL REMARKS ON CORRELATION

We are sure you have heard the phrase "correlation does not equal causation." Now it is time to remind yourself of that. Statistical software packages can calculate the correlation between any two sets of numeric variables, but they cannot tell you if one variable caused the other—no matter how high the correlation coefficient. Instead, they simply tell you if there is a relationship between two variables. However, when interpreting correlation, it is important to consider other factors that might influence the relationship. Sometimes a third variable may be driving the observed association between two variables, making it appear as though they are directly related when they are not.

This brings us to the concept of a **confounding variable**, or third variable, an extraneous variable that was not accounted for but influences both variables of interest, leading to a "confounded" result.

Consider the following example. A researcher investigates whether there is a correlation between ice cream consumption and drowning incidents. The data reveal a positive correlation—when ice cream consumption increases, so do drowning cases. Does this mean that ice cream consumption causes drownings? Should policymakers consider banning ice cream to improve public safety? Of course not. The underlying *confounding variable* in this case is *temperature*—warmer weather leads to both increased ice cream consumption and more people swimming, which in turn raises the risk of drowning. This example illustrates why identifying potential confounding variables is critical when interpreting correlation results.

When discussing correlation, it is essential to describe how changes in one variable are *associated* with changes in another rather than implying causation. Avoid phrases such as "the effect of variable X on variable Y" and statements suggesting that one variable *affects* or *influences* another. Instead, use precise language, stating that variables are *correlated* or *associated* with each other. This distinction is critical, as correlation alone does not establish a cause-and-effect relationship.

Finally, while correlation provides information about the *direction* and *strength* of a relationship, it does not determine the exact slope of the best-fitting line. In other words, we can conclude that math scores and motivation are correlated, but we cannot infer *how much* math scores typically increase for every one-point increase in motivation. The correlation coefficient (r) only tells us the direction of the relationship (positive or negative) and how tightly the individual values cluster around the line—it does not quantify the rate of change between the two variables.

It is important to note that the slope of the best-fitting line is independent of the strength of the correlation. A steep line can still correspond to a weak correlation, while a nearly flat line can reflect a strong correlation. Correlation simply describes the direction and degree of association between two variables—it does not allow for precise predictions. If you want to determine the slope and use one variable to predict another, stay tuned for the upcoming discussion on regression analysis.

CHI-SQUARE TEST FOR INDEPENDENCE

A **chi-square test for independence** is a statistical test used to determine whether there is a significant association between two categorical variables.

When learning about independent-samples t tests or correlations, a common misconception is that you can analyze just two numbers rather than two sets of numbers. However, comparing two means without additional context is not statistically meaningful. To determine whether differences between means are significant, you need to account for the spread of the data (variance) and the number of observations (N). The same principle applies to correlation—you cannot determine whether two values are correlated in isolation. Instead, you need two sets of numbers to estimate variance and covariance, which are essential for assessing the strength and direction of a relationship.

In contrast, chi-square tests allow you to analyze relationships *without requiring individual raw data points.* In a chi-square test for independence, you compare observed values to expected values, where the expected values are calculated from the sample data itself rather than being predetermined.

Because the chi-square test for independence examines the relationship between two categorical variables, it can be used as an alternative to a correlation or the independent-samples t test. When using it, you should clearly define a dependent variable and an independent variable to inform your conclusions from the analysis. For example, say you want to check the effectiveness of a persuasive video on whether college athletes should be paid. Your independent variable shows if individuals watched the video or not, and your dependent variable is a dichotomous variable indicating agreement or disagreement with a statement in favor of paying college athletes.

If you had an entire dataset with individual responses, you could use statistical software packages to run a chi-square test. However, what if all you had was what appears in Table 11.1?

TABLE 11.1 *Support for Paying College Athletes*

		SUPPORT PAYING COLLEGE ATHLETES	
		YES	NO
Watched video	**Yes**	18	3
	No	12	11

There are a few important assumptions for this test, just like other chi-square tests:

- *Random sampling*: Each independent variable is drawn from a random sample.
- *Independence of observations*: The outcome for one individual does not influence the outcome for another.
- *Mutually exclusive categories*: Each observation belongs to one, and only one, category within the contingency table.

Since all assumptions are met, we went ahead and entered the information into a statistical software package and found that χ^2 (1) = 5.692, p = 0.017. There is an association between watching the video and support for paying college athletes. The phi coefficient, which represents practical significance, was 0.36, indicating a medium effect size.

Let's take a minute to understand how we would calculate the outcome by hand. First, we must calculate the expected frequencies for each cell. To do this,

- Calculate the total for each of the rows and columns by simply adding them together.
- For each cell, compute the following:

$$f_e = \frac{row\ total \times column\ total}{N}$$

N equals the total number of participants. Table 11.2 shows you how to calculate these values for our example.

TABLE 11.2 *Calculating Expected Frequencies*

		SUPPORT PAYING COLLEGE ATHLETES		
		YES	NO	TOTAL
Watched video	**Yes**	$\frac{(21 \times 30)}{44} = 14.3$	$\frac{(21 \times 14)}{44} = 6.7$	18 + 3 = 21
	No	$\frac{(23 \times 30)}{44} = 15.7$	$\frac{(23 \times 14)}{44} = 7.3$	12 + 11 = 23
	Total	18 + 12 = 30	3 + 11 = 14	18 + 12 + 11 + 3 = 44

The degrees of freedom are calculated by multiplying the degrees of freedom for each categorical value: $df = (r - 1) \times (c - 1)$, where r = number of rows (categories of one variable) and c = number of columns (categories of the other variable). Each of the variables in our example has one degree of freedom (2 - 1) = 1 and (2 - 1) = 1. We then multiply them together for one total degree of freedom.

Next, we need to calculate the test statistic. The test statistic is calculated by:

$$\chi^2 = \sum \frac{(f_o - f_e)^2}{f_e}$$

$$\chi^2 = \frac{(18 - 14.3)^2}{14.3} + \frac{(3 - 6.7)^2}{6.7} + \frac{(12 - 15.7)^2}{15.7} + \frac{(11 - 7.3)^2}{7.3}$$

$$\chi^2 = 0.96 + 2.04 + 0.87 + 1.88$$
$$\chi^2 = 5.75$$

Our hand-calculated value is very close to the value calculated by statistical software. Comparing the obtained value to Table A6 in the appendix tells us that the p value falls somewhere between .05 and .01, since 5.75 falls between 3.84 and 6.64. Figure 11.14 shows that a χ^2 value of 5.75 exceeds the critical value and falls in the rejection region.

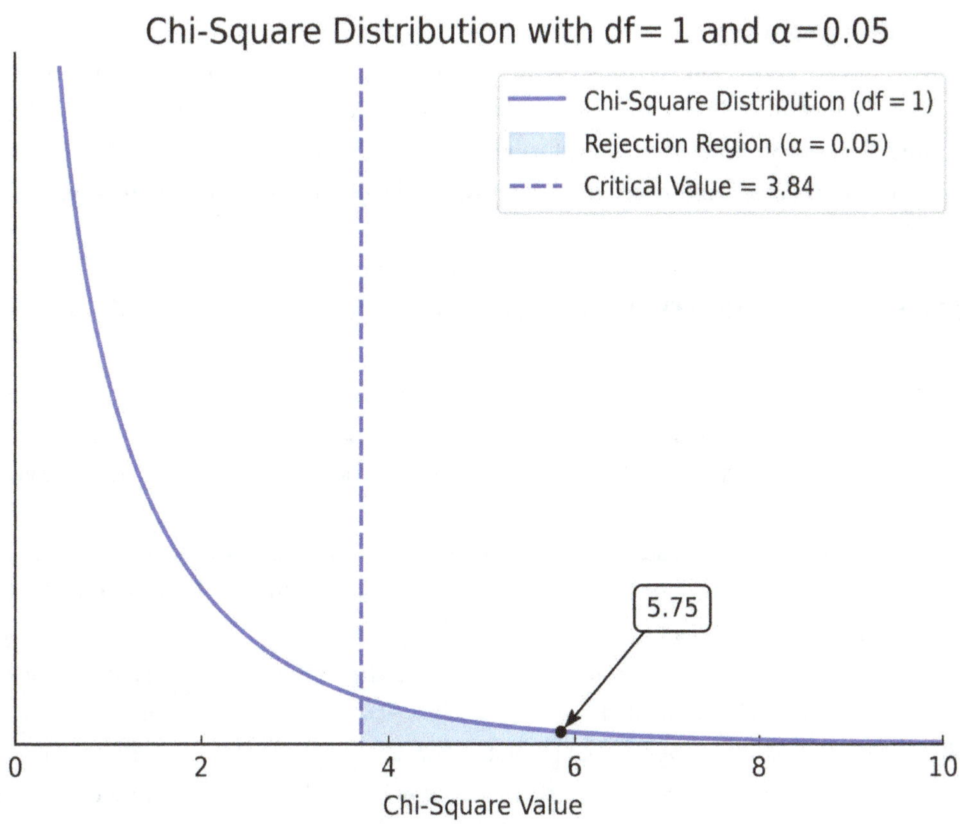

FIGURE 11.14 Observed χ^2 Compared to Critical Threshold for Rejection ($df = 1$, $\alpha = .05$)

There are two key limitations to consider when utilizing chi-square tests. First, large sample sizes can make trivial associations statistically significant. Small differences between observed (*O*) and expected (*E*) frequencies are magnified because the test statistic aggregates these differences across all cells. Second, Cochran (1954) explicitly stated that a minimum expected frequency of 5 is a reasonable threshold for the validity of the chi-square approximation in most practical cases because this helps ensure the validity of the test's results, particularly the accuracy of the *p* values. If expected cell frequencies are consistently below 5, alternative methods such as Fisher's exact test should be used.

CONCLUSION

This chapter highlights the critical role of statistical tools in understanding relationships between variables, with particular emphasis on correlational analyses and their limitations. Pearson's correlation coefficient (r) is foundational for assessing the direction and strength of associations, ranging from -1 to +1. Scatterplots and key assumptions such as linearity, homoscedasticity, and bivariate normality ensure meaningful and valid interpretations. Practical significance is illustrated through the coefficient of determination (r^2), which quantifies the variance in the dependent variable explained by the independent variable.

Furthermore, the chapter introduces the chi-square test for independence, which serves as a robust method for analyzing associations between two categorical variables. Unlike correlation or independent-samples t tests, the chi-square test allows researchers to examine associations using summary statistics without requiring the underlying dataset.

The discussion on limitations underscores the importance of recognizing outliers, range restriction, and sample size, which can significantly influence the results of correlational or chi-square analyses. Together, these tools provide complementary approaches to exploring relationships, reinforcing the distinction between correlation and causation, and preparing researchers for advanced techniques such as regression.

KEY TAKEAWAYS FOR EDUCATIONAL RESEARCHERS FROM CHAPTER 11

- Correlation measures association, not causation. Pearson's correlation coefficient (r) quantifies the strength and direction of a linear relationship between two continuous variables, but it does not imply that one variable causes the other. Researchers must avoid causal language and instead focus on interpreting the relationship as an association.
- Checking assumptions is crucial. Valid correlation analyses require meeting key assumptions such as *linearity*, *homoscedasticity*, and *bivariate normality*. A scatterplot is an essential tool for visually inspecting these assumptions before running statistical tests.
- Outliers, range restriction, and sample size impact correlation. Extreme values can distort correlation estimates, while a limited range of data (e.g., only studying high-achieving students) may weaken or even reverse a correlation. A *larger, more representative sample* would provide a more accurate estimate of the relationship.
- Chi-square tests for independence analyze categorical relationships. Unlike correlation, a chi-square test for independence is used to examine whether two categorical variables are significantly associated. The chi-square test can be an alternative to correlation when dealing with dichotomous or multicategory variables. This

method allows researchers to assess whether an observed association is statistically significant, helping policymakers and educators determine whether interventions such as different types of intervention programs are effective.

- Statistical significance does not always imply practical significance. Even a small correlation can be statistically significant in large datasets, but effect size measures such as r^2, the coefficient of determination, help determine how meaningful the relationship is in real-world educational settings.

KEY DEFINITIONS FROM CHAPTER 11

Bivariate normality is an extension of the normal distribution to two dimensions where two variables are jointly distributed in a way that their relationship forms a bivariate normal distribution.

A **chi-square test for independence** is a statistical test used to determine whether there is a significant association between two categorical variables.

The **coefficient of determination** (r^2) indicates the proportion of the variance in the dependent variable that is explained by the independent variable.

A **confounding variable**, or third variable, is an extraneous variable that was not controlled for and is the reason a particular "confounded" result is observed.

Covariance describes the extent to which two variables change (or vary) together.

Heteroscedasticity means that the spread of data points is not uniform across values of the independent variable. A heteroscedastic scatterplot often resembles a cone.

Homoscedasticity means that the spread of points above and below a line of best fit is the same along the entire line.

Linearity is the assumption that a straight line is the best way to describe a dataset instead of some type of curve.

Pearson's correlation coefficient (r) is an estimate of the population correlation ρ. It tells us the direction and the strength of the correlation.

Point-biserial correlation coefficient (r_{pb}) is used to measure the relationship between a *continuous variable* (e.g., motivation to learn math, achievement scores) and a *dichotomous variable* (e.g., pass/fail, enrolled/not enrolled, full-time/part-time).

Range restriction occurs when the variability of a variable (or variables) in a dataset is artificially limited or constrained, such that the range of values is smaller than it would be in the population. This restriction can significantly impact your results, potentially leading to biased estimates and incorrect conclusions.

Scedasticity is the extent to which data points are spread out above and below the line.

Spearman rank-order correlation coefficient (r_s), also known as Spearman's rho, measures the relationship between *two ranked (ordinal) variables.*

CHECK YOUR UNDERSTANDING

1. What is the main purpose of Pearson's correlation coefficient (r) in correlational analysis?
   a. To compare means across groups
   b. To measure the extent of causation between two variables

c. To measure the direction and strength of a relationship between two variables
d. To test for the significance of regression equations

2. Which of the following is NOT an assumption of correlational analysis?
   a. Linearity
   b. Homoscedasticity
   c. Causality
   d. Normality

3. What does a Pearson's *r* value of 0 indicate?
   a. A strong positive correlation
   b. No correlation
   c. A strong negative correlation
   d. A perfect correlation

4. Which statistical method is used when investigating a relationship between a continuous variable and a dichotomous variable?
   a. Pearson's *r*
   b. Spearman's rho
   c. Point-biserial correlation
   d. ANOVA

5. What does the assumption of homoscedasticity in correlational analysis require?
   a. The data must form a straight line.
   b. The variables must be normally distributed.
   c. The sample size must be large enough.
   d. The spread of data points must be consistent above and below the line across the entire range.

12

PREDICTING AN OUTCOME

LEARNING OBJECTIVES

By the end of this chapter, you will be able to:

- Explain the concepts of predictor and criterion variables, derive the equation of a regression line using the method of least squares, and interpret the slope and intercept within the context of data analysis;
- Identify and assess the key assumptions for multiple regression, including linearity, homoscedasticity, independence of errors, multicollinearity, and model specification, ensuring valid and reliable results in regression analyses; and
- Compute and interpret diagnostic measures such as variance inflation factors, tolerance, and the coefficient of determination (r^2 *or* R^2), and assess model fit to draw meaningful conclusions from regression analyses.

In Chapter 11, we reviewed the procedures to compute a correlation coefficient that indicates the extent to which two variables (X and Y) are related to each other. The value of r indicates the direction and strength of the relationship. Two variables change in opposite directions when r is negative, and two variables change in the same direction when r is positive. The closer r is to +1 or -1, the stronger the correlation and the more closely the two variables are related.

The strength of a correlation reflects how closely a set of data points fits a line of best fit on a scatterplot. This line is also called a *regression line*. We will now learn how to use the value of r to compute the equation of a regression line. We will then use this equation to predict values of one variable given known values of a second variable in a population before incorporating additional variables into our predictions.

FUNDAMENTALS OF LINEAR REGRESSION

Linear regression is a statistical method used to determine the equation of a regression line for a set of data points and to assess how well this equation can predict values of one variable based on known values of another variable. In linear regression, we identify two types of variables:

- **Predictor variable**, or known variable (X), is the variable with values that are known and can be used to predict values of another variable. We also call this variable an independent variable.
- **Criterion variable**, or to-be-predicted variable (Y), is the variable with unknown values that can be predicted or estimated, given known values of the predictor variable. We sometimes call this variable a dependent variable or an outcome variable.

As always, it is important to consider the data assumptions that must be met before performing any analysis. We will provide more details on some of these assumptions later, but the conditions that indicate data are a good fit for regression analysis include the following:

- *Linearity of the X-Y relationship.* The relationship between the predictor and criterion variable should be linear.
- *No range restriction.* The range of data should not be artificially limited, as this can affect the correlation and regression results.
- *Homoscedasticity.* The variance of the residuals (errors) should be consistent across all values of the predictor variable.
- *No significant outliers.* Outliers can disproportionately influence the regression line and distort results.
- *Sufficient sample size.* The sample should be large enough to ensure that estimates of the correlation coefficient (r) and regression coefficients (b) are stable and reliable.

Linear regression allows us to find the equation of a line that best describes a set of data points and test whether predictions made from this equation are statistically significant. Before we work through an example to answer these questions, let us first explain what the regression line is. Specifically, we want you to know what it means to say that a straight line has the best fit to the data.

WHAT MAKES THE REGRESSION LINE THE BEST-FITTING LINE?

Once we have determined that there is a linear pattern in a set of data points using a scatterplot, we want to find a regression line, or a straight line that has the best fit. The criterion used to determine the regression line is sum of squares (SS), or the sum of the squared distances of data points from a straight line. The line, associated with the smallest total value for SS, is the best-fitting straight line, called the regression line. Remember, we don't square the distances just to be complicated; we square the distances so that all the values are positive even if the actual value is *below* the expected value. Otherwise, one point that was very far above the line would cancel out a point that was equally far below. Instead, we want them to both count as being quite far away from the line, so we square the distance.

Suppose that we measure two variables, one predictor or independent variable plotted on the X-axis and one criterion or dependent variable plotted on the Y-axis. Figure 12.1 shows a scatterplot with five data points (A, B, C, D, and E) along with the line of best fit. The arrow from each data point to the line represents the distance from each data point to the best-fitting line. As you see, all data points deviate from the line to some extent. Data point D (below the line) has the largest deviation from the line. Data point A (above the line) has the smallest. To quantify these deviations, we square each distance to ensure that all values are positive. The sum of squares in this example is the total of these squared deviations across all five

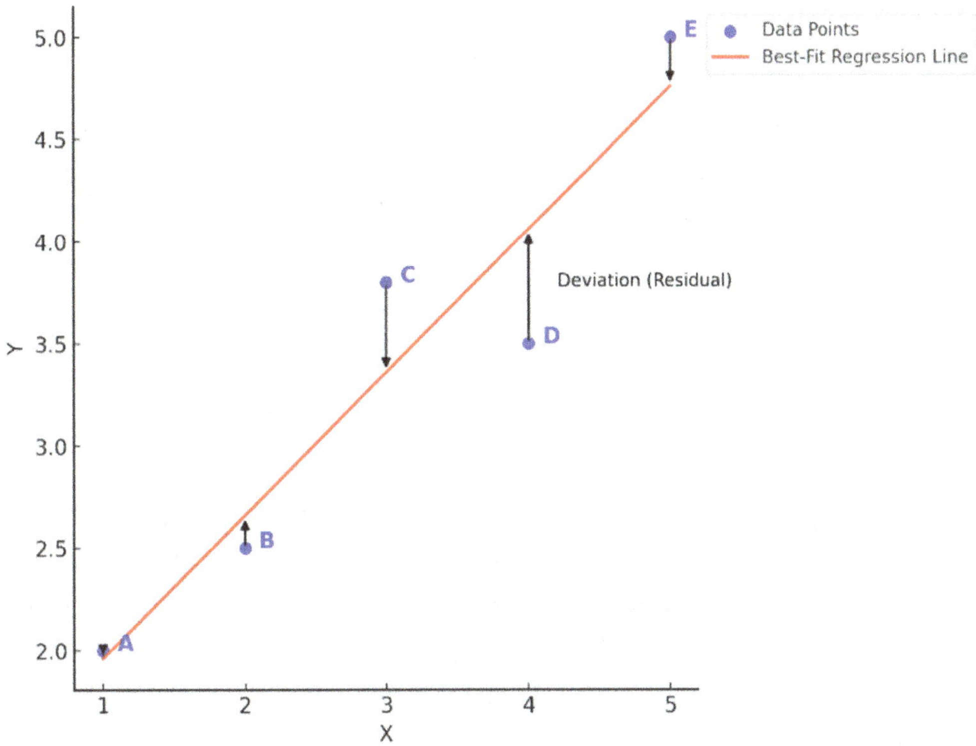

FIGURE 12.1 Concept of Sum of Squares: Deviations of Data Points from the Regression Line

data points. The regression line is defined as the one that minimizes this sum, providing the best possible fit to the data.

THE SLOPE AND INTERCEPT OF THE REGRESSION LINE

As we learned during grade school, the equation $y = mx + b$ represents the slope-intercept form of a linear equation. This equation describes a straight line on a Cartesian coordinate plane. You can think of the equation of regression as representing the same thing, but we use b to label the coefficients and a to represent the intercept. So, our new equation is $Y = bX + a$.

- The slope (b) of a straight line is used to measure the change in Y relative to the change in X. When X and Y change in the same direction, the slope is positive. When X and Y change in the opposite direction, the slope is negative.
- The intercept (a) of a straight line is the value of Y when $X = 0$. It is where a straight line crosses the Y-axis on the graph.

The method of least squares is the statistical procedure used to compute the slope (b) and the intercept (a) of the best-fitting straight line or the **ordinary least squares (OLS) regression line** to a set of data points. This line minimizes the sum of the squared differences (residuals) between the observed data points and the values predicted by the line. By minimizing these squared differences, ordinary least squares ensures that the line represents the closest possible approximation of the relationship between the variables. This is why the regression line is often referred to as the *least squares regression line.*

We will walk through how to compute the slope and intercept for the dataset by using the sum of squares. The process is tedious in particular when the sample size is large. In practice, you will use statistical software to compute the slope and intercept using the method of least squares, but it is good to understand what the output actually means.

To use the method of least squares, complete the following three steps:

- Step 1. Compute the preliminary calculations.
- Step 2. Calculate the slope (b).
- Step 3. Calculate the Y-intercept (a).

Suppose a teacher wants to examine the relationship between the number of hours students spend studying for a quiz (predictor variable: X) and their quiz scores (criterion variable: Y). To better understand this pattern, the teacher conducts a linear regression analysis to find the best-fitting line that predicts quiz scores based on study time. In this example, we use a small dataset to illustrate the three steps involved in fitting a regression line. However, in practice, regression analysis is conducted with larger datasets. The teacher gathers data from four students (Table 12.1).

TABLE 12.1 *Hours Studying and Quiz Scores*

PREDICTOR VARIABLE STUDY HOURS (X)	CRITERION VARIABLE QUIZ SCORE (Y)
0	3
1	3
2	5
3	5

STEP 1: COMPUTE PRELIMINARY CALCULATIONS

We begin by making the preliminary calculations shown in Table 12.2. The signs (+ and -) of the value that we measure for each variable are essential to making accurate computations. They tell us whether the actual

TABLE 12.2 *Computing SS_X, SS_{XY}, and Key Components for Regression Analysis*

X	Y	$X - \bar{X}$	$Y - \bar{Y}$	$(X - \bar{X})(Y - \bar{Y})$	$(X - \bar{X})^2$
0	3	-1.5	-1	1.5	2.25
1	3	-0.5	-1	0.5	0.25
2	5	0.5	1	0.5	0.25
3	5	1.5	1	1.5	2.25
$\bar{X} = 1.5$	$\bar{Y} = 4$			SS_{XY}: $\Sigma\,(X - \bar{X})(Y - \bar{Y}) = 4$	SS_X: $\Sigma\,(X - \bar{X})^2 = 5$

value is above or below the regression line. The goal of this step is to compute the sum of squares needed to calculate the slope and intercept of the best-fitting line in regression analysis.

The first two columns are observed values for X (number of hours) and Y (quiz score). But other parts of the chart may require some thinking to understand:

- $\bar{X}$: The mean of X values.
- $\bar{Y}$: The mean of Y values.
- $X - \bar{X}$: The deviation of each score (X) from the mean ($\bar{X}$).
- $Y - \bar{Y}$: The deviation of each score (Y) from the mean ($\bar{Y}$).
- $(X - \bar{X})(Y - \bar{Y})$: The product of the deviation scores for X and Y.
- $(X - \bar{X})^2$: The product of the deviation scores for X. This should look familiar from back when we learned about variance.
- $\Sigma (X - \bar{X})(Y - \bar{Y})$: The sum of products for X and Y. This is also referred to as the covariance of X and Y, or SS_{XY}.
- $\Sigma (X - \bar{X})^2$: The sum of squares for X, which is represented as SS_X.

It's easy to get lost in all the SSs, Xs, and Ys, so let's break it down differently. First, consider why it's important to understand how much X and Y deviate from their respective means. When determining the best-fitting line, a key factor is how much the values of X and Y vary. Are the data points widely spread out, or are they clustered closely together? Does X have a broad range while Y is more constrained or vice versa? Or do both variables exhibit similar variability? A scatterplot provides a visual representation of this spread, while these calculations quantify it mathematically.

Next, we introduce a column where we multiply the two deviation scores and sum them. If you recall from our discussion on correlation, the numerator of the correlation formula was SS_{XY}—this represents the *covariance*, or how much the two variables *vary together*. Finally, in the last column we calculate SS_X, which when divided by $n - 1$ gives us the *variance of X*.

STEP 2: CALCULATE THE SLOPE (b)

It is time to recall a fundamental math concept from grade school: the formula for *slope* as *rise over run*, or the *change in Y divided by the change in X*. Keep this general idea in mind as we walk through the actual calculations—it will help the process make more sense.

You can calculate the slope of an OLS regression line by using the following equation:

$$b = \frac{SS_{XY}}{SS_X} = \frac{\sum(X_i - \bar{X})(Y_i - \bar{Y})}{\sum(X_i - \bar{X})^2}$$

Go back to Table 12.2 earlier in the chapter and see what that represents. On top we have the sum of the changes in X times the changes in Y, and on the bottom we have the changes in X times the changes in X. If we cancel out changes in X from top and bottom, we have the changes in Y over the changes in X, which sounds a lot like *rise over run*. The slope of our regression line is

$$b = \frac{4}{5} = 0.8$$

STEP 3: CALCULATE THE INTERCEPT (a)

Remember, our regression line formula is $Y = bX + a$. We now have the value for b, but we still need a. If we have values for Y and X, we can plug them in and solve for a. But we can't use any specific individual's values because we don't know that they fall exactly on the line. We do know one point that is always on the regression line, however, and that is $(\bar{X}, \bar{Y})$. And we already calculated both of those values in Table 12.2. So, if we rearrange the formula for the slope of a line, we end up with:

$$a = \bar{Y} - b\bar{X}$$

$$a = 4 - [(0.8) \times (1.5)] = 2.8$$

We now have our equation for least squares regression:

$$\hat{Y}_i = 0.8X_i + 2.8,$$

where $\hat{Y}_i$ (which you actually pronounce "Y-hat") is the predicted value of Y for person i for a given value of X_i. The values of b and a are sometimes called regression coefficients.

What do we do with this equation? You might have wondered—when we discussed using a known variable X to predict an unknown variable Y—how that works if we don't already know the values of Y for new cases. The key is that we first need a set of observed data, where we already have X and Y values for multiple individuals. Using these known values, we calculate the regression equation, which then allows us to make predictions about Y for new observations based on their X values.

So, is Y known or unknown? In our initial dataset, Y is known because we have chosen to measure it. However, in future scenarios if our sample is representative of a larger population, we can use this equation to estimate Y for any given X value. For example, if we wanted to predict the outcome when X = 5, we would simply substitute 5 for X in the regression equation:

$$\hat{Y}_I = 0.8(5) + 2.8 = 6.8.$$

Based on that equation, we would predict that for X = 5, Y would equal 6.8. Figure 12.2 shows our regression line passing through the point (5, 6.8).

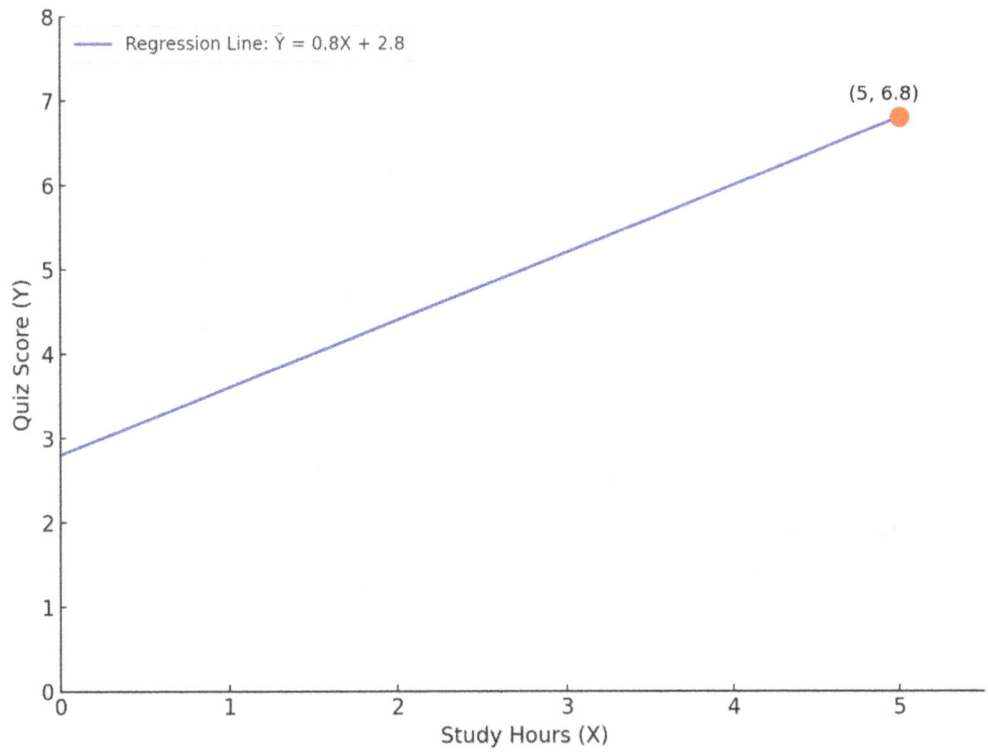

FIGURE 12.2 Regression Line with Predicted Value at X = 5

CONDUCTING REGRESSION ANALYSIS WITH SAMPLE EDUCATION DATA

We are now going to use more complicated but realistic data to explain the process of regression analysis. Imagine we want to determine whether student motivation can be used to predict math achievement. If a significant relationship exists, we could develop a *predictive model* based on this sample data. This model would allow us to estimate the math scores of future students using only their motivation scores, providing a useful tool for educators and researchers in understanding and supporting student performance.

We collected data from 100 high school juniors enrolled in an Algebra II course. At the beginning of the semester, students completed a motivation assessment (scored from 1 to 10), which measured their self-discipline, interest in math, and effort toward learning. Students took a state-mandated final exam at the end of the semester, with scores recorded on a scale of 0–100. To investigate this, we will run a simple linear regression analysis:

- *Dependent variable (outcome)*: Math final exam score
- *Independent variable (predictor)*: Motivation score

We have three more steps, but this time they are about the entire process of analyzing data with regression rather than just computing the equation of the line:

- Step 1. Construct the regression model (i.e., estimate the regression coefficients).
- Step 2. Assess the statistical adequacy of the "fit" of the regression model to the data.
- Step 3. Evaluate the practical utility of the fitted regression model.

For now, we will only consider a case of simple linear regression in which the X-Y relationship must be linear and there is only a single X. Using more than one predictor variable is known as multiple regression, which we'll get to later in this chapter.

STEP 1: CONSTRUCT THE REGRESSION MODEL

Figure 12.3 provides a scatterplot showing math and motivation scores, which suggests that a straight-line regression model might be appropriate.

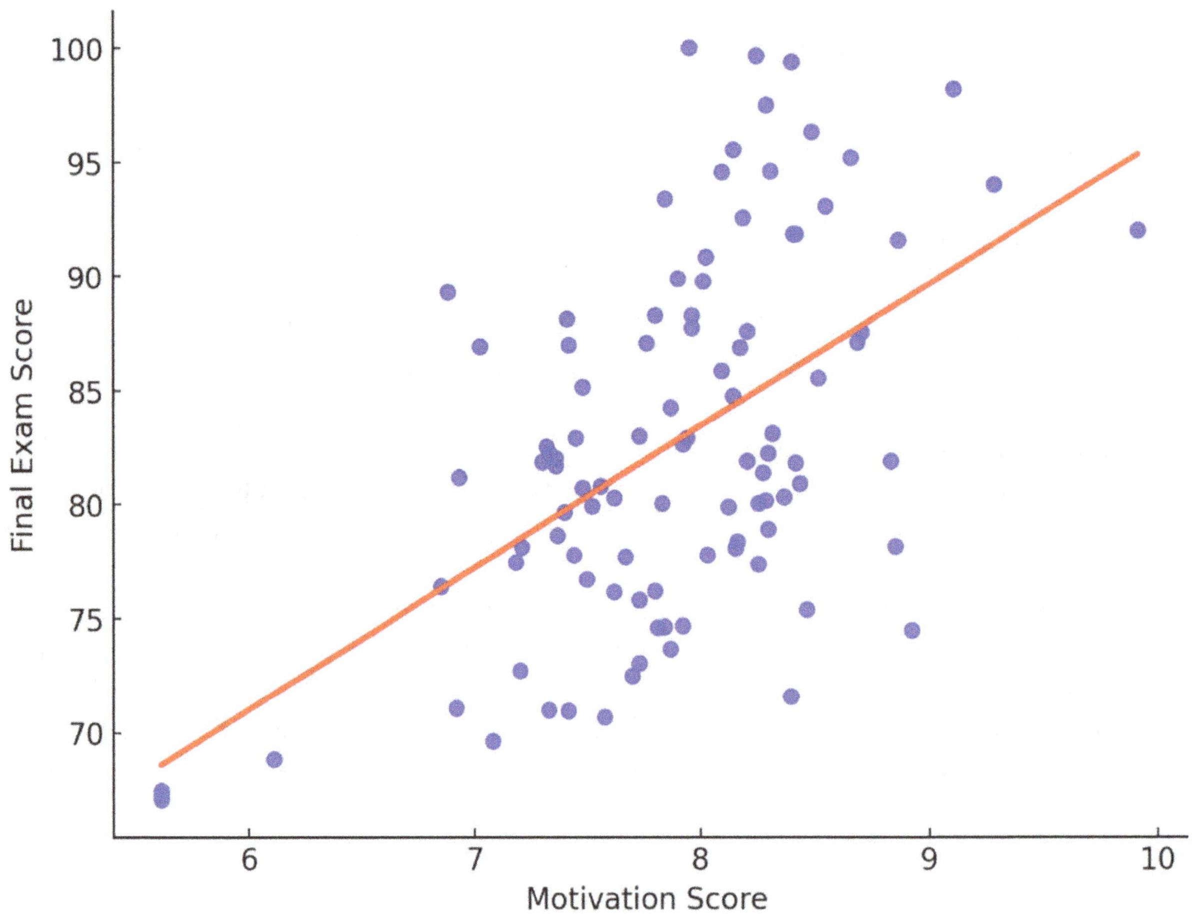

FIGURE 12.3 Scatterplot of Math Final Exam and Motivation Scores

Next, we estimate the regression coefficients using least squares regression, which finds the best-fitting straight line by minimizing the vertical distance, called *error*, between the data points in the scatterplot and the line.

$$Y_I = bX_i + a + \varepsilon_i$$

That equation should look familiar except that we've now included the error term, which we'll explore in the next section. Recall that to calculate the regression coefficients, we needed to determine the mean of X, the mean of Y, the variance of X, and the covariance between X and Y—all of which we previously computed by hand. However, when $n = 100$, performing these calculations manually becomes impractical. Instead, we rely on statistical software to generate these values, allowing us to focus on understanding and interpreting the regression analysis rather than manually computing each step.

First, let's look at descriptive statistics for math and motivation scores.

TABLE 12.3 DESCRIPTIVE STATISTICS FOR MATH AND MOTIVATION SCORES

	N	MEAN	*SD*	VARIANCE	*SS*	(SS_{XY})	CORRELATION (*r*)
Motivation score	100	7.86	0.7	0.49	48.87	303.82	0.54
Final exam score	100	82.59	8.02	64.28	6364.12		

We could now plug those values into the same formula we used before to calculate the slope:

$$b = \frac{SS_{XY}}{SS_X} = \frac{303.82}{48.87} = 6.22$$

This means that for every *one-unit increase* in motivation scores, we can expect an *average increase of 6.22 points* in final exam scores. Now, the next step is to calculate the intercept, which represents the predicted exam score when motivation is zero.

$$a = \bar{Y} - b\bar{X} = 82.59 - (6.22)(7.86) = 33.70$$

Our prediction equation is $\hat{Y}_i = 6.22X_i + 33.70$.

Statistical software often provides both standardized and unstandardized estimates for coefficients. The coefficients we just calculated are unstandardized, meaning they retain the original metric of the scores.

STEP 2: ASSESSING THE STATISTICAL ADEQUACY OF THE "FIT" OF THE REGRESSION MODEL TO THE DATA

There are two primary ways to assess the fit of a regression model (aside from statistical significance, which we will discuss later in the chapter): the accuracy of the predictions and the proportion of variation explained by the model.

PREDICTION ACCURACY

Let's take a few steps back and imagine there is no known independent variable that could be used to predict another variable. For example, suppose you wanted to predict a student's final exam score, but you don't have their motivation score. What is your best guess for the student's final exam score? Probably the overall average exam score is the best you can do.

Now suppose you do have access to the student's motivation score, and you have the regression equation we generated above. Now, your best guess is $\bar{Y}_i$. However, the prediction equation is not perfect, and there will be some amount of prediction error. Therefore, for any given value of X for the ith student, there are three sources of deviation in Y:

- $(Y_i - \bar{Y})$ = total deviation; the deviation between the observed Y and the average Y (or, $\bar{Y}$). This is the amount that you were "off" when you just used the average math score to guess this individual's score.
- $(\hat{Y}_i - \bar{Y})$ = explained deviation by regression or model; the deviation between the average of Y (or $\bar{Y}$) and the predicted $\bar{Y}$ (or $\bar{Y}$). This is the improvement of your prediction using the regression model.
- $(Y_i - \hat{Y}_i)$ = error or unexplained deviation, the deviation between the observed Y and the predicted Y (or, $\hat{Y}$). This is also known as residual variance.

Figure 12.4 labels all three of those sources of deviation, with the total on the left and then error deviation above explained deviation on the right.

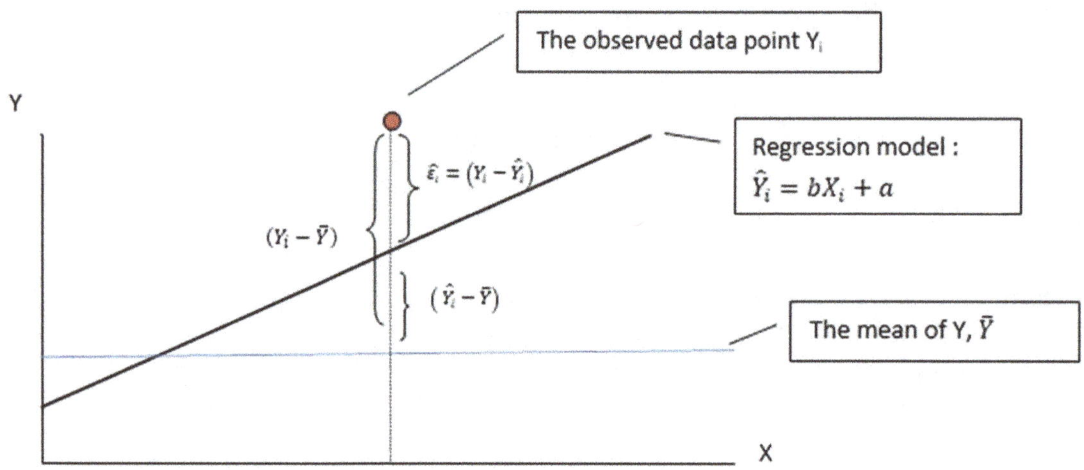

FIGURE 12.4 Breaking Down Total Variation: Explained versus Unexplained Deviation

When we use least squares regression, we're finding the regression line that minimizes the error or, in other words, minimizing the value of $\Sigma (Y_i - \hat{Y}_i)$. But remember, if we just use the actual deviations, those will add up to zero, so instead we square them. For all students in the sample, as summarized in Table 12.4,

TABLE 12.4 ***Summary of Sum pf Squares for a Simple Least Squares Regression***

	SUM OF SQUARES
SS Total	6364.12
SS Residual	4475.42
SS Regression	1888.7

$$\sum(Y_i - \bar{Y})^2 = \sum(Y_i - \hat{Y})^2 + \sum(\hat{Y}_i - \bar{Y})^2$$

$$SS_{Total} = SS_{Residual} + SS_{Regression}$$
$$6364.12 = 4475.42 + 1888.7$$

These equations are just representations of Figure 12.4. We have residual variation plus explained (or regression) variation on the right, and they equal total variation on the left. All of this is leading up to the way we measure the accuracy of our prediction: the standard error of the estimate. Remember, standard deviation from the mean indicates the average distance between each data point and the mean. Similarly, the standard error of the estimate measures the standard deviation of data points in the sample from the regression line. Standard deviation is the square root of the variance, and the standard error of the estimate is the square root of the variance for residual (error) variation.

When calculating standard deviation, we take the square root of the sum of squared deviations from the mean divided by the degrees of freedom. Mathematically, this is expressed as:

$$S_y = \sqrt{\frac{\Sigma(Y_i - \bar{Y})^2}{N-1}} = \sqrt{\frac{SS_{Total}}{N-1}}$$

This is something we covered in Chapter 3, so don't worry too much if it looks unfamiliar now. We just want you to see the similarity to the formula for the standard error of the estimate. This time we're looking at residual variation (or error variation) rather than total, and degrees of freedom are $N-2$. The formula for the standard error of the estimate is:

$$S_{y-\hat{y}} = \sqrt{\frac{\Sigma(Y_i - \hat{Y})^2}{N-k-1}} = \sqrt{\frac{SS_{Residual}}{N-2}}$$

Another name for $SS_{Residual}/(N-2)$ is mean square residual. If we take the square root of 45.67 we get 6.76, which is the standard error of the estimate for this regression equation. In other words, the average predicted math score (or $\bar{Y}$) is off by about 6.76 points. Is that a good prediction? Well, if you look back at the descriptive statistics, you can see that the standard deviation of final exam scores is about 8.02, so if we're just using the average score to predict any given individual student's score, we can expect to be off by about 8.02 points in either direction. With our regression equation we've reduced that value somewhat, and now we're only likely to be off by around 6.76 points.

PROPORTION OF VARIATION EXPLAINED, R^2 (COEFFICIENT OF DETERMINATION)

The second way to assess fit is to look at how much of the variation in the dependent variable (final exam scores) can be explained by variation in the independent variable (motivation scores). The **proportion of variance explained** in regression is a measure of how well the independent variable(s) explain the variability in the dependent variable. This measure is often expressed as the coefficient of determination (r^2 or R^2). R^2 represents the proportion of the total variance in the dependent variable (Y) that is explained by the independent variable (X) through the regression model. Remember this equation from earlier about the sources of variation?

$$SS_{Total} = SS_{Residual} + SS_{Regression}$$

When we were estimating accuracy, we were interested in the residual variation. This time, we're interested in the variation that is explained by the model.

$$R^2 = \frac{SS_{\text{Explained}}}{SS_{\text{Total}}} = \frac{1888.7}{6364.12} = .297$$

In other words, R^2 represents the ratio of explained variation to total variation in the dependent variable. This information is typically included in the standard output of a regression analysis or, at the very least, can be requested as part of the results.

R^2 for this model is 0.297, meaning that approximately 30% of the variation in final exam scores among high school juniors can be explained by differences in motivation scores. R^2 ranges from 0 to 1, where 0 means that the model explains none of the variance in Y and 1 means that it explains all the variance. A higher R^2 value suggests that the predictor variable(s) provide a stronger explanation of the variation in the dependent variable.

The next value reported is adjusted R^2, which is adjusted based on how many predictors are included in your model. This matters more when we have more than one predictor for multiple regression. In this case, the values of R^2 and adjusted R^2 are almost the same.

We will explore adjusted R^2 in more detail later in this chapter. However, it is important to note that unlike R^2, which *always* increases when more predictors are added (even if they are not meaningful), Adjusted R^2 accounts for the number of predictors and only increases *if* the new predictor improves the model more than what would be expected by chance alone. In other words, adding meaningless predictors reduces adjusted R^2, and in some cases it can even become negative, indicating that the additional predictors do not contribute useful information to the model:

$$R^2_{adjusted} = 1 - \left(\frac{(1 - R^2)\,(N - 1)}{N - p - 1}\right),$$

where N is the number of observations and p is the number of predictors.

STEP 3: EVALUATE THE PRACTICAL UTILITY OF THE FITTED REGRESSION MODEL

With the available information, we must now decide whether to use, modify, or discard the fitted regression model. While metrics such as R^2 and the standard error of the estimate can help assess the model's statistical validity, we must also determine its practical value. For example, do the results hold significance for the field? Are they publishable? Another important consideration is whether the model performs consistently across different subgroups. Often, these questions cannot be answered solely through statistics—they require a deep understanding of the existing literature, context, and practical implications. That being said, statistical significance remains an important factor to consider when deciding whether to use, refine, or discard the model.

TESTING REGRESSION SLOPES FOR STATISTICAL SIGNIFICANCE

In a simple linear regression analysis, we run two separate significance tests. One is for the regression model as a whole. This test uses the F distribution. You can also test the significance of each coefficient individually to determine if the slope (b) is significantly different from 0 and if the intercept (a) is statistically different from zero, which uses the t distribution. When your model includes only one predictor variable, the statistical significance of the overall model and the significance of the individual coefficient will always yield the same result. This is because in simple regression, the predictor's contribution to explaining variance in Y is identical to the model's overall explanatory power.

For now, we will focus on testing the individual coefficient, so let's go back to our regression equation. The slope we calculated was $b = 6.22$. This is subject to sampling error like any other statistic, so we need to test it. This means that we follow the four steps we've used before.

STEP 1: STATE THE HYPOTHESIS

The null hypothesis (H_0) states that there is no linear relationship between the independent variable (X) and the dependent variable (Y), implying that the slope of the regression line is 0. The alternative hypothesis (H_A) posits that there is a significant linear relationship between X and Y, meaning the slope of the regression line is not 0. By testing these hypotheses, we determine whether changes in X are significantly associated with changes in Y and thus whether the regression line provides meaningful predictions for the data.

H_0: $B = 0$, changes in Y are not related to changes in X.
H_A: $B \neq 0$, changes in Y are related to changes in X.

In our example,

H_0: Changes in final exam scores (Y) are NOT related to changes in motivation scores (X).
H_A: Changes in final exam scores (Y) ARE related to changes in motivation scores (X).

Or we can simply write:

$$H_0: B = 0$$
$$H_A: B \neq 0$$

Note that we use B instead of b to distinguish it from the sample statistic (b).

STEP 2: SET THE CRITERIA FOR A DECISION AND EVALUATE DATA

- We will use a 0.05 level of significance.
- *df* for the *t* test is equal to the sample size minus 2 or ($N - p - 1$).
    - The sample size minus the number of predictors in the model -1 (1 for each variable, *X* and *Y*)
    - Because $N = 100$, $df = 98$
- The critical value for this two-tailed test is ±1.98, shown in Figure 12.5.

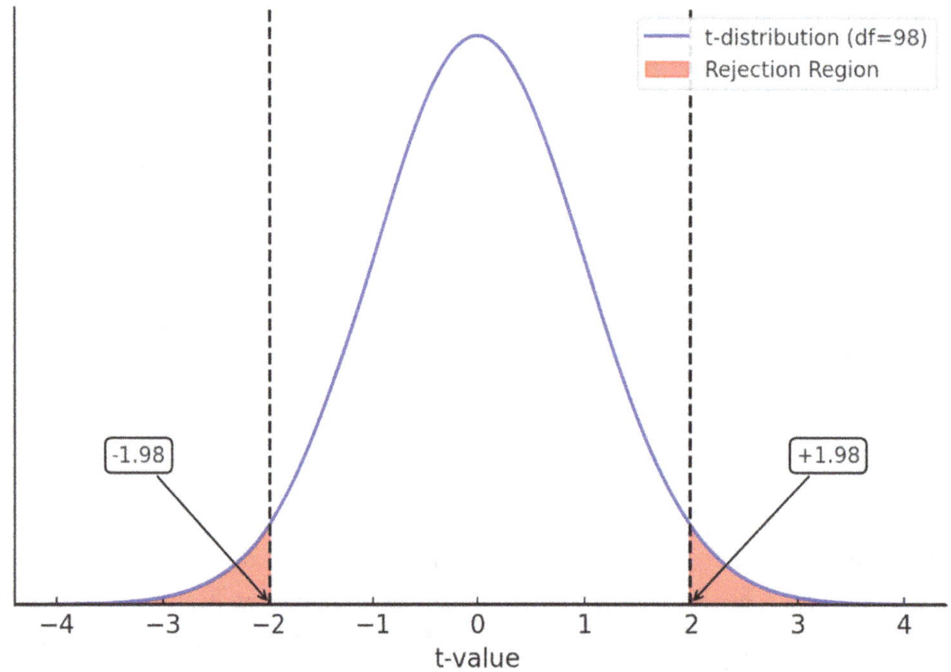

FIGURE 12.5 Critical Values and Rejection Regions for a Two-Tailed Test (α = 0.05, *df* = 98)

Linear regression relies on several important assumptions above and beyond those listed earlier in the chapter to ensure that the results are valid and reliable. The first assumption is *linearity*, meaning there should be a straight-line relationship between the dependent variable (the outcome being predicted) and each of the independent variables (the predictors). For example, if a school administrator wants to predict student test scores based on the number of hours students spent preparing for exams, there should be a roughly linear relationship between preparation time and scores. If the relationship is curved—say, where additional study time beyond a certain point shows diminishing returns—then the linearity assumption would be violated, and the regression model may not provide accurate predictions.

Another key assumption is *the independence of errors*, which means that the errors (the differences between actual and predicted values) should not be related to one another. This assumption is particularly important when analyzing data collected over time or across groups. For instance, if a principal is studying the relationship between teacher professional development hours and student achievement across several years, the test score data from one year could influence the next year's results due to shared factors such as evolving curriculum and teacher collaboration. A violation of this assumption might indicate that patterns, like trends over time, are not fully captured by the model.

The third assumption is *homoscedasticity*, which means that the errors should have a consistent spread across all predicted values. This assumption is violated if the variability in errors increases or decreases depending on the value of the predictors. For example, when examining the relationship between high school grade point average (GPA) and college readiness test scores (such as SAT and ACT), the errors might be small for students with average GPAs but larger for students with very high or very low GPAs. This could happen because other unaccounted factors, such as extracurricular involvement and test anxiety, might play a bigger role at those extremes.

Another critical assumption is *the normality of errors*, meaning the residuals (errors) should follow a normal bell-shaped distribution. This is important for conducting statistical tests and accurately interpreting the results. For example, if a university researcher is predicting student retention rates based on variables such as campus engagement and financial aid, the prediction errors should cluster around zero, with few extreme errors. If errors are skewed—perhaps because certain student subgroups (such as first-generation students) behave differently—the results might not be reliable.

STEP 3: COMPUTE THE TEST STATISTIC

TEST STATISTIC (t) AND DEGREES OF FREEDOM

Let's determine whether the slope in our sample dataset is significantly different from zero. Typically we would use statistical software for these calculations, but since we have already computed the descriptive statistics, we can work through the calculations by hand.

If you recall, calculating a t statistic always involves comparing two values—whether it's two group means or a group mean and a reference point—and dividing the difference by some form of standard error. In this case, we follow the same logic:

- We subtract zero from the estimated beta coefficient (our slope).
- We then divide this difference by the standard error of the beta coefficient (which is not the same as the standard error of the estimate discussed earlier).

This approach allows us to assess whether the slope is statistically significant and meaningfully different from zero.

Let's start with the equation for the t statistic:

$$t = \frac{b - B}{SE_b}$$

Because we expected $B = 0$ under the null hypothesis,

$$t = \frac{b - 0}{SE_b} = \frac{b}{SE_b}$$

We have already calculated the slope as $b = 6.22$, but how do we calculate the standard error of the slope? The formula is

$$SE_b = \frac{S_{y-\hat{y}}}{\sqrt{\Sigma(X_i - \bar{X})^2}},$$

where $S_{y-\hat{y}}$ is the standard error of the residuals or errors, also known as the standard error of the estimate. It is calculated as:

$$S_{y-\hat{y}} = \sqrt{\frac{\Sigma(Y_i - \hat{Y})^2}{N-2}} = \sqrt{\frac{SS_{Residual}}{N-1}},$$

$\hat{Y}_i$ represents the predicted values of Y using the regression line.
N equals the number of data points in our dataset.

$$S_{y-\hat{y}} = \sqrt{\frac{4475.42}{100-2}} = 6.76$$

Now, we have all the values we need to calculate the standard error of the slope:

$$SE_b = \frac{6.76}{\sqrt{48.87}} = 0.96$$

Compute the t statistic:

$$t = \frac{6.22}{0.96} = 6.48$$

Calculate the degrees of freedom:

$$df = 100 - 2 = 98$$

STEP 4: MAKE A DECISION

Before starting the analysis, we decided to use an alpha level of .05 for this two-tailed test. Since the obtained t statistic (6.48) is greater than the critical value (1.98), we reject the null hypothesis. Therefore, we conclude that on average, motivation scores significantly predict final exam scores. From our earlier calculations, we know that motivation scores explain approximately 30% of the variation in the dependent variable.

> APA style sample write-up: A simple linear regression was conducted to examine whether motivation scores significantly predict final exam scores. The results of the regression indicated that the model was statistically significant, $F(1, 98)^* = 41.36$, $p < .05$, and explained 29.7% of the variance in final exam scores. The unstandardized regression coefficient for motivation was $b = 6.22$, $t(98) = 6.43$, $p < .001$, indicating that for every one-unit increase in motivation score, final exam scores increased by approximately 6.22 points. These findings suggest that higher motivation scores are significantly associated with higher final exam scores.

*The omnibus F test will be explained in the next section.

MULTIPLE REGRESSION: PREDICTING AN OUTCOME WITH MULTIPLE PREDICTORS

So far we have learned the fundamentals of simple linear regression, where we use changes in one variable to predict changes in another. Now, let's extend this concept by incorporating additional predictor variables.

While there is a lot to cover, multiple regression is a powerful and widely used method in research and practice. Multiple regression allows us to better understand relationships between multiple predictors and an outcome variable. Since you will likely use and encounter multiple regression frequently in your career, developing a strong understanding of its principles is essential.

Here's the general outline for this section so you don't get lost:

1. *Getting started.* What variables should you include and how? What do you need to check before running your analysis?
2. *Assumptions and diagnostics.* What you need to check to avoid possible problems with your results.
3. *Basics of multiple regression.* Running the analysis and interpreting results, understanding the effects of each predictor separately, and reporting results.

GETTING STARTED

Multiple regression uses two or more predictors, or independent variables, to predict variation in a quantitative outcome, or dependent variable. Sometimes there are multiple factors influencing an outcome, and you want to control for one factor to understand the influence of another. You can do this by including both variables in a regression model. Here is what the statistical model for multiple regression looks like:

$$Y_i = \beta_0 + \beta_1 X_{i1} + \beta_2 X_{i2} + \ldots + \beta_p X_{ip} + \varepsilon \text{ for } i = 1, 2 \ldots, n.$$

- Y_i is the score of the ith subject ($i = 1, 2, \ldots, n$) on the dependent variable.
- β_0 represents the intercept in multiple regression, which is the predicted value of the dependent variable when all predictor variables are zero. In simple linear regression, we previously used the letter a to denote the intercept, but in multiple regression, β_0 is the standard notation.
- β_1 is the slope (or regression coefficient) associated with the predictor X_1 and reflects the expected change in Y for each one-unit increase in X, with the other predictors held constant.
- β_2 through β_p are the regression coefficient for predictors 2 through p. In this context, p refers to the number of predictors in the model and is not related to the p value used for statistical significance testing.
- ε_i is the error term.

CONSIDERATIONS FOR BUILDING A MULTIPLE REGRESSION MODEL

When you are planning to analyze data with multiple regression, you need to decide what independent variables to include. There are several factors to consider when making this decision.

Sample size. There is no strict sample size requirement for multiple regression analysis, as the ideal sample size depends on the complexity of the model. However, various rules of thumb exist. A common guideline suggests having at least 10 cases per predictor variable. For example, you should aim for a sample size of at least 50 cases if you plan to include five predictors. This estimate is only a rough guideline, as the actual sample size needed to detect a statistically significant effect depends on the effect size—or how well the independent variables predict the outcome. To determine a more precise sample size, researchers can use power analysis, which calculates the minimum number of observations needed to reliably detect an effect.

Purpose of model. Is your regression model exploratory or explanatory?

- If your research is relatively new and lacks a strong theoretical framework, your focus may be on exploring new significant predictors.
- If your study is grounded in an established research base and aims to test a specific theory, your analysis is explanatory.

In general, it is best to use theory whenever possible when designing a regression model. A strong theoretical foundation not only guides model building but also helps in interpreting and discussing results when significant predictors are identified. Without a theoretical framework, explaining your findings can be more challenging.

Parsimony. One key metric for assessing how well a regression model predicts the dependent variable is R^2. We will discuss this in more detail later, but an important consideration is that R^2 always increases as more predictors are added, regardless of whether they meaningfully contribute to the model. This can make it tempting to keep adding predictors in an attempt to improve the model. However, doing so can lead to **overfitting**, which is a situation where the model becomes overly complex and fits the training data too closely. Instead of capturing the true underlying relationship between variables, an overfitted model captures random noise or fluctuations, reducing its ability to generalize to new data.

To avoid overfitting, predictors should be selected based on theoretical justification from your literature review. Ideally, they should be known predictors of your outcome variable or variables for which you have a strong rationale for testing their influence.

DATA EXPLORATION AND PREPARATION

Once you have an idea about which predictor (independent) variables you want to include, you start your analysis by looking at the descriptive statistics for your independent variables and dependent variable. What should you check?

- *Frequencies*
  - If you have dichotomous variables (e.g., enrolled in gifted education program/not enrolled in gifted education program), check to see if you have sufficient numbers in each category. It is okay if the groups are not the same size, but we suggest having at least 10 cases in each group.
  - Before conducting a regression analysis, it is essential to check for missing data within your dataset. If a variable contains a significant amount of missing data, it can pose a challenge because regression analysis typically employs listwise deletion to address missing values. This means that any case with a missing value in one or more variables will be excluded from the analysis entirely. As a result, extensive missing data can lead to a substantial reduction in the sample size, potentially affecting the validity and generalizability of the results.

Therefore, identifying and addressing missing data through appropriate techniques—such as imputation and sensitivity analysis—can help mitigate these issues.

- *Correlations*
    - It is important to assess the correlation between the dependent variable and each independent variable. This helps determine whether a relationship exists and provides insight into the strength and direction of each predictor's association with the dependent variable. However, correlation alone does not fully determine a variable's importance in the model. Even if an independent variable shows a weak correlation with the dependent variable, it may still be theoretically significant or serve as a crucial control variable to better isolate the effects of other predictors. This highlights the importance of reviewing the existing literature on the dependent variable to make informed modeling decisions.
    - Intercorrelations among independent variables can indicate potential issues with *multicollinearity*, which occurs when predictors are highly correlated with one another. While we will explore this topic in greater detail later in this chapter, a general guideline is that a correlation of $r = 0.7$ or higher between two independent variables may signal a multicollinearity concern. If such high correlations are present, further investigation will be necessary later in the analysis to determine their impact on the regression model and whether corrective measures, such as removing or combining variables, are needed.
- *Outliers*
    - Unusual observations can disproportionately affect model fitting and inferences about your results. It is a good idea to check the minimum and maximum values for each of your predictors to see if you have any extremely unusual cases, which can sometimes be due to data entry errors. But there are also three diagnostic indices for outliers: distance, leverage, and influence.
        - *Distance:* This metric assesses how well the model fits the data by measuring the difference between the observed value and the predicted value, expressed as $(Y_i - \hat{Y}_i)$. Larger differences suggest greater random error or potential systematic issues due to outliers. A commonly used measure of distance is the studentized residual, which expresses how far a residual deviates from zero in standard deviation units.
        - *Leverage:* This measure helps identify potential outliers based on predictor variables. Leverage values close to $\frac{1}{N}$ are generally unremarkable, while values exceeding $\frac{3(p+1)}{N}$, where p is the number of the predictors, may indicate influential points that require further investigation.
        - *Influence:* Influence combines both distance and leverage to detect observations that significantly impact the regression model. One widely used measure is *Cook's D*, which evaluates how much the regression coefficients would change if a particular data point were removed. Observations with *Cook's D* values greater than 1 are often flagged as potentially influential.
    - In short, an outlier is a data point that does not follow the general trend of the dataset. An outlier has high leverage if the value on any independent variable (or combination of independent variables) is extremely high or low. A data point with high leverage has the potential to be influential (i.e., change the results of the model), but not all high leverage data points are highly influential.
- *Recoding Variables (If Necessary)*
    - Before conducting a regression analysis, ensure that your data are properly formatted and ready for use. This often involves data cleaning to transform categorical responses into numerical values suitable for analysis. For instance, if survey respondents answered "Yes" or "No" to a question, you must recode these responses as 1 and 0, respectively. Typically, the focal group (the category of interest) is coded as 1, while the reference group is

coded as 0. Properly preparing your data in this way ensures that your variables are appropriately structured for regression modeling.

- Another common mistake in regression analysis occurs when dealing with categorical variables with more than two possible responses. If your data contain a categorical variable with multiple groups, you cannot treat it as a continuous variable. Doing so would imply an order or numeric relationship that does not exist. For example, suppose you are analyzing preferred instructional style, and your dataset codes them as:
  - 1 = lecture-based
  - 2 = discussion-based
  - 3 = hands-on
  - 4 = self-paced

  You *cannot* use this as a single independent variable in regression, as it would be incorrect to assume that "self-paced" (coded as 4) is four times "lecture-based" (coded as 1). To properly include categorical variables in a regression model, you must create dummy variables or use statistical software that recognizes the variable's categorical nature and adjusts appropriately. **Dummy coding** is a method of converting categorical variables into a set of binary variables (0's and 1's) that can be used as predictors in regression and other statistical models. If you manually create the set of dummy variables, you convert each category into a separate binary variable (0 or 1) except for one category, which serves as the reference group. For instance, you could create the following three dummy variables:
  - Discussion-based (1 = discussion-based, 0 = Otherwise)
  - Hands-on (1 = hands-on, 0 = Otherwise)
  - Self-paced (1 = self-paced, 0 = Otherwise)

- The omitted category, lecture-based, becomes the reference group, meaning that the regression coefficients for other instructional styles represent their differences relative to lecture-based. Using dummy coding correctly prevents *incorrect assumptions about numerical relationships* between categories and allows for meaningful comparisons between groups. Choosing an appropriate reference category depends on the context of your study and research question. The choice of the reference group affects the interpretation of results but does not impact the overall validity of the model. Dummy variables also collectively capture the categorical variable without implying order or magnitude.

ASSUMPTIONS OF MULTIPLE REGRESSION AND DATA DIAGNOSTICS

Earlier in the chapter, we reviewed data assumptions for linear regression. The same assumptions apply for multiple regression:

- Linearity
- Independence of errors
- Homoscedasticity
- Normality of errors

Let's break the normality of errors assumption down a little more. The key here is that error should be independent and randomly distributed. This assumption is closely tied to the independence of observations, meaning that no individual data point should influence another. If observations are not independent (e.g., if data points are clustered, repeated measures are used, or there is some form of dependence between cases), the errors may also become correlated, violating the assumption of independence in regression analysis. This can lead to biased estimates and incorrect inferences. This assumption can be violated in education studies, particularly when data from students in the same school or under the same teacher share common influence, such as teaching style, curriculum, or resources. For example, if you're studying the relationship between homework completion and test scores with student samples from multiple schools, students in the same class might perform similarly due to the same instruction quality, violating the independence assumption. In such cases, advanced techniques such as hierarchical linear modeling and clustered standard errors may be necessary. If you can see patterns in your errors, or residuals, then your independent variables are not capturing all the variability they should. In general, the distribution of residuals should be normal and centered on zero, meaning that over- and underprediction occur evenly.

In addition to the prior assumptions, multiple regression introduces additional considerations due to the presence of multiple independent variables:

- Multicollinearity
- No endogeneity
- Correct model specification

Multicollinearity occurs when there is a strong linear relationship between two or more independent variables in a regression model. Ideally, independent variables should be uncorrelated, but it is common for variables to have intricate relationships that are challenging to partial out in social science research. In multiple regression, we try to isolate the unique effect of each independent variable on the dependent variable while holding all other predictors constant. However, when two independent variables are highly correlated, changing one often means changing the other in a similar way, making it difficult to determine their distinct contributions.

Imagine we are studying the effect of property tax revenue (X_1) and school district wealth (X_2) on per-student funding. These two variables are highly correlated—school districts in wealthier areas typically generate more revenue from property taxes due to higher property values. If both are included in a multiple regression model predicting per-student funding, it becomes difficult to determine whether funding differences are driven more by property tax revenue or by the overall wealth of the district, as these factors move together. To properly separate their effects, we would need data from districts where these relationships are not perfectly aligned, such as districts with high property tax revenues but relatively lower overall wealth or high overall wealth but lower property tax revenues—scenarios that are less common due to the typical structure of school finance systems.

Why does multicollinearity matter? High multicollinearity results in coefficient estimates that can swing wildly based on which other independent variables are included. They are very sensitive to small changes in the model (i.e., untrustworthy βs). You also have less precision, meaning your standard errors increase and you can't trust the p values to identify independent variables that are statistically significant. A classic sign of near-perfect multicollinearity is when you have a high R^2 but none of the variables show significant effects. However, you can still have high multicollinearity without having a high R^2.

If multicollinearity is detected, should you remove a variable? Not necessarily.

- The severity of problems increases with the degree of multicollinearity, so if the issue is only moderate, it may not require correction.
- Multicollinearity impacts only the correlated independent variables, but it is important to note that two variables may jointly show high correlation with a third even if their pairwise correlations are not strong. If multicollinearity arises among covariates but does not affect your main independent variable, you can still interpret its results reliably.
- Multicollinearity affects the coefficients of independent variables and *p* values but not the overall model predictions or fit statistics. If your primary goal is using the model to make predictions rather than interpreting individual coefficients, multicollinearity may be less of a concern.

It may be tempting to remove one of two highly correlated independent variables, such as student age and student grade level, if they are essentially measuring the same concept. In that case, it makes sense to drop one. However, if both variables are included based on theoretical justification and prior research, you should not remove one solely to reduce multicollinearity.

Why? Remember when you learned about variance and the trade-offs between bias and precision? Removing a key independent variable biases the estimates of other predictors. While removing a variable may improve statistical precision, it can also introduce omitted variable bias, making it difficult to interpret results accurately. Instead, you can report the standard errors and acknowledge potential imprecision rather than altering the model structure unnecessarily. Two diagnostics that tell you how much of a problem multicollinearity is creating for your model are *variance inflation factors* (*VIF*) and *tolerance*. We will revisit those concepts later in the chapter. For now, let's move to the next assumption.

A more complex assumption is no *endogeneity*, meaning that the predictors should not be influenced by the dependent variable or share a relationship with the error term. For instance, consider a study examining how access to Advanced Placement courses influences college enrollment rates. If higher college enrollment rates also encourage schools to expand Advanced Placement course offerings, there's a two-way relationship that violates this assumption. This requires careful study design or advanced statistical methods to address.

Another essential assumption is *correct model specification*, which means including all relevant variables while excluding unnecessary ones. Ensuring correct model specification involves theory-driven variable selection and testing models for robustness to avoid either omitting key variables or overfitting the model with irrelevant predictors.

- *Under-specification (omitted variable bias).* If a school district administrator predicts student success on state exams but fails to include key predictors such as socioeconomic status or English-language proficiency, the model will produce biased estimates.
- *Over-specification (including irrelevant variables).* Adding unnecessary variables, such as the distance a student lives from school, might not improve predictions but could increase model complexity without adding explanatory power.

BASICS OF MULTIPLE REGRESSION: HYPOTHESIS TESTING AND INTERPRETATION

As with simple linear regression, we rely on the principle of least squares to estimate the slopes in a regression model subject to the constraint that the SS_{error} or $SS_{residuals}$ is minimized:

$$SS_{residuals} = \Sigma (Yi - \hat{Y}i)^2$$

where $\hat{Y}_i$ is the predicted (fitted) value and is obtained after estimating model parameters. Let's walk through a brief example. A school district wants to understand how different factors predict average standardized test scores. The school district conducts a multiple regression analysis using student-teacher ratio, average parental education level (measured in years), school funding per student, percentage of students on free/reduced lunch, and student attendance rate. This is a very large school district composed of 56 schools.

$$\hat{Y}_i = b_0 + b_1 X_{S\text{-}Tratio} + b_2 X_{Parent\ ED} + b_3 X_{Funding} + b_4 X_{Lunch} + b_5 X_{Attendance}$$

STEP 1: STATE THE HYPOTHESIS

There are two types of hypotheses to test in multiple regression: the omnibus test and tests for individual predictors. The **omnibus hypothesis test** assesses whether the overall model is statistically significant. Specifically, the omnibus hypothesis test assesses the null hypothesis that all regression coefficients for the predictors (except the intercept) are equal to zero. This test evaluates whether at least one predictor variable has a statistically significant relationship with the dependent variable. The omnibus null hypothesis can be represented by:

$$H_0: \beta_1 = \beta_2 = \beta_3 = \ldots = \beta_p = 0$$

The alternative omnibus hypothesis is simply that at least one regression coefficient is not 0. The hypothesis can be represented as:

$$H_A: at\ least\ one\ \beta_p \neq 0$$

In our example, we have five predictors to predict standardized test scores. Thus, our hypotheses are:

$$H_0: \beta_{S\text{-}Tratio} = \beta_{Parent\ ED} = \beta_{Funding} = \beta_{Lunch} = \beta_{Attendance} = 0$$

$$H_A: at\ least\ one\ \beta_p \neq 0$$

If the omnibus test is significant, we then examine each predictor individually to identify which variables contribute meaningfully to the model.

For each predictor p, we set the following hypotheses:

H_0: $B_p = 0$, the predictor has no effect on the dependent variable.
H_A: $B_p \neq 0$, the predictor has a significant effect on the dependent variable.

Since our model includes five predictors, we conduct five separate hypothesis tests, each testing whether the respective coefficient is significantly different from zero.

STEP 2: SET THE CRITERIA FOR A DECISION AND EVALUATE DATA

- We will use a 0.05 level of significance for the omnibus F test.
- The degrees of freedom (df) for the F test in regression are determined as follows: for the regression component, the df equals the number of predictors in the model (p). For the error term, the df is calculated as the sample size (N) minus the number of predictors (p) minus 1, expressed as ($N - p - 1$). This accounts for one degree of freedom for each independent variable (X) and the dependent variable (Y). Because $N = 56$, the F ratio follows the F distribution with (5,50). The critical value at $\alpha = 0.05$ is 2.40. Figure 12.6 provides a visual representation.

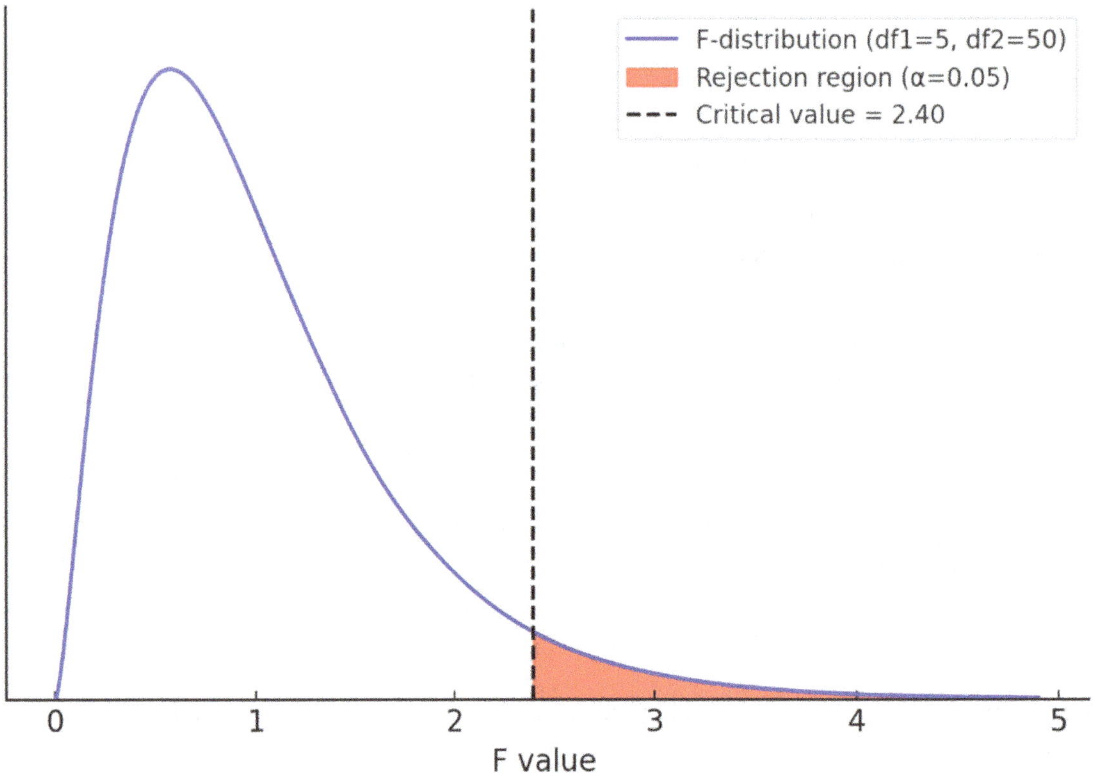

FIGURE 12.6 F Distribution with Critical Value at $\alpha = 0.05$ for $df = (5, 50)$ for Omnibus Test

When the omnibus test is significant, we proceed with hypothesis testing for individual independent variables. Using $\alpha = 0.05$ for each test, the test statistic follows a t distribution with degrees of freedom calculated as $df = N - p - 1$. In this case, with $N = 56$ observations and $p = 5$ predictors, the degrees of freedom for each t test is $df = 56 - 5 - 1 = 50$. Figure 12.7 shows this distribution.

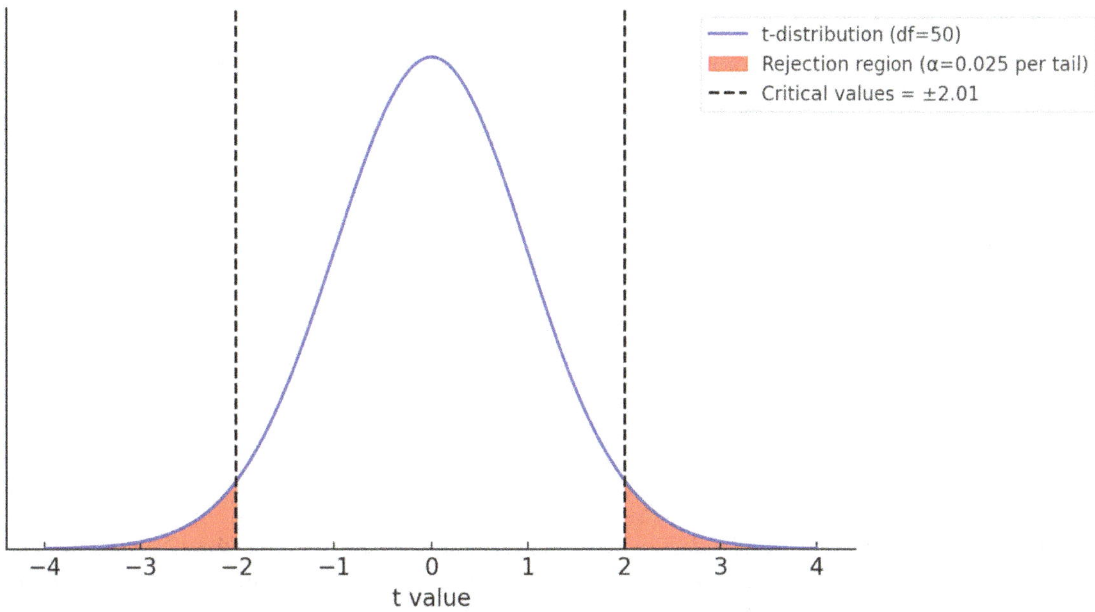

FIGURE 12.7 *t* Distribution with Critical Values and Rejection Regions for Testing Individual Predictors

While we will not cover the process of assessing whether the data meet multiple regression assumptions here, it is always essential to visualize the data through graphs and compute descriptive statistics to identify potential violations. Checking assumptions—such as linearity, normality of residuals, homoscedasticity, and multicollinearity—ensures the validity and reliability of the regression results.

STEP 3: COMPUTE THE TEST STATISTIC

Similar to ANOVA, we use the *F* test for the omnibus hypothesis. To run the *F* test, we need mean squares (MS) regression and mean squares (MS) residual values. Table 12.5 summarizes how to calculate the degrees of freedom, the sum of squares, the mean square, and the *F* statistic. Remember that p indicates the number of predictors and n indicates the sample size.

Similar to what we have already learned, simply add the $SS_{Regression}$ and SS_{Error} to get SS_{Total}. You also do the same thing for your $df_{Regression}$ and df_{Error} to get df_{Total}.

$$F = \frac{MS_{Regression}}{MS_{Residual}} \sim F_{(df_{Regression}, df_{Error})}$$

TABLE 12.5 *Regression Formulas*

	DEGREES FREEDOM (*df*)	SUM OF SQUARES (SS)	MEAN SQUARES (MS)	F
Regression (Model)	p	$\sum(\hat{Y}_i - \bar{Y})^2$	$\frac{SS_{Regression}}{df_{Model}}$	$\frac{MS_{Model}}{MS_{Error}}$
Error (Residual)	$N - p - 1$	$\sum(Y_i - \hat{Y}_i)^2$	$\frac{SS_{Error}}{df_{Error}}$	
Total	$N - 1$	$\sum(Y_i - \bar{Y})^2$	$\frac{SS_{Total}}{df_{Total}}$	

There are 5 degrees of freedom in the numerator ($df_{Regression}$), representing the 5 predictors (independent variables), and there are 50 degrees of freedom in the denominator (df_{Error}), representing 56 schools – 5 predictors – 1. Table 12.6 contains the results of a regression analysis on the data.

$$F = \frac{MS_{Regression}}{MS_{Residual}} = \frac{500}{30} = 16.67 \sim F_{(5,50)}$$

TABLE 12.6 *ANOVA Summary Table for Multiple Regression Analysis*

	DEGREES OF FREEDOM (df))	SUM OF SQUARES (SS)	MEAN SQUARES (MS)	F
Regression	*5*	2,500	500	16.67
Error (Residual)	*50*	1,500	30	
Total	*55*	4,000		

While the statistical software you use will provide the p value, let's look at Table A3 in the appendix to determine if we have a significant result. Find where five degrees of freedom in the numerator and 50 degrees of freedom in the denominator intersect, which identities a critical value of 2.40. Our F statistic of 16.76 is larger than the critical value of 2.40 at the .05 level. So, we can reject the null as $p < .05$.

Are we finished? Not so fast. It is common for those first learning multiple regression to mistakenly assume that finding a significant result when testing for the omnibus hypothesis means that all β coefficients are significantly different from 0. However, this test only tells us that at least one β coefficient is significantly different from 0.

Since we have a significant result, the next step is to run t tests on individual coefficients. If the omnibus test is not statistically significant, it does not make sense to look at the individual predictors. It was significant in this case, so we move to the next set of hypothesis tests. Recall that the null hypothesis for any given individual predictor can be represented as:

$$H_0: \beta_p = 0$$

The alternative hypothesis can be represented as:

$$H_A: \beta_p \neq 0$$

Let's suppose that we obtained the results summarized in Table 12.7 with the multiple regression analysis.

TABLE 12.7 ***Multiple Regression Results: Predictors of Academic Performance***

PREDICTOR	UNSTANDARDIZED COEFFICIENT (*b*)	SE	*t* STATISTIC	P
Student-teacher ratio	-1.5	0.3	-5.0	< .001
Parental education level	2.0	0.5	4.0	< .001
School funding	0.2	0.4	0.5	0.619
Free/reduced lunch percentage	-1.2	0.3	-4.0	< .001
Attendance rate	0.6	0.2	3.0	0.004

The unstandardized coefficient column presents the slope for each predictor, indicating the expected change in the dependent variable for a one-unit increase in the predictor, holding other variables constant. The far-right column in Table 12.7 provides the statistical significance of each coefficient, helping determine whether a predictor has a meaningful impact on the model. Before diving further into Table 12.7, it is important to discuss partial correlations, which help assess the unique contribution of each predictor while controlling for other variables in the model.

PARTIAL CORRELATIONS

Multiple regression involves accounting for the relationships among multiple variables, a process known as partialling. **Partialling** means separating the effects of one predictor from the others. The statistical measures that accomplish this are partial correlations, which help determine the independent contribution of each predictor to the dependent variable.

Suppose we have a three-variable problem with Y, X_1, and X_2. Y is the dependent variable, and X_1, and X_2 are independent variables. Both independent variables correlate with the dependent variable, but as is often the case with educational data, they also correlate with each other. We can't add the two coefficients of determination together to find the R^2 of the regression model, because we must account for the variation that's shared, or overlapping. In Figure 12.8, the circles represent the variability of data in the three variables. Because they are all related, the circles overlap.

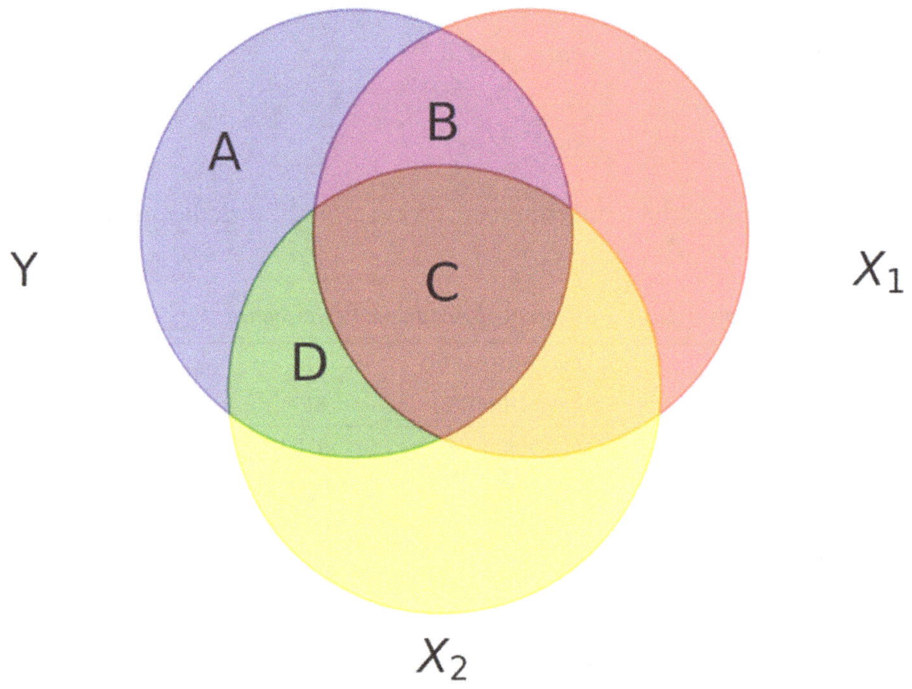

FIGURE 12.8 Variability Across Three Variables

- Area A+B+C+D represents all the variability in the dependent variable to be explained (shaded in purple).
- Area B+C+D represents R^2 for the model as a whole because they represent the variability in the dependent variable that is shared with the two predictors combined.
- Area A now represents the residual variance. If R^2 is small, the unexplained variance in the dependent variable will be relatively large.
- Area B+C represents the total shared variability between *Y* and X_1. But notice that Area C is also shared with X_2.
- Area B by itself represents the unique relationship between *Y* and X_1. This is the partial correlation that's represented in the regression results—the extent to which X_1 can predict *Y* with X_2 held constant.

To compute partial correlations, you first calculate the residuals—the errors that remain after predicting the dependent variable using each independent variable. Then, you correlate these residuals with each other to determine the unique contribution of each predictor while controlling for the others.

An important takeaway is that partial correlations are not fixed values; they depend on the specific set of variables included in the model. As the relationships among independent variables change, the partial correlations will also adjust, reflecting the shifting influence of each predictor. As those correlations change, so will the magnitude (and sometimes even the direction) of any given predictor variable on the dependent variable. Just picture the X_2 circle in Figure 12.8—if it shifted down so that Area B grew and Area C got smaller, the regression coefficient for X_1 would be larger.

Most of the time, partial correlations are smaller than the simple bivariate correlation between the independent variable and the dependent variable but not always. Sometimes one of your other predictors could actually *suppress* the relationship between the two variables, so the partial correlation is larger.

Now let's look at our regression results again in Table 12.8, this time with the partial correlation statistics included.

TABLE 12.8 *Multiple Regression Results with Partial Correlations*

PREDICTOR	UNSTANDARDIZED (*b*)	*SE*	STANDARDIZED β	PARTIAL CORRELATION	BIVARIATE CORRELATION	*t* STATISTIC	*p*
Student-teacher ratio	-1.5	0.3	-0.3	-0.58	-0.65	-5.0	< .001
Parental education level	2.0	0.5	0.1	0.49	0.55	4.0	< .001
School funding	0.2	0.4	2.0	0.03	0.15	0.5	0.619
Free/reduced lunch percentage	-1.2	0.3	-0.5	-0.49	-0.55	-4.0	< .001
Attendance rate	0.6	0.2	0.1	0.39	0.45	3.0	0.004

The bivariate correlation between standardized test scores and student-teacher ratio was -0.65, but the partial correlation is only –0.58. Removing the effects of the other predictors has "purified" the relationship. We can also see that while the original correlation between school funding and standardized test scores was 0.15, the partial correlation is even smaller, only 0.03. The correlation table just considers two variables at a time. With regression, each independent variable has a coefficient that is estimated after holding all the others constant (controlling for the other predictors).

Now we'll look in more detail at Table 12.8 focusing on the independent variable of student-teacher ratio.

- The unstandardized coefficient for reading tells us that with the effect of the other predictors held constant (partialled out), each one-point increase in the student-teacher ratio is associated with a 1.5-point decrease in standardized test score.
- This prediction is based on a sample, and we always have sampling errors. The value of 0.3 indicates the standard error of the associated regression coefficient estimate and is used for the statistical test.
- To aid in interpretating your results, it is common to report the *standardized coefficients*, which are interpreted like z scores, and can be obtained by converting the raw data to their z score counterparts (measured in standard deviations above and below the mean) before running the regression.
  - The standardized coefficients tell us which independent variables are the most powerful predictors of the dependent variable.
  - To convert from unstandardized coefficients to standardized, we use the expression:

$$\beta_p = b_p \frac{SD_Y}{SD_P},$$

where β_p represents the standardized slope associated with predictor p, b_p is the unstandardized slope for predictor p, SD_Y is the standard deviation of the dependent variable, and SD_p is the standard deviation of predictor p. Standardized slopes are typically included in computer output and reported as part of your results. Our result for the student-teacher ratio independent variable shows that with the effects of other predictors held constant, for each one standard deviation increase in student-teacher ratios, we expect standardized test scores to decrease 0.3 standard deviations. It can get a little confusing because people use either β or b to refer to regression coefficients, but if you see both β and b, the former means standardized, and the latter is unstandardized.

- We're finally ready to test the null hypothesis about this individual predictor using $\alpha = 0.05$. The table reports that the t statistic for student-teacher ratio is -5.0, and the p value shown in Table 12.8 is $p < 0.001$. This tells us that the student-teacher ratio is a significant predictor of standardized test scores.

After controlling for other predictors, the results show that the β coefficient for 4 predictors is significantly different from 0. However, we find that school funding is not, which is actually good news for the district because it spent years developing an equitable funding model to support student success.

ASSESSING MODEL FIT

Similar to what we learned earlier in the chapter, the basic measure of model-data fit is the squared multiple correlation, R^2. The statistic can be calculated using:

$$R^2 = \frac{SS_{Regression}}{SS_{Total}}$$

For our example, the coefficient of determination (R^2) is calculated as follows: $R^2 = \frac{SS_{Regression}}{SS_{Total}} = \frac{2500}{4000} = 0.625$ = 62.5%. This means that 62.5% of the variation in students' standardized test scores is explained by the set of five predictors. However, it is important to note that this dataset was created for illustrative purposes. In real-world education research, it is unlikely that such a high percentage of variance in standardized test scores would be explained by just five predictors, as many other factors influence student performance. A larger R^2 indicates a stronger model, meaning the independent variables collectively explain more of the variance in the dependent variable. However, as we have learned, R^2 *is biased* because it tends to increase as more predictors are added even if they do not meaningfully contribute to the model.

To address this issue, a less biased version, *adjusted* R^2, is reported. Adjusted R^2 accounts for the number of predictors and sample size, providing a more reliable measure of model fit. It is calculated using the following formula:

$$R^2_{adjusted} = 1 - \left(\frac{(1 - R^2)(N - 1)}{N - p - 1}\right) = 1 - \left(\frac{(1 - .625)(56 - 1)}{56 - 6 - 1}\right) = .5875$$

As you can see from the formula, adjusted R^2 depends on the sample size and the number of predictors in the model. If the sample size is large enough, the values of R^2 and adjusted R^2 should be very similar. If the sample size is small and you have a large number of predictors, the discrepancy between R^2 and adjusted R^2

will be larger. In general, we recommend using adjusted R^2 when reporting your results but encourage you to check to see what is typical for studies in your field.

When reviewing regression results, it is important to assess whether multicollinearity may be influencing the model estimates. To do this, we examine two key diagnostics: *VIF* and *tolerance* (*the reciprocal of VIF*). These metrics help determine the extent to which predictor variables are correlated with each other. High VIF values and low tolerance levels indicate potential multicollinearity, which can distort coefficient estimates and reduce the stability of the model. Evaluating these diagnostics ensures that the model specification is appropriate and helps identify whether adjustments, such as removing or combining variables, are necessary.

VARIANCE INFLATION FACTORS (VIFS)

The **variance inflation factor (VIF)** quantifies how much the variance of a regression coefficient is inflated due to multicollinearity among predictor variables. We can use multiple regression to calculate the degree of multicollinearity among independent variables by making each independent variable a dependent variable predicted by all the others. We end up with R^2 values for each of those regression models, indicating, just like with any other regression model, the percentage of the variance in one independent variable that is explained by the other independent variables. The VIF calculation utilizes these R^2 values to diagnose the extent of multicollinearity using the following formula:

$$VIF_p = \frac{1}{1-R_p^2}$$

The subscript p indicates the independent variable, and you calculate a VIF for each independent variable. When R^2 equals zero, there is no multicollinearity because the set of independent variables could not explain any of the variability in the remaining independent variable. In that case the fraction would be 1/1, so the lowest possible VIF is 1. But if you look at the equation, as R^2 gets bigger, the denominator gets smaller and the VIF increases.

There are no hard rules about acceptable VIF values. However, Kutner et al. (2013) state that "a maximum VIF value in excess of 10 is frequently taken as an indication that multicollinearity may be unduly influencing the least squares estimates" (p. 409). Models with VIF < 5 further reduces the inflation of standard errors and therefore reduces the likelihood of Type II errors.

TOLERANCE

Tolerance is a statistical measure that indicates how much a particular predictor variable contributes to the overall variance in the dependent variable while accounting for the effects of other predictors. Tolerance is closely related to multicollinearity and serves as the reciprocal of the VIF.

- Tolerance = $(1 - R_p^2)$
- Tolerance = 1/VIF

Table 12.9 shows the relationship of VIF and tolerance to r_p and R_p^2. There is no new information here, as tolerance and VIF are based on the same calculations. They are just expressed in different forms, and you may see reference to one or the other, so it helps to be aware of how they are related. If you only had two predictors in your model, then r_p would be the same as the bivariate correlation between them. While sets of independent variables may correlate differently, that column can give you a sense of likely multicollinearity problems based on your preliminary review of the correlation table.

TABLE 12.9 *VIF, Tolerance, r_p and R_p^2*

VIF	r_p	R_p^2	TOLERANCE
1	0.00	0.00	1.00
1.05	0.22	0.05	0.95
1.10	0.30	0.09	0.91
1.20	0.41	0.17	0.83
1.30	0.48	0.23	0.77
1.40	0.53	0.28	0.71
1.50	0.58	0.34	0.67
2.00	0.71	0.50	0.50
5.00	0.89	0.79	0.20
10.00	0.95	0.90	0.10
∞	1.00	1.00	0.00

A Tolerance value below 0.1 indicates a serious multicollinearity problem, suggesting that a predictor shares a substantial amount of variance with other independent variables. It is important to remember that VIF and tolerance are measures of the extent to which regression coefficients may be affected by multicollinearity. Higher VIF values and lower tolerance values signal greater risk, potentially distorting coefficient estimates and reducing the reliability of the model.

Including more control variables or predictors in nonexperimental or quasi-experimental designs is usually a good thing as long as you are not adding so many predictors that you're overfitting the model. In theory, multicollinearity is not a problem if VIFs are lower than published rules of thumb (i.e., VIF < 10) or, more conservatively, VIF < 5. However, these rules of thumb do not protect against substantive changes in regression coefficient magnitude, so you can't only rely on VIF and tolerance values.

While addressing multicollinearity, it is important to balance reducing redundancy with maintaining theoretically important predictors. Simply removing variables based on VIF values alone can introduce omitted variable bias, potentially distorting the findings. Instead, these values should be considered alongside theoretical justification and empirical patterns in the data to ensure both statistical validity and substantive relevance.

REPORTING RESULTS OF A MULTIPLE REGRESSION ANALYSIS

There is no standard format to write up results, but we recommend including the following information:

- Preliminary descriptive results
    - The number of cases used in your analysis

    - Correlations among dependent and independent variables, indicating if any variables have very low correlations
    - Whether or not the data meet the assumptions for multiple regression
        - Indicate whether or not there is a potential problem with multicollinearity
        - Indicate if any outliers exist
- Regression results
    - Model fit (R^2 and adjusted R^2)
    - Coefficient estimates and standard errors
    - t statistic, df, and significance level of each predictor

APA-style sample write-up: A multiple regression analysis was conducted to examine the relationship between student-teacher ratio, parental education level, school funding per student, free/reduced lunch percentage, and student attendance rate with standardized test scores. The regression model included five predictors and was based on data from 56 schools.

The overall model was statistically significant, $F(5, 50) = 16.67$, $p < .05$, indicating that at least one of the predictor variables significantly contributes to explaining the variance in standardized test scores. The model accounted for 62.5% of the variance in test scores, as indicated by R^2. The adjusted R^2, which corrects for potential bias due to the number of predictors, was .588.

The results indicate that four of the five predictors significantly contribute to explaining standardized test scores. Specifically, student-teacher ratio, parental education level, free/reduced lunch percentage, and attendance rate were all statistically significant predictors ($p < .05$), while school funding per student was not significant ($p = .619$).

- The student-teacher ratio had a significant negative association with standardized test scores ($b = -1.5$, $SE = 0.3$, $p < .01$), meaning that for every one-unit increase in student-teacher ratio, test scores decreased by 1.5 points, holding all other variables constant. The standardized coefficient ($\beta = -0.3$) indicates that one standard deviation increase in student-teacher ratio corresponds to a 0.3 standard deviation decrease in test scores.
- Average parental education level was positively associated with test scores ($b = 2.0$, $SE = 0.5$, $p < .01$), indicating that an increase in parents' education level corresponded to an increase in test scores by 2 points per year of additional education.
- Free/reduced lunch percentage was negatively associated with test scores ($b = -1.2$, $SE = 0.3$, $p < .01$), suggesting that as the percentage of students on free/reduced lunch increases, test scores decrease.
- Attendance rate was positively associated with test scores ($b = 0.6$, $SE = 0.2$, $p < .01$), indicating that students in schools with higher attendance rates tend to perform better.
- School funding per student was not a significant predictor of test scores ($b = 0.2$, $SE = 0.4$, $p = .619$), indicating that after accounting for the other variables, funding alone did not explain additional variance in test scores. This finding aligns with the district's efforts to implement an equitable funding model, suggesting that other factors may have a more direct impact on student performance.

LOGISTIC REGRESSION: PREDICTING CATEGORICAL OUTCOMES

Up to this point we have focused on linear regression, which predicts a continuous dependent variable (Y) using one or more independent variables (X). However, many research questions involve categorical outcomes rather than continuous ones. For example, instead of predicting a student's final exam score, we might want to determine whether a student passes or fails. In such cases logistic regression is the appropriate statistical technique, as it models the probability of an outcome occurring based on predictor variables.

Logistic regression is a type of regression analysis used when the dependent variable is categorical, meaning it consists of two or more categories. Unlike linear regression, which assumes a continuous outcome, logistic regression models the probability that a given outcome occurs. The most common form is *binary logistic regression*, which predicts an outcome with two categories such as pass/fail, admitted/not admitted, or employed/unemployed. When there are more than two categories, we can use *multinomial logistic regression* (for unordered categories) or *ordinal logistic regression* (for ordered categories).

In linear regression, we model the relationship between X and Y using the equation: $Y = bX + a$, where b is the slope and a is the intercept. However, this equation is inappropriate for categorical outcomes because it can yield values outside the range of 0 to 1, which are not valid probabilities. Instead, logistic regression models the *log odds* of an event occurring using the logistic function: $\log\left(\frac{p}{1-p}\right) = bX + a$,
where:

- p is the probability of the event occurring;
- $\frac{p}{1-p}$ represents the odds of the event occurring;
- $\log\left(\frac{p}{1-p}\right)$ is the logit (log odds);
- b is the regression coefficient, which represents the change in log odds of the outcome for a one-unit change in X; and
- a is the intercept.

To obtain the probability of the event occurring, we use the inverse logistic function:

$$p = \frac{e^{(bx + a)}}{1 + e^{(bx + a)}}$$

This transformation ensures that predicted probabilities remain within the valid range of 0 to 1.

INTERPRETING LOGISTIC REGRESSION COEFFICIENTS

Because logistic regression models log odds rather than raw probabilities, interpreting coefficients requires some additional steps:

- A positive coefficient ($b > 0$) means that as X increases, the probability of the event occurring increases.
- A negative coefficient ($b < 0$) means that as X increases, the probability of the event occurring decreases.
- The odds ratio is the exponentiated coefficient e^b. An odds ratio greater than 1 suggests that an increase in X increases the odds of the event occurring, while an odds ratio less than 1 suggests a decrease in the odds.

Let's walk through a quick example. Suppose we want to predict whether a student is admitted to college. As discussed earlier, dummy coding is used to include categorical variables in regression models by converting them into a series of binary (0 or 1) variables. Each dummy variable represents one of the categories of the original categorical variable, allowing the regression model to quantify the effects of the categorical predictor. This time let's use dummy coding for our outcome variable, where 1 = *admitted* and 0 = *not admitted* based on one's SAT score (X). The logistic regression equation might look like this:

$$\log\left(\frac{p}{1-p}\right) = 0.005(SAT) - 3.5$$

This equation tells us that for each one-point increase in SAT score, the log odds of admission increase by 0.005. To convert this into an odds ratio, use:

$$e^{0.005} \sim 1.005$$

This means that for every one-point increase in SAT score, the odds of admission increase by approximately 0.5%.

ASSUMPTIONS OF LOGISTIC REGRESSION

While logistic regression does not assume normality or homoscedasticity like linear regression, it does require the following:

- *Binary or categorical dependent variable.* The outcome must be categorical.
- *Independent observations.* Each observation must be independent of others.
- *No perfect multicollinearity.* Predictors should not be too highly correlated.
- *Linearity in the logit.* The independent variables should have a linear relationship with the log odds of the outcome.

For education researchers, logistic regression is particularly useful in analyzing outcomes related to student success, such as graduation likelihood, college acceptance, and course completion rates. By modeling the probability of categorical outcomes, education researchers can identify key predictors of student achievement and inform policy decisions. For instance, understanding how standardized test scores, socioeconomic status, and attendance patterns impact student retention can help educators develop targeted interventions. Logistic regression provides a powerful method to translate complex educational data into actionable insights, ultimately contributing to more informed decision-making in educational settings.

CONCLUSION

Understanding and applying regression analysis is a critical skill for interpreting relationships between variables and making data-driven predictions. This chapter has provided a comprehensive exploration of both simple and multiple regression techniques, emphasizing their theoretical underpinnings, assumptions, and practical applications. By mastering these concepts, researchers and practitioners can build models that not only identify meaningful relationships but also provide robust predictions. Regression analysis is a powerful tool that, when used responsibly and with a solid theoretical foundation, enables a deeper understanding of complex phenomena. As you continue to practice these techniques, always remember to prioritize sound research design and thoughtful interpretation of results to ensure meaningful and actionable insights.

KEY TAKEAWAYS FOR EDUCATIONAL RESEARCHERS FROM CHAPTER 12

- Ensuring regression assumptions to improve education data analysis. Before conducting a regression analysis in education research, it is crucial to verify that the assumptions of the model (such as linearity, independence of errors, homoscedasticity, and multicollinearity) are met. For instance, when analyzing student achievement, failing to check these assumptions can lead to incorrect conclusions about the impact of factors such as teacher effectiveness and school funding. If errors are correlated (e.g., students within the same classroom share similar unmeasured characteristics), standard errors will be underestimated, leading to misleading significance tests.
- Addressing multicollinearity to accurately identify key educational predictors. When independent variables are highly correlated, it becomes difficult to determine their unique contributions to outcomes. In education research, factors such as socioeconomic status and parental education are often correlated, making it essential to assess variance inflation factors (VIF) and tolerance to detect multicollinearity. A high VIF may indicate that two or more predictors are measuring similar underlying constructs, requiring researchers to consider alternative modeling strategies.
- Interpreting omnibus and individual tests to understand educational interventions. The F test in regression helps determine whether the overall model explains a significant amount of variance in outcomes such as student test scores and graduation rates. However, even if the overall model is significant, this does not mean that every predictor is important. Education researchers must follow up with t tests to identify which specific factors (such as teacher-student ratios, school funding, and instructional methods) are significantly related to student success. This ensures that resources and interventions are directed toward the most impactful variables.
- Using adjusted R^2 to avoid overfitting in educational models. While a high R^2 suggests a strong model, it can be misleading in education research if too many predictors are included, leading to overfitting. For example, if a researcher includes every possible student characteristic in a model predicting standardized test scores, the model may explain a large proportion of variance but may not generalize to other student populations. The adjusted R^2 accounts for the number of predictors and the sample size, helping researchers avoid inflated model performance and ensure real-world applicability.
- Distinguishing between standardized and unstandardized coefficients for policy and practice. In education research, unstandardized coefficients (b) provide real-world interpretations, such as "for every additional student per teacher in a classroom, student test scores decrease by 1.5 points." This is useful for policymakers when making funding and curriculum decisions. On the other hand, standardized coefficients (β) allow comparisons across different factors, such as whether attendance rates and teacher qualifications have a stronger influence

on student achievement. Using both types of coefficients helps translate statistical findings into actionable policy recommendations.

- Using regression as a tool to inform, not as a definitive answer. Even when a regression model produces statistically significant results, education researchers must critically evaluate the findings in context. A significant predictor does not necessarily imply causation—just because school funding is correlated with student performance does not mean that increasing budgets alone will improve outcomes. Additionally, omitted variables, such as classroom climate and extracurricular involvement, may introduce bias. Researchers should integrate regression results with educational theory, qualitative insights, and policy considerations to develop more holistic recommendations for improving student learning and success.

KEY DEFINITIONS FOR CHAPTER 12

Criterion variable, or to-be-predicted variable (Y), is the variable with unknown values that can be predicted or estimated, given known values of the predictor variable. We sometimes call this variable a dependent variable or an outcome variable.

Dummy coding is a method of converting categorical variables into a set of binary variables (0's and 1's) that can be used as predictors in regression and other statistical models.

Linear regression is a statistical procedure used to determine an equation of a regression line for a set of data points and to determine the extent to which the regression equation can be used to predict values of one variable, given known values of a second variable in a population.

Logistic regression is a type of regression analysis used when the dependent variable is categorical, meaning it consists of two or more categories.

Multicollinearity occurs when there is a strong linear relationship between two or more independent variables in a regression model.

Multiple regression uses two or more predictors, or independent variables, to predict variation in a quantitative outcome, or dependent variable.

The **omnibus hypothesis test** assesses whether the overall model is statistically significant. Specifically, the omnibus hypothesis test assesses the null hypothesis that all regression coefficients for the predictors (except the intercept) are equal to zero.

The **ordinary least squares regression line** is a statistical method used to find the best-fitting straight line through a set of data points. This line minimizes the sum of the squared differences (residuals) between the observed data points and the values predicted by the line.

Overfitting in the context of multiple regression occurs when the model is too complex and fits the training data too closely, capturing noise or random fluctuations rather than the true underlying relationship between the variables in the population.

Partialling means separating the effects of one predictor from the others.

A predictor variable, or a known variable (X), is the variable with values that are known and can be used to predict values of another variable. We also call this variable an independent variable.

The **proportion of variance explained** in regression is a measure of how well the independent variable(s) explain the variability in the dependent variable. This measure is often expressed as the coefficient of determination (r^2 *or* R^2).

Tolerance is a statistical measure that indicates how much a particular predictor variable contributes to the overall variance in the dependent variable while accounting for the effects of other predictors. Tolerance is closely related to multicollinearity and serves as the reciprocal of the variance inflation factor.

The **variance inflation factor (VIF)** quantifies how much the variance of a regression coefficient is inflated due to multicollinearity among predictor variables.

CHECK YOUR UNDERSTANDING

1. What does the slope (*b*) in a simple linear regression equation represent?
   a. The predicted value of Y when $X = 0$
   b. The change in X relative to Y
   c. The change in Y for a one-unit change in X
   d. The coefficient of determination

2. What does R^2 represent in simple linear regression analysis?
   a. The correlation between independent variables
   b. The proportion of variance in Y explained by X
   c. The slope of the regression line
   d. The standard error of the residuals

3. What does tolerance measure in regression analysis?
   a. The extent of multicollinearity among independent variables
   b. The proportion of variance in the dependent variable explained by predictors
   c. The normality of residuals
   d. The degree of model fit

4. What happens if a model includes highly correlated independent variables?
   a. The dependent variable becomes unreliable.
   b. The coefficients of the independent variables become unstable.
   c. The model fit (R^2 will always decrease).
   d. The assumption of linearity is violated.

5. What does homoscedasticity mean in regression analysis?
   a. The residuals have a consistent spread across all predicted values.
   b. The independent variables are not correlated.
   c. The predictors have no relationship with the error term.
   d. The regression coefficients are stable across models.

13

ANALYZING CHANGE AND ADJUSTING FOR COVARIATES

LEARNING OBJECTIVES

By the end of this chapter, you will be able to:

- Explain the fundamentals of repeated-measures ANOVA, including its application in testing differences across multiple conditions or time points for the same participants, and recognize the advantages of within-subjects designs in reducing error variance;
- Identify the assumptions underlying repeated-measures ANOVA, particularly the concept of sphericity, and apply appropriate corrections (e.g., Greenhouse-Geisser or Huynh-Feldt adjustments) when these assumptions are violated to ensure valid statistical conclusions; and
- Explain the purpose and application of analysis of covariance (ANCOVA), including how it controls for covariates to reduce error variance, and interpret ANCOVA results in the context of experimental and nonexperimental research.

A **repeated-measures ANOVA** is used to determine whether there are significant differences between three or more related groups or measurements taken from the same participants under different conditions, at different time points, or across repeated trials. Similar to how a one-way between-subjects ANOVA extends the independent-samples *t* test, the one-way repeated-measures ANOVA extends the repeated-measures *t* test (also known as the related-samples *t* test and the dependent-samples *t* test).

The term *one-way* indicates that only one factor is being tested, while *within-subjects* signifies that the same set of participants is measured across all conditions, eliminating between-subject variability and increasing statistical power.

An example of a longitudinal study would be measuring a group of college students at one point during each of their four undergraduate years to track changes in a dependent variable over time. This design is considered longitudinal because the same individuals are measured repeatedly across different time points. A repeated-measures ANOVA would be appropriate here.

In contrast, if students were only measured once and grouped based on their current year in college (e.g., freshman, sophomore, junior, senior), the study would be classified as a cross-sectional design rather than longitudinal. In this case, a repeated-measures ANOVA would not be appropriate, as each student would only contribute a single measurement rather than repeated observations over time.

Throughout this chapter we will continue to discuss differences between groups, but it is important to remember that in a one-way repeated-measures ANOVA, we are examining the same individuals measured at different time points or under different conditions. Instead of thinking in terms of distinct groups of people, it is more accurate to consider them as groups of scores on a dependent variable, collected from the same participants. In a one-way repeated-measures ANOVA, there is only one group of individuals, and their responses are analyzed across multiple measurements.

SOURCES OF VARIATION IN THE ONE-WAY REPEATED-MEASURES ANOVA

We already learned how to analyze group mean differences with the one-way *between-subjects* ANOVA, when data are collected from a different group of participants for each condition. We apply the same idea to test our hypotheses with the one-way *within-subjects* ANOVA. Either way, the test statistic has the same basic format:

$$F = \frac{MS_{BG}}{MS_E}.$$

In a repeated-measures ANOVA, we analyze the variance associated with group membership, where a "group" represents a point in time or experimental condition. We also examine the variance that is not associated with group membership, known as error variance.

Unlike a one-way between-subjects ANOVA, a repeated-measures ANOVA introduces an additional source of variation: within-subject variability. This requires a slight adjustment in how we calculate degrees of freedom. While the between-groups variation remains the same, the error variance is divided into two components:

- Variance due to individual differences (accounting for variability within participants), and
- Residual error variance (remaining unexplained variation).

This separation allows for a more precise estimation of effects while controlling for individual differences.

We will guide you through understanding each type of variation in a repeated-measures ANOVA, but we will not be performing any manual calculations. If you want to understand how to calculate the test statistics by hand, websites such as Laerd Statistics (https://statistics.laerd.com/statistical-guides/repeated-measures-anova-statistical-guide-2.php) provides a clear example.

BETWEEN-GROUPS VARIATION (TREATMENT EFFECT)

This component represents the variation between different conditions or time points in a repeated-measures ANOVA. You should already be familiar with this concept:

- Calculate the mean for each group (condition or time point).
- Subtract each group mean from the overall grand mean.
- Square each difference and sum them up.

The resulting value is the sum of squares between groups (SS_{BG}), which quantifies how much variability in the dependent variable is due to differences between conditions rather than random error.

WITHIN-GROUPS VARIATION (ERROR VARIATION)

The process of calculating within-groups variation begins similarly to between-groups variation:

- Compute the mean for each group (condition or time point).
- Subtract this mean from each individual's score.
- Square the differences and sum them up to obtain the sum of squares within groups (SS_{WG}).

However, in a repeated-measures ANOVA, we go one step further by splitting the error variance into two components, based on the two primary reasons why individual scores may differ within a group:

1. *Measurement error.* No assessment is perfectly reliable—there may be sampling error, inconsistencies in measurement, or external factors affecting performance (e.g., a participant having a bad day). These random fluctuations introduce noise into the data, contributing to the error term.
2. *Individual differences.* Some people naturally perform better or worse on a dependent variable, regardless of the time point or condition. These stable individual differences can influence results and complicate interpretations in between-subjects designs.

The second reason is a common challenge with education research. If we are assessing the results of an intervention and we have three different classrooms of students, it is hard to know if different outcomes are due to our intervention or if the students are just different. But here is the magic of a repeated-measures ANOVA: this time, they are *not* different. They are the same people in every condition or at every time period. Any differences in ability, motivation, or anything else are the same across groups. Each person brings the same differences to each point in time or condition. We can quantify those differences and then remove them from the error term.

How do we do that? To remove individual differences from the error term, we calculate the sum of squares between persons (SS_{BP}) using the following steps:

- Calculate each individual's mean score across all time points or conditions.
- Subtract this individual mean from the overall grand mean (the mean across all participants and conditions).
- Square each difference and sum them up.

The result is the sum of squares between persons (SS_{BP}), which quantifies variability due to stable individual differences. By accounting for and removing this variation, a repeated-measures ANOVA isolates the true effect of the independent variable, improving statistical power and reducing error.

DEGREES OF FREEDOM IN THE REPEATED-MEASURES ANOVA

We are going to keep reminding you of this, but everything we do with an ANOVA comes back to the concept of variance, and you should remember that variance = $\frac{\textit{sum of squares}}{\textit{degrees of freedom (df)}}$. In other words, to get from multiple *sum of squares* values to actual variance estimates, we need to divide them by the degrees of freedom. With a between-subjects ANOVA we have two separate degrees of freedom, but now with repeated-measures ANOVA we have three.

It is easier to understand this with an example (Table 13.1). Imagine we have a study with 20 participants (n = 20) being tested at three points in time (k = 3), which means we have 60 total data points.

The two rows in bold in Table 13.1 are the degrees of freedom you use to find the critical value for the F test, and they are the ones you include when reporting results. Please note that the degrees of freedom listed here are only applicable when the sphericity assumption, which we will explain later in the chapter, is satisfied.

TABLE 13.1 *Degrees of Freedom for Example*

SOURCE OF VARIATION	FORMULA	*df* VALUE
Total	$kn - 1$	59
Between groups (conditions)	$\mathbf{k - 1}$	**2**
Between persons (participants/subjects)	$n - 1$	19
Within groups (error)	$\mathbf{(k - 1)(n - 1)}$	**38**

CALCULATING A REPEATED-MEASURES ANOVA

It's time to revisit fractions to understand why a within-subjects ANOVA generally has greater *statistical power* than a between-subjects ANOVA and when it does not.

We begin with the familiar formula for the F statistic:

$$F_{obt} = \frac{MS_{BG}}{MS_E}.$$

Table 13.2 summarizes how these values are calculated, though it refers to "conditions" rather than "between-groups" for clarity. Understanding these components will help explain why within-subjects designs often result in greater power compared to between-subjects designs and when that advantage might not hold.

Remember that a mean square is the same as a variance. To find MS_{BG}, we find the SS_{BG} and divide by df_{BG}, which is k - 1. This is the value for the numerator of our fraction. The bigger this value is, the more likely we are to get a significant result.

MS_E is a little more complicated because we have to divide it into two parts. We first find the total SS_E, and then we figure out how much of that is due to differences between people, which is the $SS_{Subjects}$. We subtract out the part of the error that is due to people being different, and we are left with the SS_E that we truly consider error. We divide this by the df_E, which is (k - 1)(n - 1), and we have MS_E for the denominator of the fraction. The smaller this value is, the more likely we are to get a significant result.

TABLE 13.2 REPEATED-MEASURES ANOVA TABLE FORMULAS

SOURCE OF VARIATION	SS	df*	MS	F
Between groups (conditions)	$SS_{BG} = \Sigma n_i[\bar{x}_i - \bar{x}_{grand}]^2$	$k - 1$	$\frac{SS_{BG}}{df_{BG}}$	$\frac{MS_{BG}}{MS_E}$
Between persons (subjects)	$SS_{Subjects} = \Sigma k(\bar{x}_{Subject} - \bar{x}_{grand})^2$	$n - 1$	$\frac{SS_{Subjects}}{df_{Subjects}}$	
Within groups (error)	$SS_E = SS_{Total} - SS_{BG} - SS_{subjects}$	$(k - 1)(n - 1)$	$\frac{SS_E}{df_E}$	
Total	$SS_{Total} = \Sigma[x_{ij} - \bar{x}_{grand}]^2$	$kn - 1$		

Here is where we talk about power and why the repeated measures ANOVA usually has more power than a between-subjects ANOVA. If the MS_E value is smaller, our fraction gets bigger, right? So how are MS_E values different in the two types of ANOVAs?

- The big difference is that with a repeated-measures ANOVA, we can remove a major component of the error variance. All the error that is due to people being different gets removed from the equation altogether. This tends to make the error variance much smaller in a repeated-measures ANOVA, and therefore we have more statistical power.
- The other difference, however, is in the degrees of freedom. In a between-subjects ANOVA if we had three separate groups of 20 participants each, our total sample size would be 60. The degrees of freedom for error would be calculated as $df_E = n - k = 60 - 3 = 57$. In the example repeated-measures ANOVA, we have a total of 20 participants, and the degrees of freedom for error are calculated as: $df_E = (k - 1)(n - 1) = 38$. Although the sum of squares for error (SS_E) is smaller in a repeated-measures design—because we remove variance due to individual differences—the MS_E is calculated by dividing SS_E by the smaller degrees of freedom (38 instead of 57). This trade-off slightly offsets the advantage of reducing SS_E but usually does not eliminate the overall gain in power.

Whether or not we end up with a more powerful test depends on if the reduction in SS_E more than compensates for the reduction in degrees of freedom for the error term. It does most of the time because we can usually take out quite a bit of error. If participants behave consistently across all time points or conditions, meaning they differ from one another in the same way each time, then a repeated-measures ANOVA allows us to remove a substantial amount of error. To understand this, recall how we determine how much error can be removed due to *between-person differences*. We do this by subtracting each individual's average score across all conditions from the overall grand mean.

- If a participant is *consistently above* the mean across all conditions, their *overall average* will also be above the mean, allowing us to *partial out* that difference completely.
- Conversely, if a participant scores *above the mean in one condition, below the mean in another, and close to the mean in a third*, their *overall average* will be very close to the grand mean. In this case, there is little individual variation to remove.

Thus, the more *consistent* individuals are in their responses across conditions, the greater the reduction in error variance, increasing the statistical power of the analysis.

ASSUMPTIONS OF A REPEATED-MEASURES ANOVA

Most of the assumptions for a repeated-measures ANOVA should look familiar.

- *Normality.* We assume the dependent variable is normally distributed in the population and the sample of individuals in our study. This assumption is particularly important when the sample size is small.
- *Independence within groups.* Within each group, participants are independently observed within the group, and the score of one person will not be related to the score of another person in the group.
- *Homogeneity of variance.* The variance of the dependent variable in each group (time period/condition) is the same.
- *Homogeneity of covariance.* We assume that participants' scores in each group are related because the same participants are observed across the groups/conditions. We also must assume that the extent to which they are related is consistent from one pair of groups to another. Removing the between-persons error from the total error only makes sense if the difference between an individual's average score and the grand mean is representative of all the groups.

These last two assumptions are together referred to as sphericity. In short, **sphericity** refers to the condition where the variances of the differences between all combinations of related groups (levels of the independent variable) are equal. Violating the assumption of sphericity artificially increases the value of between-group differences, which makes it more likely that we will commit a Type I error and reject the null when there is no true difference in the population.

CONDUCTING A REPEATED-MEASURES ANOVA

We will once again use data from the vocabulary recall study, displayed in Table 13.3, examining the effect of learning strategies to walk through an example. However, this time, we will assume that the same group of students ($n = 10$) experienced *all four learning strategies* rather than being assigned to only one. This repeated-measures design allows us to analyze how each student's recall performance varies across the different strategies while controlling for individual differences.

STEP 1: DETERMINE THE NULL AND ALTERNATIVE HYPOTHESES

The hypotheses for a one-way repeated measures ANOVA are the same as those for the one-way ANOVA, which are as follows:

For k groups,

H_0: $\mu_1 = \mu_2 = \ldots = \mu_k$ (all group means are equal)

H_A: *not all the* μ_k *are equal* (at least one group mean differs from the others)

In our example, $k = 4$ and thus:

$H_0: \mu_1 = \mu_2 = \mu_3 = \mu_4$ or $H_0: \mu_{Repetition} = \mu_{Visual} = \mu_{TechAid} = \mu_{Control}$
H_A: *not all the* μ_k *are equal* (at least one group mean differs from the others)

TABLE 13.3 *Vocabulary Recall Across Four Learning Strategies*

REPETITION	VISUALIZATION	TECHNOLOGY-AIDED LEARNING	CONTROL
13	17	17	10
12	15	18	12
12	15	16	9
11	17	16	12
9	18	20	4
11	18	21	11
11	14	18	9
14	15	20	8
13	17	19	9
8	18	17	5
$\bar{X}$=11.4	$\bar{X}$=16.4	$\bar{X}$=18.2	$\bar{X}$=8.9

STEP 2: SET THE CRITERIA FOR A DECISION

Let's set the alpha level at 0.05. In our example, $n = 10$ with four conditions, so the sampling distribution of F ratio follows F distribution with $df_{BG} = 3$, $df_E = (k - 1)(n - 1) = 3 \times 9 = 27$ when the sphericity assumption holds. We reject the null hypothesis when the obtained F statistic is larger than 2.96. Later, we will discuss how the *dfs* are adjusted for evaluating the F statistic when the sphericity assumption is violated to control the Type I error rate, but for now see Figure 13.1 on the next page for a visual representation of F distribution under the assumption is met.

STEP 3: CHECK ASSUMPTIONS AND COMPUTE THE TEST STATISTIC

We won't go through all the data assumptions, as you should already be familiar with some of them. Instead, we will focus on evaluating the sphericity assumption. We can begin this evaluation with a descriptive analysis by examining the variance-covariance matrix.

Table 13.4 shows the variance and covariance matrix. The diagonal elements of the matrix, shaded in grey, represent the *variances* of each condition. If these values are substantially different, it may indicate a violation of sphericity. By looking at these values, we see that the variances among four conditions range from 2.27 to 7.21. The variance for the control condition (7.21) is much larger than the others, suggesting that it may contribute to a potential violation of sphericity.

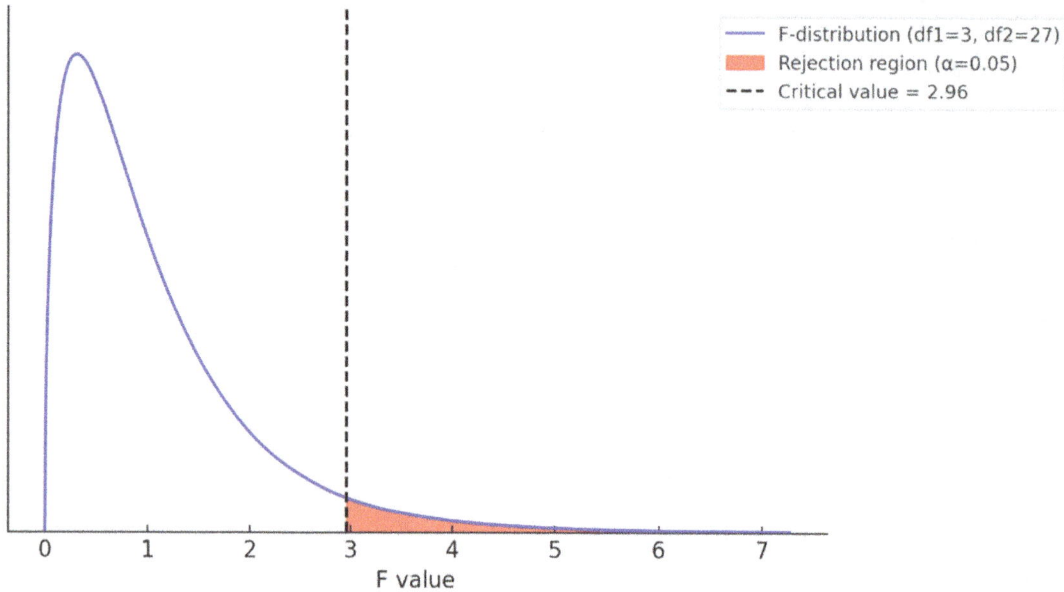

FIGURE 13.1 *F* Distribution with Critical Value for Repeated Measures ANOVA Under Sphericity (df = 3, 27)

TABLE 13.4 ***Variance and Covariance Matrix***

LEARNING STRATEGY	REPETITION	VISUALIZATION	TECHNOLOGY-AIDED LEARNING	CONTROL
Repetition	3.38	-1.4	0.24	2.71
Visualization	-1.4	2.27	0.58	-1.29
Technology-aided learning	0.24	0.58	3.07	-0.98
Control	2.71	-1.29	-0.98	7.21

Now, let's examine the off-diagonal elements. Since the upper and lower triangles contain identical values, we will focus on the upper triangle. These values represent the covariances between different conditions. If the covariances vary substantially, it may indicate a violation of the sphericity assumption. In this case, the covariances range from -1.4 to 2.70, demonstrating considerable variation, particularly since some are positive while others are negative. This pattern suggests a potential violation of the sphericity assumption.

Next, we can formally test this assumption with Mauchly's test for sphericity. The null and alternative hypotheses of Mauchly's tests are as follows:

H_0: The variances of the differences between conditions are equal.

H_A: The variances of the differences between conditions are not equal. (At least one variance differs from the others.)

To meet this assumption, we aim to fail to reject the null hypothesis. Table 13.5 summarizes the result of Mauchly's test, which can be obtained from common statistical software.

TABLE 13.5 *Mauchly's Test Results*

MAUCHLY'S W	APPROX. X^2	*df*	*p*	GREENHOUSE-GEISSER ε	HUYNH-FELDT ε	LOWER BOUND ε
0.445	6.247	5	0.286	0.666	0.856	0.333

Using $\alpha = 0.05$, we fail to reject the null hypothesis. Although we observed some inconsistencies in variances and covariances among the four groups, these inconsistencies are not severe enough to violate the assumption. Therefore, we conclude that the sphericity assumption is satisfied in our example. The second half of Table 13.5 (on the right-hand side) tells us more about the violation, assuming one exists. The epsilon (ε) values are estimates of the degree of sphericity in the population. If they are close to 1.0, then we expect that the sphericity assumption has been met. However, if they are substantially less than 1.0, this indicates that the sphericity assumption is violated. The lower the epsilon value is, the worse the violation.

What happens if we violate the assumption of sphericity and obtain a significant result for Mauchly's test ($p < 0.05$)? Fortunately, when this assumption is violated, we can adjust the degrees of freedom for the omnibus *F* test to compensate. This adjustment is made using epsilon (ε), which is applied to both the numerator and denominator degrees of freedom by multiplying them by the epsilon value. Since higher degrees of freedom increase statistical power, reducing them (by multiplying by a value less than 1.0) makes it more difficult to detect a significant result.

Conceptually, this adjustment acknowledges that our assumptions have not been fully met, so we apply a more *conservative approach* before rejecting the null hypothesis. The more severe the violation the *lower the epsilon value*, leading to a greater reduction in degrees of freedom, making it even harder to obtain statistical significance. There are three commonly used methods for calculating epsilon, each based on different criteria, and we must decide which one to use. Similar to how we learned about multiple post hoc test options in Chapter 10, the three epsilon values differ in how conservative/careful they are:

- The Huynh-Feldt correction is the least conservative.
- Lower-bound is the most conservative.
- The Greenhouse-Geisser correction is somewhere in between.

Which one should you use? Typically, when epsilons are smaller than 0.75, we suggest using the Greenhouse-Geisser value. If they are above 0.75, it is safe to use the Huynh-Feldt correction. Fortunately, you do not need to do the math yourself, because common statistical software will conduct the calculations for you and those for our current example are displayed in Table 13.6.

First, observe the various degrees of freedom in the numerator for the repeated conditions. If sphericity is assumed, we use the standard calculation of $k - 1$, which in this case equals 3. However, when we apply different epsilon (ε) values to adjust for violations of sphericity, the degrees of freedom decrease. At the lower bound, where $\varepsilon = 0.333$, the adjusted degrees of freedom become $3 \times 0.333 = 1.0$.

Why does the *F* statistic remain unchanged? If you refer back to Table 13.1, you'll see that degrees of freedom are multiplied by the same epsilon value in both the numerator and the denominator. This proportional adjustment ensures that the calculated *F* statistic stays the same and that only the degrees of freedom change, affecting the significance threshold. While statistical software handles these calculations automatically, it is important to understand how and why these adjustments are made when sphericity is violated.

TABLE 13.6 *Summary of Sphericity Corrections for Within-Subjects Effects*

CASES	SPHERICITY CORRECTION	SUM OF SQUARES	*df*	MEAN SQUARE	F	*p*	η^2
RM Factor	**None (Sphericity Assumed)**	**558.675**	**3**	**186.225**	**46.524**	**<.001**	**0.838**
	Greenhouse-Geisser	558.675	1.998	279.687	46.524	<.001	0.838
	Huynh-Feldt	558.675	2.567	217.642	46.524	<.001	0.838
	Lower-bound	558.675	1	558.675	46.524	< .001	0.838
Residuals	**None (Sphericity Assumed)**	**108.075**	**27**	**4.003**			
	Greenhouse-Geisser	108.075	17.978	6.012			
	Huynh-Feldt	108.075	23.102	4.678			
	Lower-bound	108.075	9	12.008			

STEP 4: FIND THE *p* VALUE, DRAW A CONCLUSION, AND REPORT YOUR CONCLUSION

Since the data meet the sphericity assumption, we use the result with no adjustment for degrees of freedom and ignore the rest of the rows. As it turns out, our *F* statistic is much higher than the critical value of 2.96, so we can reject the null hypothesis. We can conclude that there is a significant difference in means across recall conditions, $F(3, 27) = 46.524, p < .001$.

MEASURING EFFECT SIZE

Once we have determined that group means vary significantly in the population, we next determine the size of this effect using either η^2 or ω^2. In a between-subjects ANOVA, we divided SS_{BG} by SS_{Total} to get η^2 and then corrected that slightly for ω^2. But remember, with a repeated-measures (or within-groups) ANOVA, we are partialling out some of the total variation because it is due to between-persons differences that are the same across groups. We distinguish these new measures of effect size from the original η^2 and ω^2 by referring to them as partial η^2 and partial ω^2.

To compute partial eta squared (η_p^2), we remove $SS_{Subjects}$ from SS_{Total}. There are two ways to do this:

1. We can subtract SS_{Subjects} from SS_{total}:

$$\eta_p^2 = \frac{SS_{BG}}{SS_{Total} - SS_{Subjects}}$$

Or

2. We can add $SS_{between\ groups}$ to SS_{error}:

$$\eta_p^2 = \frac{SS_{BG}}{SS_{BG} + SS_E}$$

Remember that the original denominator was SS_{Total}. In this type of ANOVA, total variation is made up of variation between groups, variation between persons, and error variation. So, we can take the whole and subtract out the part we don't want to include (Option 1), or we can just add up the other two parts (Option 2).

You will get to the same place, but which one you choose depends on which values are easier to calculate. If software is doing the math for you, then it doesn't matter.

In our example, it is easier to use Option 2. Plug in the numbers and we get:

$$\eta_p^2 = \frac{558.675}{558.675 + 108.075} = 0.837$$

So, partial eta squared for this analysis is 0.837, meaning that 83.7% of the variance in vocabulary recall can be explained by learning strategies.

We also previously learned that η^2 tends to overestimate the effect, and the more conservative measure of the effect size is ω^2. For the repeated measures ANOVA, we use partial omega squared, ω_p^2. We have the same two basic options for calculating partial omega squared as with eta squared, but we will just use the second one since it is easier with our data:

$$\omega_p^2 = \frac{SS_{BG} - (df_{BG} * MS_E)}{SS_{BG} + SS_E + MS_E}$$

Notice that when the sphericity assumption is violated, we use df_{BG} in the numerator rather than k - 1. If sphericity is assumed, these two values are the same.

$$\omega_p^2 = \frac{558.675 - (3 * 4.003)}{558.675 + 108.075 + 4.003} = 0.815$$

We conclude that 81.5% of the variation in vocabulary recall is explained by the learning strategies applied.

POST HOC COMPARISONS

As we previously learned, post hoc tests are computed following a significant ANOVA to determine which pair or pairs of group means significantly differ. Since we have four groups, we need to make multiple pairwise comparisons using a post hoc test to determine which means are significantly different. A good starting point is to examine the means plots in Figure 13.2 to identify any patterns in the mean differences. Visualizing the data in this way can provide an initial sense of trends before conducting post hoc tests.

Remember, if we conduct multiple pairwise comparisons using repeated-measures t tests, the likelihood of committing a Type I error increases. The Bonferroni procedure is best adapted for use following a significant repeated-measures ANOVA. This adjusts the alpha level for each pairwise comparison by dividing the familywise alpha by the total number of pairwise comparisons. In our case there are six unique pairs:

- Repetition vs Visualization
- Repetition vs Technology-Aided Learning
- Repetition vs Control
- Visualization vs Technology-Aided Learning
- Visualization vs Control
- Technology-Aided Learning vs Control

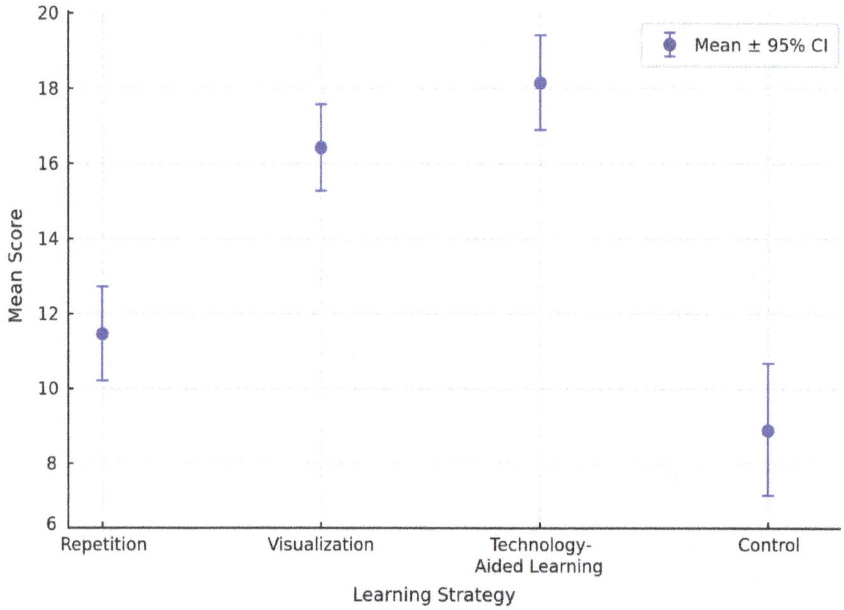

FIGURE 13.2 Mean Recall Scores by Learning Strategy with 95% Confidence Intervals

$$testwise\ alpha = \frac{familywise\ alpha}{total\ number\ of\ pairwise\ comparisons} = \frac{0.05}{6} = 0.0083$$

This means that each pairwise comparison is evaluated at $\alpha = 0.0083$, which is quite conservative. In fact, it would be too conservative if the number of groups were much higher, but we don't have the same number of options in terms of other post hoc tests with a repeated-measures ANOVA because they don't use the related-samples t test.

When we run the analysis using statistical software, Table 13.7 tells us that both the repetition and control groups are significantly different from both the visualization and technology-aided strategies. However,

TABLE 13.7 ***Post Hoc Analysis: Pairwise Comparisons of Learning Strategies with Bonferroni Adjustment***

			95% CI FOR MEAN DIFFERENCE				
		MEAN DIFFERENCE	LOWER	UPPER	SE	t	p_{bonf}
Repetition	Visualization	-5.000	-7.547	-2.453	0.895	-5.588	<.001***
	Technology-aided	-6.800	-9.347	-4.253	0.895	-7.600	<.001***
	Control	2.500	-0.047	5.047	0.895	2.794	0.057
Visualization	Technology-aided	-1.800	-4.347	0.747	0.895	-2.012	0.326
	Control	7.500	4.953	10.047	0.895	8.382	<.001***
Technology-aided	Control	9.300	6.753	11.847	0.895	10.394	<.001***

* $p < .05$, *** $p < .001$

Note. p value and confidence intervals adjusted for comparing a family of 6 estimates (confidence intervals corrected using the Bonferroni method).

APA sample write-up: A study was conducted to examine the effect of learning strategy on the number of words recalled by 10 participants across four different learning strategies. The assumption of sphericity was assessed using Mauchly's test, and since the assumption was met, no corrections to degrees of freedom were applied. The results of the one-way repeated measures ANOVA showed a significant main effect of learning strategy, $F(3, 27) = 46.52$, $p < .001$, $\omega^2 = .815$, indicating that the number of recalled words differed significantly across the four learning strategies. The partial omega squared value indicated a large effect, suggesting that a substantial proportion of the variance in vocabulary recall can be attributed to differences in learning strategy. Pairwise comparisons using Bonferroni-adjusted *t* tests revealed that both the repetition group and the control group were significantly different from both the visualization strategy and the technology-aided learning strategy ($p < .05$). However, the repetition and Control groups were not significantly different from each other after Bonferroni adjustment.

there is no statistically significant difference between the repetition and control groups or between visualization and technology-aided strategies.

ANALYSIS OF COVARIANCE (ANCOVA)

Analysis of covariance (ANCOVA) is a statistical technique used to evaluate whether there are significant differences between groups on a dependent variable while statistically controlling for the effects of one or more continuous variables (called covariates). ANCOVA offers a combination of multiple regression and analysis of variance. We believe that the best place to start is to consider the relationship between ANOVA, ANCOVA, and multiple regression, as these are all forms of the general linear model (Table 13.8). The difference has to do with what independent variables you are using.

Notice that all three require a dependent variable that is continuous and normally distributed. A traditional ANOVA has categorical grouping variables (e.g., treatment/control, education level) for the independent variable(s). A traditional multiple regression has continuous predictors for independent variables. You can also

TABLE 13.8 *Comparing ANOVA, ANCOVA, and Multiple Regression*

VARIABLE	ANOVA		ANCOVA		MULTIPLE REGRESSION
Dependent variable	Continuous normally distributed variable	⟷	Continuous normally distributed variable	⟷	Continuous normally distributed variable
Independent variable	Categorical (grouping) variable(s)		Categorical (grouping) variables + Continuous variables (i.e., Covariates)		Continuous variables (interval or ordinal) + Dummy coded variables

include categorical variables in regression by using dummy variables, which is a way of turning a categorical variable into something that can be treated like it is continuous in the regression model.

WHY USE ANCOVA?

We use ANCOVA if our primary focus is on the categorical grouping variable, but we want to include a continuous variable as a covariate. First, we need to understand what is meant by "covariate" here. **A covariate** is a continuous variable that is included in the analysis to account for variability in the dependent variable that is not directly related to the independent variable(s). By controlling for the covariate, ANCOVA provides a clearer understanding of the relationship between the independent variable(s) and the dependent variable. Along with being a continuous variable, an appropriate covariate is:

- Correlated with your dependent variable;
- Not highly correlated with/predictive of your independent variable, and
- Not something you're interested in interpreting as part of your analysis.

Remember that with any ANOVA, we are comparing the variance explained by our grouping variable(s) to error variance. Anything we can do to reduce the error variance gives our analysis greater power to detect a significant effect. ANCOVA allows us to subtract out part of the error variation that can be explained by a covariate. You are using the covariate to reduce the noise (error variation), so the differences associated with your actual independent variable(s) are easier to detect.

Now we should think about what would make a covariate an effective noise-reducer, and to do that we need to think about sources of variation. Figure 13.3 represents the total variation in our dependent variable. This is broken into two sources: variation due to our model, which potentially includes both main effects and interactions, and variation due to error.

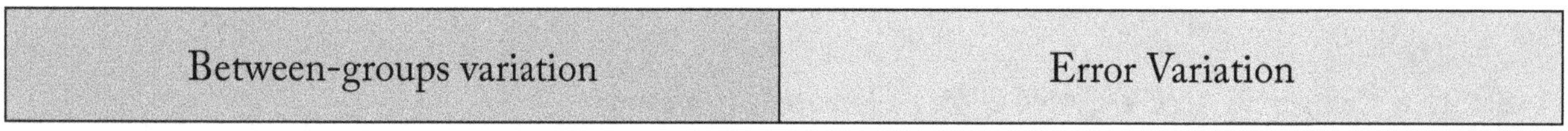

FIGURE 13.3 Total Variance in Dependent Variable

If we include a covariate in the model, we want it to take away from the error variation *without also reducing the between-groups variation*. This is why we want a covariate that is correlated with the dependent variable but not with the independent variable. Figure 13.4 represents our ideal scenario:

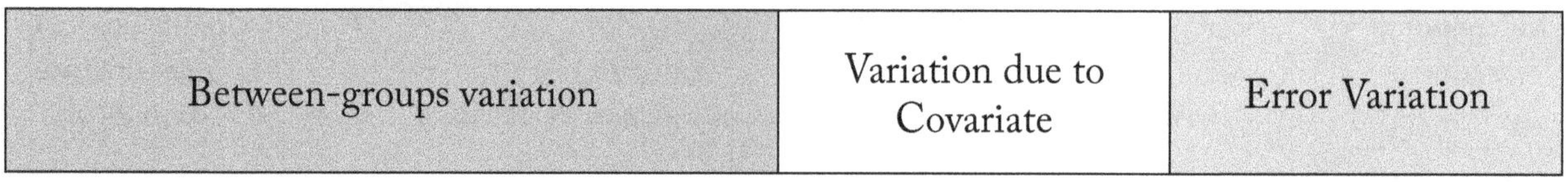

FIGURE 13.4 Ideal Partitioning of Total Variance with an Effective Covariate

In Figure 13.4 the covariate is highly correlated with our dependent variable, but not at all with the independent variables. It explains quite a bit of the error variation without removing any of the variation that we could explain with our independent variable. This combination is hard to come by in real life because more often covariates are related to each other, so we end up with something like Figure 13.5:

Between-groups variation	Variation due to Covariate	Error Variation

FIGURE 13.5 Partitioning of Variance When Covariate Shares Variance with Independent Variable

In Figure 13.5, our covariate is associated with both the independent variable and the dependent variable. While we remove some of that noise (error variation), we also remove some of our between-groups variation.

Consider two examples. First, imagine an experiment where first-year college students are randomly assigned to either a control group or one of two intervention groups designed to improve study skills. Our goal is to determine whether participation in an intervention group is associated with an increase in first-semester grade point average (GPA), our dependent variable. However, other factors may also influence GPA. Even though students were randomly assigned, each group consists of different individuals who may have varying study habits and academic abilities. These preexisting differences are unrelated to the intervention but add noise to the data, making it harder to detect meaningful effects. By using ANCOVA, we can include high school GPAs as a covariate. This allows us to control for prior academic achievement, removing any variation in first-semester GPA that can be explained by differences in high school GPAs. As a result, the analysis focuses more directly on the impact of the intervention, increasing the likelihood of detecting a true effect.

Now, consider a different scenario where students were not randomly assigned to the control or intervention groups. Instead, university administrators used a "risk" score to determine placement: students with the highest risk scores were assigned to an intensive intervention, those with moderate risk scores received a less intensive intervention, and students with lower risk scores received no intervention. In this case, high school GPA likely contributed to the risk score, meaning that including it as a covariate in an ANCOVA would reduce not only error variation but also the variation between groups, which would undermine the analysis. This setup would be inappropriate for ANCOVA, as the covariate (high school GPA) is directly related to how students were assigned to groups rather than being an independent factor. This example highlights a key requirement for selecting an appropriate covariate: it should be correlated with the dependent variable but not with the independent variable(s).

Now, consider the last bullet point about appropriate covariates. If the goal were to study the relationship between high school GPAs and first-semester college GPAs, then high school GPAs should be treated as an independent variable rather than a covariate. However, in this analysis we are not interested in explaining that well-established relationship. Instead, high school GPA is included only as a covariate to reduce error variation and increase statistical power when testing for differences in first-semester GPAs based on the intervention groups.

ASSUMPTIONS FOR ANCOVA

ANCOVA has the same assumptions as ANOVA: normality, independence, random sampling, and homogeneity of variance/covariance depending on if you're using between-subjects or within-subjects factors. There are also two new assumptions.

INDEPENDENCE OF THE COVARIATE AND THE TREATMENT EFFECT

In a traditional ANCOVA, the covariate is associated with the dependent variable but not the independent variable. A violation of this assumption is the problem we just described. If your covariate influences the independent variable as well as the dependent variable, it won't work as designed because it will reduce variation associated with your treatment/independent variable as well as error variation.

To check this, you can use a *t* test or a one-way ANOVA, depending on if your independent variable has two or more levels, to see if group membership predicts scores on the covariate. Going back to the prior example where students were randomly assigned to intervention or control groups, if we planned to use high school GPA as a covariate, we could start with a one-way ANOVA where intervention group was the independent variable and high school GPA was the dependent variable. If group membership did not significantly predict high school GPA, then that would be a good sign that the covariate was independent of the independent variable. There are exceptions to this assumption depending on why you are using ANCOVA, which we will get to. The next assumption, however, is important.

HOMOGENEITY OF REGRESSION SLOPES

This assumption is also referred to as a lack of interaction between the effect of the covariate and group membership. It probably sounds similar to the first assumption, but they are actually different. When we use a covariate, we fit a regression line to the entire dataset, ignoring which group an individual belongs to. In our example, we assume that there is a common constant relationship between high school GPA and first-semester college GPA. If it turns out that the relationship between the covariate and the dependent variable is different depending on group membership, then this overall regression line is inaccurate.

ANOTHER USE OF ANCOVA: ADJUSTED MEANS

In an ideal scenario, all analyses would be based on randomized controlled trials (RCTs), allowing researchers to make causal claims and use covariates solely to increase statistical power. However, in the real world, education researchers often rely on nonexperimental data or conduct correlational and observational studies. Without random assignment, groups may differ on multiple factors, some of which are related to both the independent variable and the dependent variable; these are known as confounding variables. One common approach to addressing these differences is to include confounding variables as covariates in an ANCOVA.

However, it is important to recognize the limitations of this method. Since it is impossible to identify, measure, and include all potential confounders, ANCOVA cannot fully substitute for ANOVA with random assignment. ANCOVA can, however, help approximate the results of an experiment by statistically adjusting for known confounding variables.

When using ANCOVA in nonexperimental research, it is crucial to avoid making causal claims, as you cannot definitively conclude that independent variable X causes dependent variable Y. Instead, the results should be interpreted as associational, acknowledging the potential influence of unmeasured confounders. So, what would be an example of this kind of research question? Imagine we wanted to examine differences in GPA by gender, but prior research suggests that motivation levels also differ between male and female students. Since motivation could influence GPA, motivation serves as a potential confounding variable in our analysis.

To account for this, we could include motivation as a covariate in an ANCOVA, allowing us to examine the relationship between gender and GPA while controlling for differences in motivation (Figure 13.6).

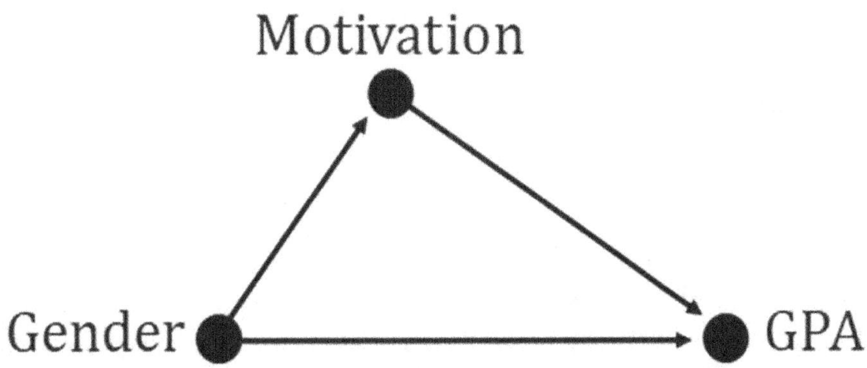

FIGURE 13.6 Motivation as a Confounding Variable in the Relationship between Gender and GPA

We will explore this concept further using a hypothetical dataset with 12 cases to illustrate how ANCOVA adjusts for confounding variables. If we only looked at the simple mean difference in GPA by gender, we would overestimate the effect. Instead, we are interested in adjusted mean differences after controlling for the covariate of motivation. To explore how ANCOVA can be used to find adjusted means and test the homogeneity of regression slopes, it is helpful to think of ANCOVA as a special type of multiple regression. It is considered special because our grouping variable (gender) is represented using dummy coding, a technique we learned in Chapter 12.

Since we only have two levels (male and female), we need only one dummy variable, with males coded as 0 and females coded as 1. This allows us to include ge nder as a predictor in the regression model while ensuring that differences between groups are appropriately quantified. To build on this, we will first examine a simple regression model using gender as the independent variable and GPA as the dependent variable in Table 13.9. Our focus will be on interpreting the regression coefficients (b), which are typically provided in statistical software output tables.

TABLE 13.9 ***Simple Regression Results***

PREDICTOR	*b*	*SE*	*t*	*p*
Intercept	2.450	0.154	15.95	<0.01
Gender	0.550	0.217	2.53	0.03

Therefore, our prediction equation is as follows:

$$\hat{Y}_i = 2.45 + 0.55(X_i)$$

If we want to know the mean GPA for each group, we just substitute the appropriate value for X and solve the equations.

For male students, $X = 0$.

$$Y\text{ - male} = 2.45 + 0.55\ (0) = 2.45$$

For female students, $X = 1$.

$$Y\text{ - female} = 2.45 + 0.55(1) = 3.00$$

Now we have the *unadjusted* mean GPAs for each group. But remember, we have reason to believe that motivation is a confounding variable that is associated with both gender and GPA. We can test whether our covariate (motivation) is independent of our grouping variable (gender) with a simple two–independent-samples *t* test (two-tailed, $\alpha = 0.05$). We used a statistical software package to run the analysis and confirmed that motivation scores are significantly different between male and female students, $t(10) = -4.82, p < .05$.

If we had randomly assigned students to two groups and were only trying to reduce error variation, then motivation would not be a good covariate to use. However, in this case we are using ANCOVA to estimate adjusted means rather than to reduce error variation. We know that motivation is associated with our independent variable, and we are using it as a covariate to try to control for the effect of motivation on GPA to get a more accurate estimate of the effect of gender on GPA.

To examine the relationship between X and Y, controlling for C (our covariate), we use a multiple regression equation using both X and C to predict Y. This is the ANCOVA model (in regression form), which can be written as $\hat{Y}_i = a + b_1 X_i + b_2 C_i$

Table 13.10 summarizes the regression coefficients.

TABLE 13.10 ***ANCOVA Results***

PREDICTOR	*b*	*SE*	*t*	*p*
Intercept	1.547	0.771	2.006	.076
Gender	0.163	0.388	0.420	.684
Motivation	.097	.081	1.194	.263

Therefore, our new prediction equation is as follows:

$$\hat{Y}_i = 1.55 + 0.163\,(X_i) + 0.097\,(C_i)$$

This time, the output tables tell us that neither the overall model nor gender on its own is a significant predictor of GPA. The relationship between X and Y has changed after controlling for C. Controlling for the covariate means that if all students had the same motivation scores, what differences in GPA would we see based on gender? To find the adjusted mean GPA by group, we substitute the grand mean of C (11.33) in the equation along with substituting in the appropriate values of X.

For male students, $X = 0$.

$$\bar{Y}_{male} = 1.55 + 0.163\,(0) + 0.097\,(11.33) = 2.65$$

For female students, $X = 1$.

$$\bar{Y}_{female} = 1.55 + 0.163\,(1) + 0.097\,(11.33) = 2.81$$

You can see that while female students still have higher adjusted mean GPAs than male students, the gap is narrower. This is consistent with the results of the regression analysis showing no significant difference in GPA by gender after controlling for motivation. We are no longer rejecting the null hypothesis, so it is possible that there is actually no difference in the population. We will get back to the actual ANCOVA analysis soon, but first a little more about regression so we can investigate the issue of homogenous regression slopes.

HOMOGENEITY OF REGRESSION SLOPES

The equations we just created were for adjusted mean values by group, and they used the grand mean for the covariate. It is possible to create separate prediction equations for each group that could include any value of the covariate. If we look back at the general prediction equation, we can substitute in appropriate values for X (either 0 or 1) to get these:

Male students:

$$\hat{Y}_{male\ i} = 1.55 + 0.097(C_i)$$

Female students:

$$\hat{Y}_{female\ i} = 1.71 + 0.097(C_i)$$

These two equations are different only in the value of the intercept; the slopes are the same. This means that the regression lines for the covariate are parallel, but the slope for male students is shifted a little lower on the Y-axis. Figure 13.7 shows the regression lines, where the dotted line represents the equation for female students and the solid line represents the equation for male students.

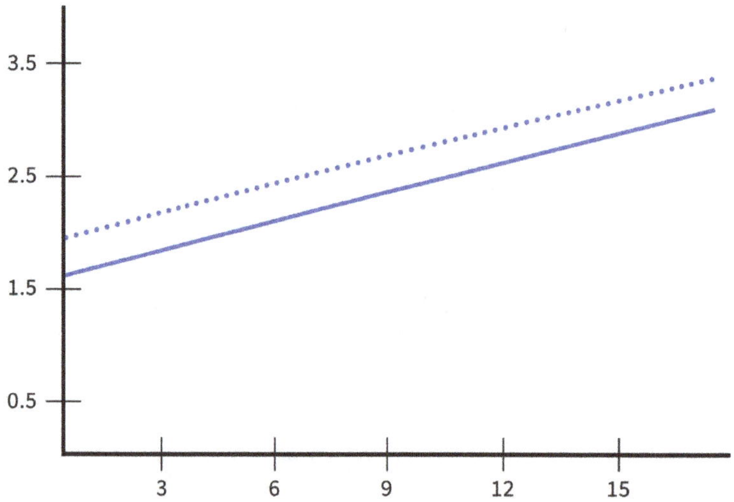

FIGURE 13.7 Parallel Regression Lines for Male and Female Students with Motivation as a Covariate

This is an example of how we can violate the first ANCOVA assumption but not the second, which is much more important. We deliberately chose a covariate that we knew was associated with our independent variable because we wanted to control for motivation to find adjusted mean GPA by gender. But even though motivation was associated with gender in this dataset, we are still assuming that the effect of motivation on GPA is not different based on gender.

What would it look like if we violated this assumption? If there was an interaction between our covariate and our independent variable such that the effect of motivation on GPA varied by gender, then the actual regression lines might look more like those displayed in Figure 13.8. If higher levels of the covariate led to higher scores on the dependent variable for one group but lower scores for another group, then this covariate would be meaningless. When this kind of interaction is present, we can't make a general statement about the effect of the covariate or about adjusted mean values by group.

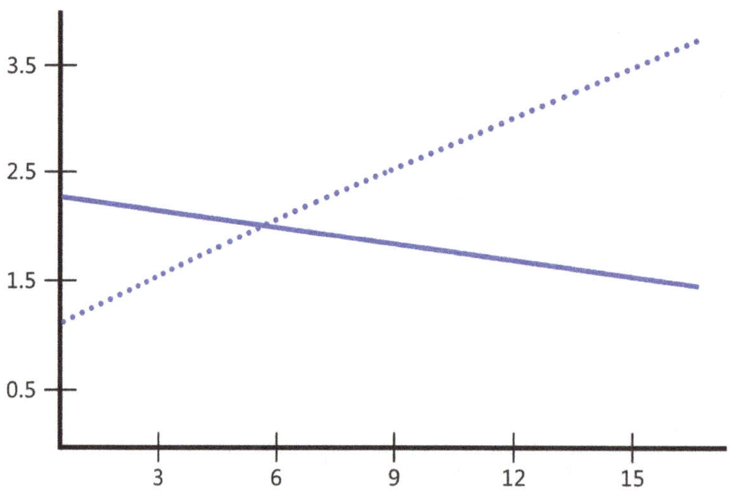

FIGURE 13.8 Violation of Homogeneity of Regression Slopes in ANCOVA

TESTING THE INTERACTION

To find out if there is an interaction between X (our grouping variable) and C (our covariate) and test the homogeneity of regression slopes, we need to add an interaction term to the regression equation and test it for significance. The interaction term is formed by multiplying the raw scores for the independent variable by the raw scores for the covariate:

$$X_i \times C_i$$

This leads to the full model we use to test the interaction term:

$$\hat{Y}_i = a + b_1 X_i + b_2 C_i + b_3 (X_i * C_i)$$

There are a few things you need to keep in mind about that equation.

1. You can't just test the interaction effect on its own. You need to include both the independent variable and the covariate in the model separately or else the interaction might show up as significant only due to the main effects of the two variables.
2. We hope to find that the interaction is not significant. Our null hypothesis is that $\beta_3 = 0$.
3. If we do manage to retain the null, we need to rerun the model without the interaction effect, because the other coefficients will change once we remove it.
4. If the interaction term is significant, then we need to keep it in the model. We will no longer have an ANCOVA model; instead we have a multiple regression model with an interaction.

MORE COMPLEX ANCOVAS

The example we've been using is the simplest possible ANCOVA, with only one independent variable that has only two levels. If you are conducting a one-way ANCOVA with a factor that has more than two levels, a factorial ANCOVA with multiple factors, or an ANCOVA with multiple covariates, the fundamental principles remain the same. However, these more complex models require additional steps upfront to thoroughly test assumptions before proceeding with the analysis.

GROUPING VARIABLE WITH > 2 LEVELS

Imagine if we wanted to use an independent variable with three levels rather than two, such as education level: high school graduates, some college experience, and college graduates. The main difference is that now we must make sure that the effect of our covariate is the same across all three groups, which means testing *two* interaction terms.

Since there are three groups, we need two dummy variables. Let's use high school graduates as the reference group, which means there is one dummy variable (X_1) to represent "some college" and a second dummy variable (X_2) to represent "college graduate." Each dummy variable is coded 0/1, so Table 13.11 shows how individuals in each group would be coded on the two variables.

TABLE 13.11 *Dummy Coding for Multiple Groups*

EDUCATION LEVEL	X_1	X_2
HS Graduate	0	0
Some College	1	0
College Graduate	0	1

Our overall prediction model will be:

$$\hat{Y}_i = a + b_1 X_{1i} + b_2 X_{2i} + b_3 C_i,$$

where b_1 represents the difference in Y between individuals with some college and high school graduates, b_2 represents the difference in Y between college graduates and high school graduates, and a indicates the adjusted mean of the high school graduate reference group.

To conduct the ANCOVA analysis, we have to check the homogeneity of regression slopes by testing interaction effects. If there are three categories in the independent variable and two dummy variables we also have two interaction terms (i.e., each dummy variable × the covariate), and then we test the full model with both interaction terms to see if either one was significant:

$$\hat{Y}_i = a + b_1 X_{1i} + b_2 X_{2i} + b_3 C_i + (b_4 X_{1i} * C_i) + (b_5 X_{2i} * C_i)$$

In this case, the hypotheses we're testing are:

H_0: $\beta_4 = \beta_5 = 0$
H_A: At least one of the beta coefficients is not zero.

If both interaction terms are not significant, you can use the covariate. The only other difference from our earlier example is that if there are significant group differences, you will need to conduct post hoc tests to find out which groups are significantly different from each other.

FACTORIAL ANCOVA

As your analysis becomes more complex, so does the work required to test assumptions. When conducting an ANCOVA with multiple factors or covariates, you need to address two key considerations:

- *The correlation between the covariate and the dependent variable.* The covariate must be correlated with the dependent variable within each combination of groups of the independent variables. For example, in a 2×2 ANCOVA, you must check this correlation separately for all four groups (cells).
- *Testing for interaction effects.* Here, the focus is on whether the covariate interacts with the treatment conditions, not the typical interaction between independent variables. Just as in cases where a factor has more than two levels, you need to test the interaction between the covariate and each dummy variable for every factor.

Note that in a 2×2 ANCOVA even though there are four groups, you only have one dummy variable, meaning there are only two interaction effects to check.

MULTIPLE COVARIATES

If you include two covariates in your ANCOVA, you must perform all the same assumption tests as you would for a single covariate—but twice (once for each covariate). While adding covariates can help control for additional factors, it is important to be cautious about overfitting the model. Every additional variable reduces your degrees of freedom, so including a weak covariate—one that does little to reduce error variation—can actually decrease the statistical power of your analysis rather than improve it. As always, the selection of covariates should be guided by theory and prior research to ensure that they meaningfully contribute to the model rather than simply increasing complexity.

SUMMARY

You are probably going to have to read this chapter more than once before it all makes sense, so here is a quick summary to reorient yourself.

- There are two main reasons you would use a covariate.
    - If you're using a classic experimental design with random assignment to groups, you would use ANCOVA to reduce error variance and therefore have greater power to detect a significant difference.
    - If you are using nonexperimental data, you might use ANCOVA to adjust group means by controlling for a confounding variable.
    - In either case, you are not interested in interpreting the results of the covariate; it is only there to clarify the relationship between the independent variable and the dependent variable.
- There are two new assumptions for ANCOVA.
    - *Independence of the covariate and the treatment effect.* Ideally, your covariate is related to your dependent variable but *not* to your independent variable. You test this assumption by using either an independent-samples t test (if your independent variable has two groups) or a one-way ANOVA (if you have more than two groups), with the covariate as your dependent variable. If the result is not significant, you can proceed because your independent variable does *not* predict the covariate.
    - *Homogeneity of regression slopes.* We need the covariate to have the same effect on the dependent variable for every level or group of your independent variable. We test this by computing a new variable to represent the interaction between the independent variable and the covariate. Then we run a multiple regression model with the independent variable, the covariate, and the interaction term as predictors of the dependent variable. If the interaction term is *not* significant, we can proceed.

CONCLUSION

This chapter explored the foundational concepts and practical applications of repeated-measures ANOVA, a powerful statistical tool for analyzing differences across multiple conditions or time points within the same group of participants. By leveraging the repeated-measures design, researchers can control for individual differences, thereby reducing error variance and increasing statistical power compared to between-subjects designs.

Central to the analysis is the ability to partition variance into components attributable to between-groups effects, within-groups error, and between-persons differences, which allows for a more nuanced understanding of the data. Additionally, the chapter highlighted the importance of understanding longitudinal versus cross-sectional designs and emphasized that repeated-measures ANOVA is uniquely suited to studies involving repeated observations on the same individuals over time or under different conditions.

A critical element of this analysis is ensuring that the assumption of sphericity is met or properly addressed. Violations of sphericity, which occur when variances of the differences between all combinations of related groups are unequal, can inflate Type I error rates, leading to invalid conclusions. To address this, researchers are equipped with tools such as Mauchly's test of sphericity and corrections such as the Greenhouse-Geisser and Huynh-Feldt adjustments. By understanding these concepts and incorporating them into their analyses, researchers can confidently interpret the results of repeated-measures ANOVA, making it an invaluable method for studies that track changes over time or across experimental conditions. Through statistical software and adherence to methodological rigor, this chapter equips readers with the knowledge to effectively utilize this robust analytical approach.

In addition to repeated-measures ANOVA, this chapter highlighted the role of analysis of covariance (ANCOVA) as a valuable statistical tool for controlling confounding variables when analyzing group differences. Unlike standard ANOVA, ANCOVA incorporates covariates—continuous variables that help reduce error variance—allowing for a more precise evaluation of treatment effects. This is particularly useful in education research, where factors such as prior achievement, motivation, and socioeconomic status may influence outcomes. By statistically adjusting for these preexisting differences, ANCOVA enhances the validity of comparisons, especially in quasi-experimental designs where randomization is not feasible. However, researchers must ensure that covariates meet key assumptions, such as independence from the treatment effect and homogeneity of regression slopes, to avoid misleading conclusions. When used appropriately, ANCOVA strengthens the ability to draw meaningful inferences, making it an essential complement to repeated-measures ANOVA in educational research.

KEY TAKEAWAYS FOR EDUCATIONAL RESEARCHERS FROM CHAPTER 13

- The use of repeated-measures design in education research is crucial for examining changes in learning outcomes. Repeated-measures ANOVA is particularly useful in education research because it allows for the study of changes in learning outcomes, student engagement, and skill development over time. For example, a researcher could track the effects of different instructional strategies on student performance across multiple testing periods. By using the same participants in each condition, this design controls for individual differences, making it a powerful tool for assessing interventions and longitudinal trends in educational settings.
- Controlling for individual differences enhances statistical power. A key advantage of repeated-measures ANOVA is its ability to remove variability due to individual differences, leading to increased statistical power. In education research, where student performance can vary due to prior knowledge, motivation, or learning styles, this method helps isolate the actual effect of a key independent variable from these potential confounding effects. For instance, in a study comparing different teaching techniques, repeated-measures ANOVA ensures that each student's unique characteristics do not confound the results, allowing for a clearer interpretation of instructional effectiveness.

- When analyzing repeated measures data, ensuring that the assumption of sphericity is met is critical. A violation of sphericity occurs when the variances of the differences between time points or conditions are unequal, which can lead to inflated Type I error rates. Researchers must check for this assumption using Mauchly's test for sphericity and apply corrections such as Greenhouse-Geisser or Huynh-Feldt adjustments if necessary. For example, if a study is evaluating student progress across multiple semesters, a potential violation of the sphericity assumption may arise when students' score variances increase over the years (e.g., early semester scores are relatively stable, but later semesters show more fluctuation) or when the variance of score differences between Semester 1 and Semester 2 is significantly smaller than that between Semester 3 and Semester 4.
- ANCOVA enhances comparisons by controlling for covariates. When studying the impact of an intervention, ANCOVA adjusts for preexisting differences (e.g., prior knowledge, socioeconomic status, or motivation levels) to isolate the true effect of the intervention. This is essential in quasi-experimental designs commonly used in education research where randomization isn't feasible.
- Education researchers should use ANCOVA thoughtfully in nonexperimental studies. When analyzing observational or survey data, ANCOVA can help control for confounding variables such as student motivation or prior achievement. However, researchers must avoid making causal claims unless random assignment was used—otherwise, differences may reflect preexisting group characteristics rather than true treatment effects.

KEY DEFINITIONS FROM CHAPTER 13

Analysis of covariance (ANCOVA) is a statistical technique used to evaluate whether there are significant differences between groups on a dependent variable while statistically controlling for the effects of one or more continuous variables (called covariates).

A **covariate** is a continuous variable that is included in the analysis to account for variability in the dependent variable that is not directly related to the independent variable(s). By controlling for the covariate, ANCOVA provides a clearer understanding of the relationship between the independent variable(s) and the dependent variable.

A **repeated-measures ANOVA** is used to determine whether there are significant differences between three or more related groups or measurements taken from the same participants under different conditions, at different time points, or across repeated trials.

Sphericity refers to the condition where the variances of the differences between all combinations of related groups (levels of the independent variable) are equal.

CHECK YOUR UNDERSTANDING

1. What is the key advantage of using a repeated-measures ANOVA?
   a. It eliminates the need for a dependent variable.
   b. It reduces error variance by accounting for individual differences.
   c. It allows for the inclusion of multiple dependent variables.
   d. It does not require any statistical assumptions.

2. What does the term "sphericity" refer to in the context of repeated-measures ANOVA?
   a. Equal variances within each group
   b. Equal sample sizes across groups
   c. Equal variances of differences between all combinations of related groups
   d. Equal means across all groups

3. What is the recommended post hoc procedure for a repeated-measures ANOVA to control for Type I error?
   a. Tukey's HSD
   b. Bonferroni correction
   c. Fisher's LSD
   d. Scheffé test

4. What does a significant result in a repeated-measures ANOVA indicate?
   a. All group means are equal.
   b. The dependent variable is unrelated to the conditions tested.
   c. The assumptions of sphericity have been violated.
   d. There is a difference in at least one pair of group means.

5. What does "partial eta squared" (η_p^2) represent in repeated-measures ANOVA?
   a. The total variation in the dependent variable
   b. The proportion of total variance explained by between-groups effects after accounting for between-persons variance
   c. The difference between the group means
   d. The degree of correlation between independent and dependent variables

6. What is the role of a covariate in ANCOVA?
   a. To control for variability in the dependent variable that is unrelated to the independent variable(s)
   b. To act as an additional independent variable
   c. To replace the independent variable in the model
   d. To reduce between-group variance

14

ADVANCED QUANTITATIVE METHODS

LEARNING OBJECTIVES

By the end of this chapter, you will be able to:

- Recognize advanced statistical techniques used. such as multilevel modeling, structural equation modeling, and meta-analysis, to address complex research questions in educational settings, accounting for nested data structures and latent constructs;
- Describe the appropriateness, strengths, and limitations of various advanced quantitative methods, ensuring alignment with research questions, theoretical frameworks, and the unique characteristics of educational data; and
- Differentiate between advanced analytical techniques and determine when each is most suitable for handling hierarchical data, modeling relationships among latent variables, or synthesizing research findings.

Educational research is a dynamic field that frequently challenges researchers to delve deeper into complex questions about teaching, learning, and policy. Designed for those seeking to expand their methodological tool kit beyond foundational techniques such as multiple regression and ANOVA and rigorously investigate educational phenomena, this chapter introduces key advanced methods, such as multilevel modeling (MLM), factor analysis, structural equation modeling (SEM), and meta-analysis. These methods are often used to explore important questions in education because they are particularly suited to the complexities inherent in education research, allowing for analyses that account for group-level predictors, interactions, and longitudinal effects. Researchers who are eager to learn more about these methods should consult additional resources, as this chapter provides only an introduction. It is also important to note that these techniques represent only a selection of the tools available. Given the nested and multifaceted nature of educational data, where understanding group-level predictors, interactions, and longitudinal effects is critical, researchers are encouraged to explore additional methods beyond those covered here.

Each method discussed in this chapter is contextualized within the unique challenges and opportunities of education research, with a focus on practical application and theoretical alignment. We hope that knowledge of these advanced methods ensures that your contributions are both innovative and impactful, driving meaningful progress in education.

MULTILEVEL MODELING

One key assumption in most of the methods we covered throughout this book is independence of observations. The assumption means that no individual outcome is dependent on the outcomes of other individuals in the sample. Does this sound like how things really work in education?

Education data are inherently nested: students are nested in classes, which are nested in schools, which are nested in communities. Because of these shared contexts, students in the same groups (e.g., same classroom or teacher) typically have more in common with one another than with students outside the group. MLM, also known as hierarchical linear modeling, is a statistical technique designed to analyze data that are organized at more than one level (i.e., nested or hierarchical). One key reason to use MLM is that it helps avoid violating the independence assumption found in standard regression methods. In standard (single-level) regression that we learned in Chapter 12, we assume that all data points are independent; however, with students clustered within classrooms and schools, this assumption does not hold. MLM accounts for the nested nature of the data when estimating relationships, providing more accurate results and standard errors. Note that although MLM can also be used for longitudinal investigations, this chapter focuses on cross-sectional MLM, which is appropriate for data collected at a single point in time, to explain the basic concepts.

Another major reason to use MLM is that you can ask new types of research questions, such as the following:

- *Impact of contextual effects (predictors) on individual outcomes.* For example, if you are interested in the effect of teacher characteristics on a student-level outcome, those characteristics are really classroom-level variables even though we sometimes treat them as if they were student-level characteristics in analysis.
- *Exploration of group variation in the individual-level independent variable–dependent variable relationship.* For example, in a study across schools, each school might show a unique relationship between students' socioeconomic background (the independent variable) and their achievement (the dependent variable). If you plot a line for each school, you might see that these slopes differ substantially across schools, as shown in Figure 14.1. In such a case, an analysis that ignores the nesting effect can oversimplify our understanding of educational phenomena and may yield invalid results by ignoring the possible contextual impact in which students are embedded.
- *Exploration of cross-sectional interactions.* This examines the degree to which a predictor at one level (e.g., a classroom- or school-level characteristic) modifies the relationship between two variables at another level. For example, you might investigate the extent to which teacher experience (a school-level variable) moderates the impact of student motivation (an individual-level independent variable) on academic performance (an individual-level outcome). This is often called a cross-level interaction because it involves variables from different levels of the data hierarchy. Such an interaction is difficult to capture accurately with single-level regression analysis.

When you read about MLMs, you will see the phrases "means as outcomes" and "slopes as outcomes."

- "Means as outcomes" looks for differences in group means as the dependent variable in an MLM. For example, the average achievement by school can be modeled with group-level predictors, such as the composition of the

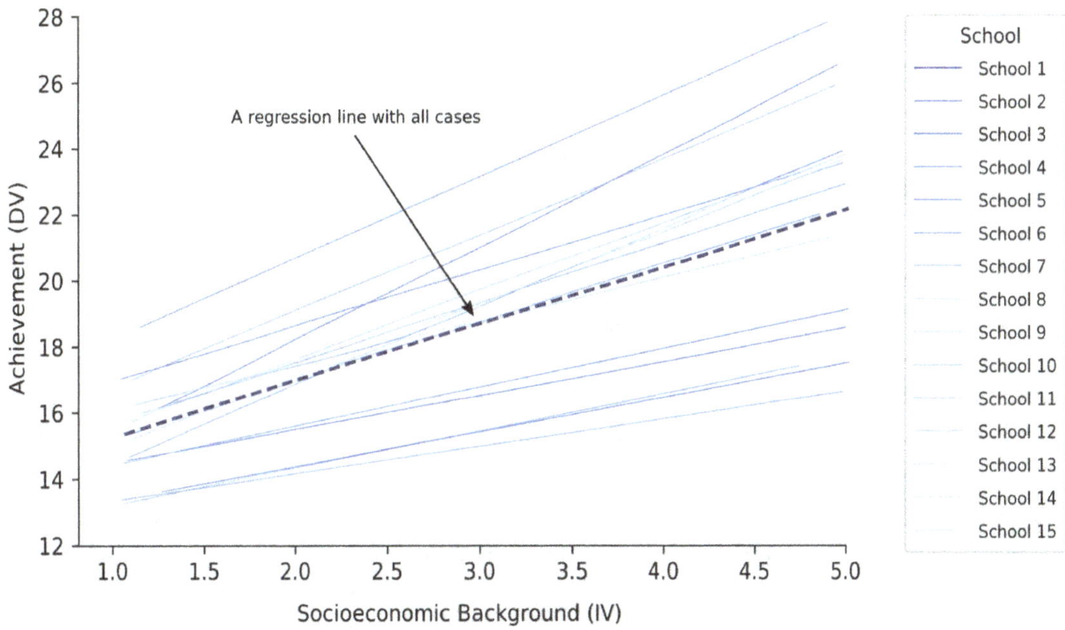

FIGURE 14.1 Multilevel Model Illustration: Random Slopes for Socioeconomic Background by School

student body or teacher experience, explaining variation in school means. Additionally, student-level predictors can be included to refine individual-level variance all within one model.

- "Slopes as outcomes" looks for differences in the effect of the student predictor on outcomes based on the group. This examines whether group-level predictors moderate the relationship between student-level predictors and outcomes, which is known as a cross-level interaction, mentioned earlier.

One of the first steps you take with MLM is to figure out how much of the variation in outcomes is due to individual-level differences and how much is due to group-level differences, a measure known as intraclass correlation. The greater the proportion of variation attributed to group differences, the more important it is to use MLM. You can use MLM within an experimental design if there is random assignment to groups or in observational studies based on either a natural grouping of individuals or a clustered sampling design.

One challenge with MLM is the need for a large sample size, particularly when you are trying to include any group-level variables. This is because at the group levels, the number of groups (rather than individuals) serves as your sample size for part of the analysis. A common rule of thumbs suggests a minimum of 20–30 groups with at least 20–30 individuals per group (Huang, 2018; Maas & Hox, 2005). However, the exact sample size requirement depends on the specific focus of the analysis and should ideally be determined through power analysis rather than relying solely on general rules of thumb. Regardless of the precise sample size needed at each level, MLM generally requires a large dataset to ensure reliable estimates and sufficient statistical power.

FACTOR ANALYSIS AND STRUCTURAL EQUATION MODELING

Factor analysis and SEM are closely related statistical techniques that help researchers understand complex relationships between observed and latent variables. Factor analysis serves as the foundation for SEM, as both methods focus on measuring latent constructs that are not directly observable. In educational research, latent constructs (psychological constructs and traits), such as student engagement, academic motivation, self-efficacy, and teaching effectiveness, cannot be measured directly but can be inferred through multiple survey items or assessments. Factor analysis is used to identify underlying structures within a dataset, grouping items based on shared variance. For example, a student engagement survey might contain a set of items about participation and effort, each of which measures a student's level of different types of engagement behavior. Instead of analyzing these items individually as variables, factor analysis groups these items into a single "engagement" factor, allowing researchers to estimate an overall engagement level based on the responses to these items.

However, when these latent constructs are incorporated into a broader framework that examines how they relate to each other, the result is SEM. SEM extends factor analysis by allowing researchers to test theoretical models that specify how unobservable constructs such as engagement, motivation, and achievement influence one another. For example, does academic motivation lead to higher engagement, which in turn enhances achievement? By modeling these relationships while accounting for measurement error, SEM provides a more precise way to test educational theories and evaluate complex learning processes. Now, let's examine each method in more detail.

Factor analysis is a statistical method commonly used in measurement research to assess the validity and structure of a construct by identifying underlying structures or patterns within a set of observed variables. It is often applied to determine whether survey or test items effectively measure the intended construct and to reduce the dimensionality of data by grouping variables into clusters (or "factors") based on shared variance. These factors often represent latent constructs such as intelligence, motivation, and satisfaction that cannot be directly measured but are inferred from multiple observed indicators (e.g., a set of numeric variables). These indicators are typically measured on a continuous scale or a Likert-type scale rather than being dichotomous (e.g., yes/no responses). To reliably measure a latent construct, researchers generally include at least three observed indicators per construct in their model.

There are two main types of factor analysis:

Exploratory factor analysis (EFA). EFA is used when researchers do not have a predefined structure in mind and aim to *explore* potential factor groupings within the data. This is often applied in early-stage research to identify how variables cluster together and to get insights into refining a measurement of latent constructs. In education research, EFA is frequently applied for the following purposes:

- Developing and refining psychological and educational assessments. For example, EFA can help identify the underlying dimensions of a new psychological scale designed to measure student motivation. Researchers can use EFA to analyze responses from the new scale to determine how items naturally group together and whether they effectively capture different aspects of motivation. The results of EFA can provide insights into revising the scale by refining or eliminating items that do not align well with the intended construct.
- Identifying the hidden patterns in responses to a set of questions. For example, researchers might use EFA to examine responses to a classroom climate survey, uncovering what factors emerge in students' perceptions, such as teacher support, peer collaboration, and engagement as distinct components.
- Reducing the number of variables in large datasets. EFA can serve as a data reduction technique by identifying redundant or highly correlated variables and summarizing them into a smaller set of meaningful factors. For example, in a large-scale education study that collects data on students' study habits, data may be collected on

variables such as time spent on different subjects, use of learning resources, time management strategies, and preferred study environments. EFA might reveal that these items load onto just two underlying factors: time investment and study strategies. This exploratory approach simplifies the dataset while preserving key information about student learning behaviors.

Confirmatory factor analysis (CFA). CFA is used when researchers have a *theoretical model* specifying how variables should load onto factors and seek to test whether the data fit this predefined structure. CFA is commonly applied in validation studies to evaluate the extent to which a measurement model appropriately represents the intended constructs. In education research, CFA is frequently applied for the following purposes:

- Testing the structure of psychological and educational assessments. For example, researchers can examine whether a measurement model provides evidence supporting the intended interpretation of scores, an essential aspect of validity. For instance, if motivation consists of intrinsic motivation, extrinsic motivation, and self-efficacy as distinct but related factors, CFA can be conducted to statistically test whether this three-factor structure fits the data. If the CFA results indicate good model fit, researchers gain supporting evidence that the scale effectively captures the theorized dimensions of motivation.
- Assessing measurement invariance across different student groups. CFA can be used to evaluate whether a scale functions consistently across diverse populations, such as by grade level, or cultural background. For instance, a researcher may use CFA to examine whether an academic stress scale measures the same construct equally across middle school and high school students. This ensures that differences in academic stress scores reflect actual differences in students' stress levels rather than differences in how students at different grade levels interpret or respond to the scale items.

Structural Equation Modeling (SEM) is an advanced statistical technique that integrates factor analysis and multiple regression to examine complex relationships among both observed and latent variables. SEM is used when researchers have a theoretical model that specifies how latent constructs influence observed responses and how these constructs relate to each other. The model is informed by theory and/or prior research, and SEM allows researchers to test how well the data fit this predefined model.

SEM offers several advantages over linear regression, including the following:

- The capability to analyze complex relationships among latent constructs. For example, SEM can have more than one dependent variable (both observed and unobserved) and can incorporate mediating variables at the same time. Unlike traditional regression, SEM distinguishes between exogenous and endogenous predictors. An endogenous variable is one that predicts the dependent variable but is itself predicted by another independent variable. Therefore, SEM enables researchers to explore a more comprehensive representation of causal relationships.
- The ability to deal with complex constructs that are difficult to measure directly or require multiple indicators while also allowing for the specification of measurement errors, resulting in a more accurate model.

When conducting SEM, sample size considerations are crucial. A commonly used guideline suggests having 5 to 10 participants per estimated parameter in the model. However, the required sample size depends on aspects such as model complexity, the number of estimated parameters, and other data characteristics. In general, a minimum of 200 participants is recommended for stable estimation, but very simple models with fewer parameters may be feasible with around 150 participants. For more complex models, larger samples (e.g., 400+) may be necessary to ensure reliable results and sufficient statistical power. It is important to note that recommendations may vary across the literature, and best practice is to conduct a power analysis to determine an appropriate sample size based on the specific model and research context.

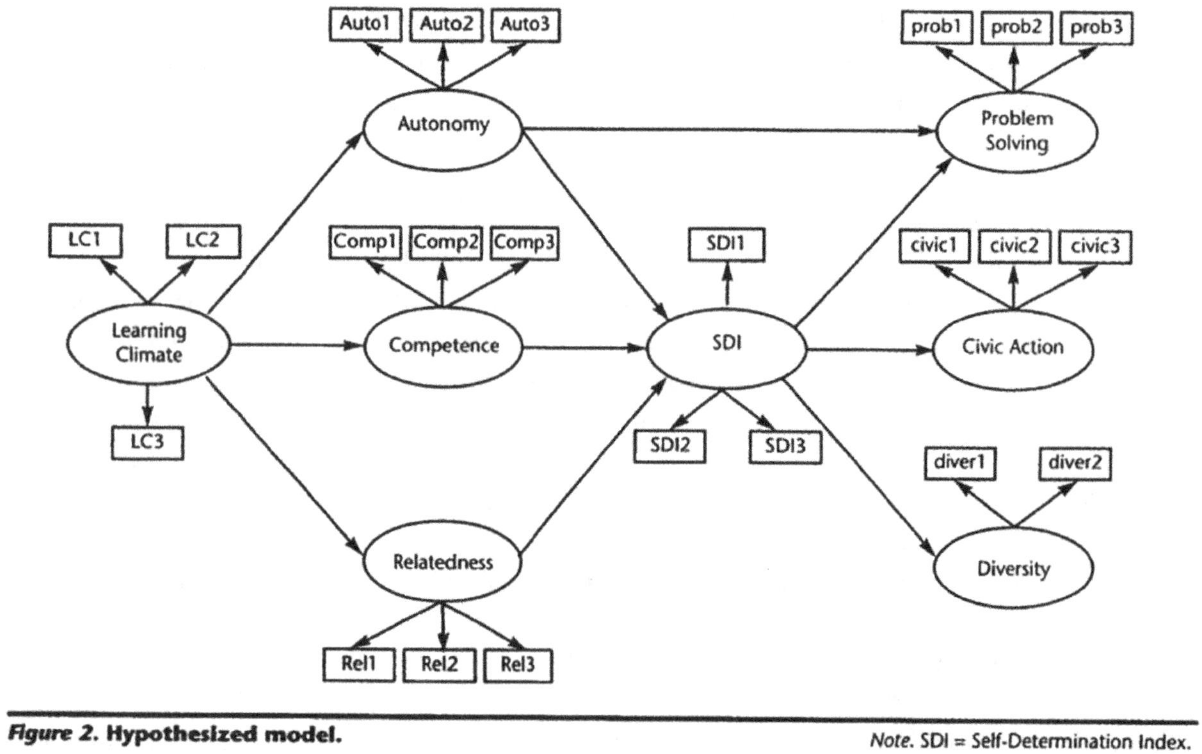

FIGURE 14.2 Hypothesized Structural Equation Model of Self-Determination Theory. Source: Levesque-Bristol, C., Knapp, T. D., & Fisher, B. J. (2011). The effectiveness of service-learning: It's not always what you think.. *Journal of Experiential Education*, 33(3), 208–224.

Figure 14.2 is from a study on self-determination theory by Levesque-Bristol et al. (2011). Each oval is a latent construct, such as autonomy or motivation, that predicts answers to survey items, represented by rectangles. The authors expect that learning climate affects student motivation, operationalized using the Self-Determination Index (SDI), a composite score that qualifies the extent to which student's motivation is self-determined. The SDI is computed by weighting different types of motivation (e.g., intrinsic, extrinsic, and amotivation) based on their position along the self-determination continuum. The model proposes that learning climate influences students' motivation through their perception of autonomy, competence, and relatedness. Levesque-Bristol et al. then suggest that more self-determined motivation (i.e., higher SDI scores) leads to positive outcomes such as problem solving (which is also directly affected by autonomy), civic action, and diversity—and each of these constructs is represented by specific items (or groups of items).

In SEM, the *measurement model* and the *structural model* serve distinct but complementary purposes in analyzing relationships among variables. The measurement model focuses on the relationship between latent constructs (unobserved variables) and their observed indicators (measured variables) and defines how well the observed variables represent the latent constructs by assessing factor loadings and measurement errors. In the example represented in Figure 14.2, a latent construct such as "autonomy" was measured by three observed variables (Auto1–Auto3), which is evaluated by factor analysis to ensure that indicators reliably measure the intendent construct. The structural model focuses on the relationships between latent constructs, specifying how one construct influences another, and examines direct and indirect relationships between variables, allowing researchers to test hypotheses about causal or theoretical relationships among constructs. Unlike the measurement model, the structural model does not focus on how latent variables are measured but rather on how they interact with one another. In Figure 14.2, the structural relationships are represented by arrows connecting one latent construct to another. For example, the path from learning climate to autonomy to SDI indicates that learning climate indirectly impacts SDI, with autonomy serving as a mediator. Significance tests on these paths examine the hypothesis, and SEM simultaneously tests multiple paths within the model.

META-ANALYSIS

Meta-analysis is a special term for quantitative synthesis, which is a type of systematic review of the literature. When there are a lot of studies on a topic but they have inconsistent results, we can get into "vote counting"—where you simply tally how many show a significant result versus a nonsignificant result, with the majority appearing to "win." However, such an approach overlooks important differences between studies that may explain the variability in findings. For example,

- Some interventions are more effective than others.
- The way outcomes are measured (operationalization) may cause variation in results.
- Differences in cultural contexts, target populations, sampling techniques, sample characteristics, and sampling error can contribute to inconsistent findings.
- Some studies are more methodologically rigorous than others, affecting the quality of their results.

Given these potential sources of variation, we want to ask to what extent do these differences have an impact on the reported findings?

Meta-analysis differs from narrative review (literature review), as it takes a *systematic, statistical approach* to synthesizing research findings across multiple studies. Unlike narrative review, which summarizes findings qualitatively, meta-analysis synthesizes numerical effect sizes from systematically selected relevant studies to provide a quantitative estimate of the effect of interest. By synthesizing quantitative findings from multiple studies, meta-analysis enhances statistical power, making its conclusions more reliable and reproducible, less prone to bias, and more generalizable than those from a single study or a narrative review. Additionally, a key advantage of meta-analysis is its ability to identify influential sources of variability in reported results. When individual studies produce inconsistent findings, meta-analysis can reveal hidden patterns and account for differences across studies. In other words, researchers can use meta-analysis to examine the extent to which study characteristics, such as sample size, methodology, and demographics, influence effect sizes. Finally, meta-analysis helps identify gaps in literature through a systematic search and review of existing studies on a given topic. Collectively, meta-analysis provides deeper insights into evidence-based practices and improves our understanding of the research landscape within a selected field.

In a meta-analysis, researchers set three main research questions regarding the following:

- *Central tendency.* The common effect (i.e., does X affect Y, and how strong is the effect?).
- *Variability.* Measuring the variation in effect sizes (i.e., is the average effect of X on Y consistent across studies or is there a significant degree of variability in the effects?).
- *Prediction.* Explaining why the variation occurs (i.e., a search for moderator variables; what variable changes the effect of X on Y?).

The outcome of interest in a meta-analysis is effect size, which demonstrates the direction and magnitude of effects. Common effect size used in a meta-analysis in education includes the following:

- Means comparisons (Cohen's *d*)
- Odds ratios
- Risk ratios
- Correlation coefficients (Pearson's *r*)

As it is a quantitative synthesis, meta-analysis is appropriate when studies are empirical instead of theoretical, have quantitative results, have findings that can be configured in a comparable form (e.g., Cohen's *d*, correlation, odds ratios), examine the same constructs and relationships, and are comparable given the question at hand. Some will claim that meta-analysis compares apples and oranges, meaning you are looking for a common effect among studies that are not really comparable. This concern often arises when studies differ significantly in terms of populations, interventions, methodologies, or outcome measures. To address this issue, we must be specific in defining our inclusion criteria of studies in a meta-analysis. This means carefully selecting studies that align with the research question by setting clear guidelines for outcomes (definition of effect sizes), study design (e.g., experimental vs. observational studies), population characteristics (e.g., age groups, educational settings), measurement methods (e.g., how the outcome variable is operationalized), and intervention details (if applicable).

One of the most common questions in meta-analysis concerns the number of studies required. In theory, a meta-analysis can be conducted with just two studies, but in this case it would only address central tendency (i.e., estimating an overall effect). To explore variation in effect sizes, at least three studies are needed, and to investigate potential sources of variation, a minimum of four studies is required. By now, you understand that determining an appropriate sample size for a statistical analysis depends on multiple aspects, and there is no single definitive answer (other than conducting the power analysis). The same applies when determining the sufficient number of studies for a meta-analysis, as it depends on factors such as the degree of expected variation in findings across studies. Additionally, the feasibility of conducting a meta-analysis is highly topic-dependent. For example, in emerging research areas, such as artificial intelligence applications in formal learning, research is rapidly expanding, but there may not yet be enough empirical studies available to support a robust meta-analysis. Finally, conducting a meta-analysis is often time-intensive, requiring careful study selection, data extraction, and statistical analysis. It is highly beneficial to work within a research team to improve efficiency and establish interrater reliability in study selection and coding, maintaining transparency in the research process. A collaborative approach ensures a more rigorous and systematic review process while reducing individual workload.

CAUSAL ANALYSIS

In social science research we are interested in identifying causal effects, but conducting randomized controlled trial experiments with random assignments is either difficult, impractical, or unethical in many cases. This limitation often forces us to rely on observational data, particularly in education research. As a remedy, one common approach researchers take is to use longitudinal data to establish temporal precedence (mentioned for causal relationships in Chapter 1), suggesting that a potential cause precedes an associated outcome. Researchers then include as many control variables as possible into a regression model, hoping to isolate the effect of interest. However, despite these efforts, the results still come with a critical limitation: correlation does not imply causation. Typically, we acknowledge this limitation by adding a cautionary note about not assuming causal effect, and that is all we can do within a traditional regression framework.

To go beyond this, researchers turn to approaches that fall under what is referred to as a counterfactual framework, using a potential outcomes model. The fundamental challenge in causal inference is that we can never observe both potential outcomes for an individual (i.e., even in an experiment we can only observe one outcome per individual—either they receive the treatment or they don't). We can't measure the treatment effect by observing outcomes under both conditions. In an ideal experiment, we would compare the same person in both conditions to measure the true treatment effect, but this is impossible in reality.

Causal analysis methods attempt to approximate this missing counterfactual, essentially estimating what would have happened if the same individual had (or had not) received the treatment. Techniques such as matching methods (e.g., propensity score matching), instrumental variables, difference-in-differences (DiD), and regression discontinuity (RD) designs are commonly used to simulate experimental conditions in observational data. These approaches help strengthen causal claims by reducing selection bias and confounding factors, although they still rely on strong assumptions. We introduce each of these methods briefly.

PROPENSITY MODELS

A **propensity model** is used to estimate the likelihood (or propensity) of an individual receiving a treatment or being exposed to a particular condition, based on observed characteristics. It is widely applied in observational studies to reduce bias and simulate the conditions of randomized experiments when random assignment is not feasible. The idea with propensity models is to understand a treatment effect by incorporating an individual's propensity (or likelihood) to experience the treatment.

For example, a researcher wanted to understand the effect of using Advanced Placement (AP) credit on subsequent college course grades. However, students who use AP credits are not randomly assigned to do so, and we know that some students (e.g., students with higher AP exam scores, SAT scores, stronger academic preparation, etc.) are more likely to use AP credit than others. Because of this selection bias, one can't just conduct a straight-up comparison of AP users and AP nonusers. If we were to do a straight-up comparison, any observed differences in college course grades might be due to preexisting academic ability rather than the use of AP credit itself. In other words, AP users might have performed better in college courses anyway, regardless of whether they used AP credit. To get a more accurate estimate of the effect of AP credit use, the researcher had to account for students' propensity to use their AP credit.

There are multiple approaches you can take to account for selection bias with a propensity model. One common method that researchers have traditionally used is propensity score matching—trying to find individuals in the control group who are similar (or "matched") to individuals in the treatment group based on relevant characteristics. This can be a simple stratified matching (e.g., choose a few key variables and match within a range, such as SAT scores within 100 points + free/reduced lunch yes/no). However, it can also get much more complicated if you use logistic regression to incorporate many different pretreatment variables to estimate a propensity score and then match study participants by their propensity scores. The latter ensures a more comprehensive balance across multiple variables.

Another approach is propensity weighting. Individuals are weighted based on their propensity scores such that all the pretreatment variables are balanced between groups. Instead of discarding unmatched participants, this method assigns weights to individuals based on their propensity scores. The goal is to adjust for imbalances so that the treatment and control groups are statistically equivalent in terms of pretreatment characteristics. Basically, participants who are less common in a group are given more importance, helping to make sure that the treatment and control groups are as similar as possible before analyzing the results. Once you have a matched/balanced treatment and control group, you then conduct your outcomes analysis, such as comparing mean differences, running regression models, and estimating causal effects with greater confidence.

Figure 14.3 has a red dot for every variable included in a propensity model—everything the researcher thought would possibly predict propensity to use AP credit. The left side shows the standardized mean differences on those variables between treatment and control groups before weighting, and the right side is after weighting.

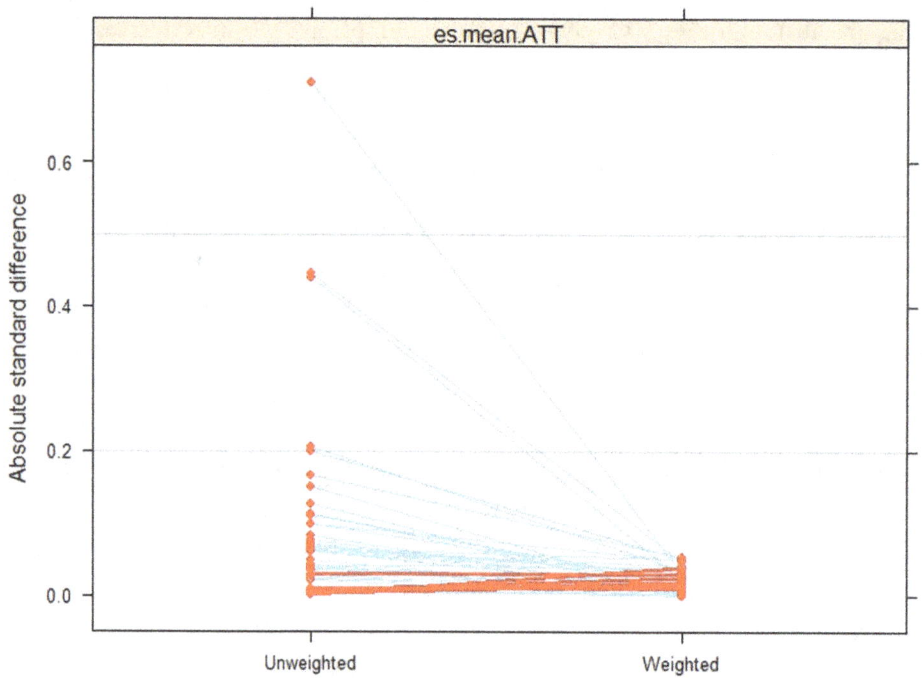

FIGURE 14.3 Standardized Mean Differences for Covariates before and after Weighting for Average Treatment Effect on the Treated

INSTRUMENTAL VARIABLES

A common challenge in causal analysis is isolating the true effect of your treatment variable or independent variable, because there are likely to be confounding variables that influence both the treatment and the outcome, which can make it hard to determine whether the relationship you observe is truly causal or just correlational. For example, if you're studying the effect of education attainment on income, you can't just compare people with different education levels because ability, motivation, and other hard-to-measure traits are likely to have a relationship with both the independent variable (education attainment) and the dependent variable (income). If these factors are not accounted for, you can't be sure whether education is truly driving income differences or if other factors are at play.

The idea of using instrumental variables is that you find a third variable that:

- Is strongly associated with your treatment variable (i.e., independent variable: education attainment).
- Does not directly affect your outcome variable (dependent variable: income) except through its effect on the treatment.
- Is independent of whatever is causing the link between the treatment and the error term of our outcome. In other words, it does not share the same hidden influence that affects both the treatment and outcome.

For various reasons that are too complicated to go into here, only the first of those assumptions is testable. The other two you just need to argue based on theory/prior research. Once you have a valid instrument, you use regression to figure out the effect of the instrument on the treatment and the instrument on the outcome and then use those values to generate your estimate of the "pure" effect of the treatment on the outcome. In simpler terms, an instrumental variable acts like a *natural randomizer*, helping to separate out the true effect of the treatment from other influencing factors.

Here is one example: Your research question is whether attending Catholic schools increases test scores, but factors influencing the decision to attend a Catholic school may also have an effect on test scores, and you can't control for all of them. So instead you use a lottery so that some number of students are able to attend the school for free. The lottery is random, so it can't have a direct effect on test scores, but it is absolutely related to whether or not a student chooses to attend Catholic school. The lottery is now your instrumental variable that you can use to figure out the effect of attending Catholic school on test scores.

Instrumental variables work great in theory, but finding a variable that meets all the assumptions can be difficult. Often researchers will use random events (e.g., natural disasters), but those don't always provide a clear and/or interesting population to study.

INTERRUPTED TIME SERIES

In most counterfactual frameworks, we must imagine what would happen to the same group of people in two different states or scenarios, one where they received the treatment and one where they did not. However, **interrupted time series (ITS)** provide a unique advantage because we can observe the same people across multiple points in time, both before and after our treatment of interest is introduced. The key assumption in ITS is that the trend in the outcome variable would have remained the same if the treatment would not have

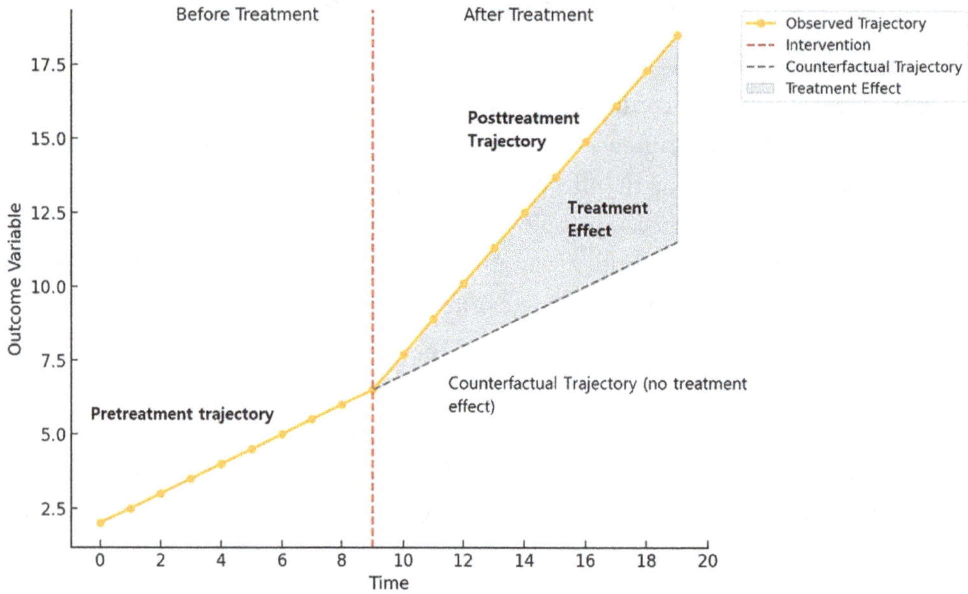

FIGURE 14.4 Interrupted Time Series Analysis of Treatment Effect over Time

occurred. In other words, we assume that any observed change in our outcome variable over time is due to the treatment effect. An example of this concept is displayed in Figure 14.4.

If the trend suddenly increases, decreases, or changes direction after the intervention, we attribute this to the treatment effect as long as no other major external factors (e.g., policy changes, economic shifts) explain the change. Therefore, it is critical to justify that your counterfactual trajectory, what would have happened without the treatment, is correct. If there exist other explanations for the posttreatment trajectory, they must be carefully considered. One way to support your case is by examining other groups/individuals who did not receive treatment to see if there were similar changes around that time even without the treatment. The better case you can make about the counterfactual trajectory, the stronger causal inferences you can draw from your analysis. ITS is powerful because it allows us to use real pretreatment data rather than relying on a hypothetical counterfactual. However, its validity depends on the assumption that nothing else significantly changed at the same time as the treatment, which could also explain shifts in the outcome.

REGRESSION DISCONTINUITY

Regression Discontinuity (RD) is a quasi-experimental research design used to estimate the causal effect of an intervention by assigning a cutoff point on a continuous variable and comparing observations just above and below that threshold. It is particularly useful when random assignment is not possible but there exists a rule-based allocation of treatment. In most of the examples we have discussed, treatment is a binary variable; that is, individuals either receive the treatment or do not. However, there are cases where treatment is a function of a continuous variable, and there are individuals just above and below whatever the cutoff point is. Consider the following scenarios:

- The legal drinking age in most states is 21, and individuals can be just under or just above 21 years of age.
- A company could decide to lay off workers who have been employed for less than two years, and there will be individuals who fall just above or below that threshold.

The key idea is to compare cases near the threshold, just above and just below the cutoff, because they are likely to be very similar in all respects except for receiving the treatment. Individuals just above are considered the treatment group, and those just below serve as the counterfactual for the treatment group. For example, AP exams are initially graded with raw scores, so students could end up with a percentage correct ranging from 0 to 100. Then cutoff values are established, so there is a sharp boundary between students who earn a 2 (a group that does not receive AP credit) and a 3 (a group that received AP credit) on the exam even though their raw scores could be almost identical. There was an interesting study by Smith et al. (2017) that wanted to understand the effect of receiving AP credit on college outcomes. They looked at whatever AP exam score was required for credit at a given institution (e.g., 3, 4, or 5) and compared outcomes of students whose raw scores were just above and below the cutoff for that AP exam score.

One important caveat is that your estimate of the effect only applies to individuals near the discontinuous region. For example, if you are looking at the effect of setting a legal drinking age on death rates or some other measure of health and safety, whatever effect you identify only applies to individuals who are close to 21 years old. Thus, while RD design is a powerful tool, its findings are only generalizable to individuals near the discontinuity point and may not apply to the entire population.

DIFFERENCE-IN-DIFFERENCES

Difference-in-Differences (DiD) is a quasi-experimental research design used to estimate the causal effect of an intervention or policy by comparing the changes in outcomes over time between a treatment group and a control group. DiD helps control for time-invariant confounders that could affect both groups equally. This concept is similar to the ITS method except that we are looking at treatment groups versus individuals. The first difference is the difference you observe in the treatment group before and after the intervention. The second difference is the gap between the control group and the counterfactual for the treatment group.

This method can handle what are known as time-invariant confounders (personal characteristics that are stable over time). As you see in Figure 14.5, there were differences between the treatment and control groups prior to the intervention, but that is okay because we can account for them. However, if there are time-variant confounders, we still have a problem because those could be an alternate explanation for why the trajectory of the treatment group changed.

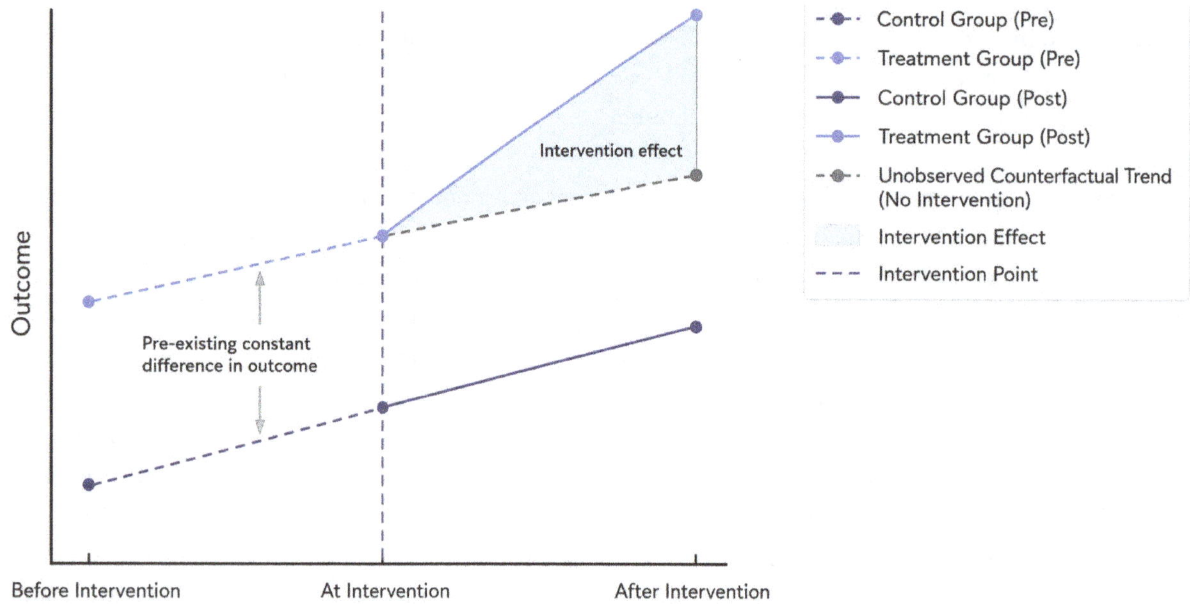

FIGURE 14.5 Difference-in-Differences (DiD)

FIXED EFFECTS

With **fixed effects approach**, instead of comparing outcomes between individuals (cases), you compare changes in outcomes within the same individual over time. This allows you to control for time-invariant factors—things that do not change over time but might still influence the outcome. For example, what changes do we see in outcomes given a change in treatment status (e.g., policy change, marriage/divorce)? The idea is that you have an outcome that varies across cases (different individuals) and over time (different time points for the same person). The model accounts for:

- Components that are time-invariant (all the factors not included in the model that are related to the outcome but don't change over time).
  - Characteristics specific to each individual that do not fluctuate across time periods (e.g., personality, genetics, early childhood environment).
  - Since they are constant, fixed effects remove their influence, preventing them from biasing the results.
- Components that are time-variant (idiosyncratic factors that do vary over time).
  - Factors that fluctuate over time and can influence the outcome (e.g., employment status, income, stress levels).
  - The fixed effects model helps estimate how changes in these time-varying factors relate to changes in the outcome.

There are several different approaches to fixed effects, but a common one is called de-meaning variables. Here is how it works:

- You first find the mean for each component of the equation for each person. For example, say you are studying the effect of marriage on happiness. If your outcome is a happiness score, you average the values of happiness over 10 time periods. Then you average the treatment indicator across each time period (if the person got married halfway through the study, their average treatment indicator would be 0.5). You also get the averages for time-invariant variables over time.

- You then take that equation (within-person averages over time) and subtract it from the baseline model. The intercept goes to zero, but so do time-invariant variables because subtracting the mean from something that doesn't vary over time takes it to zero.
- What we end up with is a de-meaned outcome that equals the de-meaned treatment indicator plus a time-varying error term.

The math gets a bit complicated, but the key takeaway is that all the factors that affect the outcome but remain constant over time are removed from the model, which eliminates many potential confounding effects that could bias the results. As long as the treatment indicator is not correlated with the time-varying error term, we can reasonably conclude that the resulting change in outcome is a result of the treatment.

Of course, there are many more methods you could learn to answer all kinds of different research questions. If you are reading this far, here are two general principles we hope you take away from this book:

1. Statistical methods are tools to help you understand phenomena, but they are only tools. You are the one to decide which tool to select and to use it wisely. Statistical software systems will provide you with some numerical outputs or fancy-looking graphs without evaluating the nature of the data. You are the one who should make sense of the data before analyzing it, and you are the one who needs to make sense of what the output means.
2. We didn't talk about any specific education or psychological theories or theoretical frameworks in this book, but that doesn't mean theory isn't important in quantitative research. You should base your research questions and the variables you include in your model on theory and/or prior research. If your sample size is large enough, you can almost always get a statistically significant result, but along the way you might overfit your model so that you are really just predicting noise in your data. As long as you remain guided by prior theory and research, you will be in good shape even if you don't end up with a p value less than 0.05.

KEY TAKEAWAYS FOR EDUCATIONAL RESEARCHERS FROM CHAPTER 14

- Education data are complex and often hierarchical, requiring careful consideration in analysis. Students are often nested within classrooms, schools, and districts, creating dependencies that must be accounted for. Additionally, measurement challenges such as survey bias and test-score variability can affect results. Researchers must select appropriate techniques to address these complexities and ensure reliable conclusions.
- Advanced methods are valuable tools, but their effectiveness depends on thoughtful application. By aligning methods with research questions, considering data complexity, balancing rigor with practicality, strengthening research design, and committing to continuous learning, education researchers can enhance the validity, reliability, and impact of their work.
- Statistical rigor must be balanced with practicality. Many advanced methods require large sample sizes, high-quality data, and careful model specifications. While complex models may offer greater insights, they also

increase the risk of overfitting and may not always be necessary. Researchers should focus on methods that enhance interpretability and applicability to educational settings.

- Strong research design is critical for meaningful causal inference. Statistical methods alone do not establish causality. Instead, study designs must minimize selection bias, control for confounders, and leverage natural experiments when possible. Combining quantitative analysis with qualitative insights can often provide a more comprehensive understanding of educational issues.
- Continuous learning and collaboration strengthen research impact. As methodologies evolve, education researchers should engage with methodologists, statisticians, and interdisciplinary teams to refine their approaches to addressing important research questions meaningfully.

KEY DEFINITIONS FROM CHAPTER 14

Difference-in-differences (DiD) is a quasi-experimental research design used to estimate the causal effect of an intervention or policy by comparing the changes in outcomes over time between a treatment group and a control group.

Factor analysis is a measurement method used to identify underlying structures or patterns within a set of observed variables and is commonly applied to reduce the dimensionality of data by grouping variables into clusters (or "factors") based on shared variance.

With **fixed effects approach**, instead of comparing outcomes between individuals (cases), you compare changes in outcomes within the same individual over time. This allows you to control for time-invariant factors—things that do not change over time but might still influence the outcome.

Interrupted time series (ITS) is a quasi-experimental research design used to evaluate the impact of an intervention, policy change, or event by analyzing trends in a time series before and after the intervention. It is particularly useful when a randomized controlled trial is not feasible.

Meta-analysis is a special name for a quantitative synthesis, which is a type of systematic review of the literature.

Multilevel modeling (MLM), also known as hierarchical linear modeling, is a statistical technique used to analyze data that is organized at more than one level, often referred to as nested or hierarchical data.

A **propensity model** is used to estimate the likelihood (or propensity) of an individual receiving a treatment or being exposed to a particular condition, based on observed characteristics.

Regression discontinuity (RD) is a quasi-experimental research design used to estimate the causal effect of an intervention by assigning a cutoff point on a continuous variable and comparing observations just above and below that threshold. RD is particularly useful when random assignment is not possible but there exists a rule-based allocation of treatment.

Structural equation modeling (SEM) is an advanced statistical technique that combines factor analysis and multiple regression to analyze complex relationships among observed and latent variables.

CHECK YOUR UNDERSTANDING

1. Which of the following is NOT a purpose of meta-analysis?
   a. Identifying the central tendency or common effect across studies
   b. Exploring variability in effect sizes
   c. Developing new experimental theories
   d. Explaining why variation in effect sizes occurs

2. A propensity model is typically used to
   a. Measure the direct effect of a treatment variable
   b. Estimate the likelihood of receiving a treatment based on observed characteristics
   c. Compare independent variables across different models
   d. Simplify complex data into manageable clusters

3. What is a key requirement for using multilevel modeling effectively?
   a. A minimum sample size of 50 participants
   b. Randomized controlled trials only
   c. Large datasets with at least 20–30 groups of 20–30 individuals per group
   d. A focus on theoretical over empirical studies

4. Which of the following is a key benefit of structural equation modeling over traditional regression analysis?
   a. It requires fewer participants to yield accurate results.
   b. It eliminates the need for latent variables.
   c. It allows the inclusion of multiple dependent variables and mediating variables.
   d. It reduces data complexity by eliminating endogenous predictors.

5. What is the minimum number of studies generally required to conduct a meta-analysis that explores variation?
   a. Two studies
   b. Three studies
   c. Four studies
   d. Twenty studies

APPENDIX A

TABLE A1. *Standard Normal Cumulative Probability*

z	0.00	0.01	0.02	0.03	0.04	0.05	0.06	0.07	0.08	0.09
-3.4	.0003	.0003	.0003	.0003	.0003	.0003	.0003	.0003	.0003	.0002
-3.3	.0005	.0005	.0005	.0004	.0004	.0004	.0004	.0004	.0004	.0003
-3.2	.0007	.0007	.0006	.0006	.0006	.0006	.0006	.0005	.0005	.0005
-3.1	.0010	.0009	.0009	.0009	.0008	.0008	.0008	.0008	.0007	.0007
-3.0	.0013	.0013	.0013	.0012	.0012	.0011	.0011	.0011	.0010	.0010
-2.9	.0019	.0018	.0018	.0017	.0016	.0016	.0015	.0015	.0014	.0014
-2.8	.0026	.0025	.0024	.0023	.0023	.0022	.0021	.0021	.0020	.0019
-2.7	.0035	.0034	.0033	.0032	.0031	.0030	.0029	.0028	.0027	.0026
-2.6	.0047	.0045	.0044	.0043	.0041	.0040	.0039	.0038	.0037	.0036
-2.5	.0062	.0060	.0059	.0057	.0055	.0054	.0052	.0051	.0049	.0048
-2.4	.0082	.0080	.0078	.0075	.0073	.0071	.0069	.0068	.0066	.0064
-2.3	.0107	.0104	.0102	.0099	.0096	.0094	.0091	.0089	.0087	.0084
-2.2	.0139	.0136	.0132	.0129	.0125	.0122	.0119	.0116	.0113	.0110
-2.1	.0179	.0174	.0170	.0166	.0162	.0159	.0154	.0150	.0146	.0143
-2.0	.0228	.0222	.0217	.0212	.0207	.0202	.0197	.0192	.0188	.0183
-1.9	.0287	.0281	.0274	.0268	.0262	.0256	.0250	.0244	.0239	.0233
-1.8	.0359	.0351	.0344	.0336	.0329	.0322	.0314	.0307	.0301	.0294
-1.7	.0446	.0436	.0427	.0418	.0409	.0401	.0392	.0384	.0375	.0367
-1.6	.0548	.0537	.0526	.0516	.0505	.0495	.0485	.0475	.0465	.0455
-1.5	.0668	.0655	.0643	.0630	.0618	.0606	.0594	.0582	.0571	.0559
-1.4	.0808	.0793	.0778	.0764	.0749	.0735	.0721	.0708	.0694	.0681
-1.3	.0968	.0951	.0934	.0918	.0901	.0885	.0869	.0853	.0838	.0823
-1.2	.1151	.1131	.1112	.1093	.1075	.1056	.1038	.1020	.1003	.0985
-1.1	.1357	.1335	.1314	.1292	.1271	.1251	.1230	.1210	.1190	.1170
-1.0	.1587	.1562	.1539	.1515	.1492	.1469	.1446	.1423	.1401	.1379
-0.9	.1841	.1814	.1788	.1762	.1736	.1711	.1685	.1660	.1635	.1611
-0.8	.2119	.2090	.2061	.2033	.2005	.1977	.1949	.1922	.1894	.1867

Standard Normal Cumulative Probability *(Continued)*

-0.7	.2420	.2389	.2358	.2327	.2296	.2266	.2236	.2206	.2177	.2148
-0.6	.2743	.2709	.2676	.2643	.2611	.2578	.2546	.2514	.2483	.2451
-0.5	.3085	.3050	.3015	.2981	.2946	.2912	.2877	.2843	.2810	.2776
-0.4	.3446	.3409	.3372	.3336	.3300	.3264	.3228	.3192	.3156	.3121
-0.3	.3821	.3783	.3745	.3707	.3669	.3632	.3594	.3557	.3520	.3483
-0.2	.4207	.4168	.4129	.4090	.4052	.4031	.3974	.3936	.3897	.3859
-0.1	.4602	.4562	.4522	.4483	.4443	.4404	.4364	.4325	.4286	.4247
-0.0	.5000	.4960	.4920	.4880	.4840	.4801	.4761	.4721	.4681	.4641
0.0	.5000	.5040	.5080	.5120	.5160	.5199	.5239	.5279	.5319	.5359
0.1	.5398	.5438	.5478	.5517	.5557	.5596	.5636	.5675	.5714	.5753
0.2	.5793	.5832	.5871	.5910	.5948	.5987	.6020	.6064	.6103	.6141
0.3	.6179	.6217	.6255	.6293	.6331	.6368	.6406	.6443	.6480	.6517
0.4	.6554	.6591	.6628	.6664	.6700	.6736	.6772	.6808	.6844	.6879
0.5	.6915	.6950	.6985	.7019	.7054	.7088	.7123	.7157	.7190	.7224
0.6	.7257	.7291	.7324	.7357	.7389	.7422	.7454	.7486	.7517	.7549
0.7	.7580	.7611	.7642	.7673	.7704	.7734	.7764	.7794	.7823	.7852
0.8	.7881	.7910	.7939	.7967	.7995	.8023	.8051	.8078	.8106	.8133
0.9	.8159	.8186	.8212	.8238	.8264	.8289	.8315	.8340	.8365	.8389
1.0	.8413	.8438	.8461	.8485	.8508	.8531	.8554	.8577	.8599	.8621
1.1	.8643	.8665	.8686	.8708	.8729	.8749	.8770	.8790	.8810	.8830
1.2	.8849	.8869	.8888	.8907	.8925	.8944	.8962	.8980	.8997	.9015
1.3	.9032	.9049	.9066	.9082	.9099	.9115	.9131	.9147	.9162	.9177
1.4	.9192	.9207	.9222	.9236	.9251	.9265	.9279	.9292	.9306	.9319
1.5	.9332	.9345	.9357	.9370	.9382	.9394	.9406	.9418	.9429	.9441
1.6	.9452	.9463	.9474	.9484	.9495	.9505	.9515	.9525	.9535	.9545
1.7	.9554	.9564	.9573	.9582	.9591	.9599	.9608	.9616	.9625	.9633
1.8	.9641	.9649	.9656	.9664	.9671	.9678	.9686	.9693	.9699	.9706
1.9	.9713	.9719	.9726	.9732	.9738	.9744	.9750	.9756	.9761	.9767
2.0	.9772	.9778	.9783	.9788	.9793	.9798	.9803	.9808	.9812	.9817
2.1	.9821	.9826	.9830	.9834	.9838	.9842	.9846	.9850	.9854	.9857
2.2	.9861	.9864	.9868	.9871	.9875	.9878	.9881	.9884	.9887	.9890
2.3	.9893	.9896	.9898	.9901	.9904	.9906	.9909	.9911	.9913	.9916
2.4	.9918	.9920	.9922	.9925	.9927	.9929	.9931	.9932	.9934	.9936

2.5	.9938	.9940	.9941	.9943	.9945	.9946	.9948	.9949	.9951	.9952
2.6	.9953	.9955	.9956	.9957	.9959	.9960	.9961	.9962	.9963	.9964
2.7	.9965	.9966	.9967	.9968	.9969	.9970	.9971	.9972	.9973	.9974
2.8	.9974	.9975	.9976	.9977	.9977	.9978	.9979	.9979	.9980	.9981
2.9	.9981	.9982	.9982	.9983	.9984	.9984	.9985	.9985	.9986	.9986
3.0	.9987	.9987	.9987	.9988	.9988	.9989	.9989	.9989	.9990	.9990
3.1	.9990	.9991	.9991	.9991	.9992	.9992	.9992	.9992	.9993	.9993
3.2	.9993	.9993	.9994	.9994	.9994	.9994	.9994	.9995	.9995	.9995
3.3	.9995	.9995	.9995	.9996	.9996	.9996	.9996	.9996	.9996	.9997
3.4	.9997	.9997	.9997	.9997	.9997	.9997	.9997	.9997	.9997	.9998

TABLE A2. *Critical Values of t Distribution*

	One-Tail Probability			
	.05	**.025**	**.01**	**.005**
	Two-Tail Probability			
df	**.10**	**.05**	**.02**	**.01**
1	6.314	12.706	31.821	63.657
2	2.920	4.303	6.965	9.925
3	2.353	3.182	4.541	5.841
4	2.132	2.776	3.747	4.604
5	2.015	2.571	3.365	4.032
6	1.943	2.447	3.143	3.707
7	1.895	2.365	2.998	3.499
8	1.860	2.306	2.896	3.355
9	1.833	2.282	2.821	3.250
10	1.812	2.228	2.764	3.169
20	1.725	2.086	2.528	2.845
30	1.697	2.042	2.457	2.750
40	1.684	2.021	2.423	2.704
50	1.676	2.009	2.403	2.678
100	1.660	1.984	2.364	2.626

TABLE A3. *Critical Values of F Distribution*

Degrees Freedom in Denominator	P	1	5	10	20	30	40	50
		DEGREES FREEDOM IN NUMERATOR						
1	.05	161.45	230.16	241.88	248.01	250.10	251.14	251.77
	.01	4,052.2	5,763.6	6,055.8	6,208.7	6,260.6	6,286.8	6,302.5
	.001	405,284	576,405	605,621	620,908	626,099	628,712	630,285
5	.05	6.61	5.05	4.74	4.52	4.50	4.46	4.44
	.01	16.26	10.97	10.05	9.55	9.38	9.29	9.24
	.001	47.18	29.75	26.92	25.39	24.87	24.60	24.44
10	.05	4.96	3.33	2.98	2.77	2.70	2.66	2.64
	.01	10.04	5.64	4.85	4.41	4.25	4.17	4.12
	.001	19.69	9.58	7.92	7.01	6.68	6.52	6.42
20	.05	4.35	2.71	2.35	2.12	2.04	1.99	1.97
	.01	8.10	4.10	3.37	2.94	2.78	2.69	2.64
	.001	14.59	6.32	4.95	4.17	3.88	3.74	3.64
30	.05	4.17	2.53	2.16	1.93	1.84	1.79	1.76
	.01	7.56	3.70	2.98	2.55	2.39	2.30	2.25
	.001	13.29	5.53	4.24	3.49	3.22	3.07	2.98
40	.05	4.08	2.45	2.08	1.84	1.74	1.69	1.66
	.01	7.31	3.51	2.80	2.37	2.20	2.11	2.06
	.001	12.61	5.13	3.87	3.14	2.87	2.73	2.64
50	.05	4.03	2.40	2.03	1.79	1.69	1.63	1.60
	.01	7.17	3.41	2.70	2.27	2.10	2.01	1.95
	.001	12.22	4.90	3.67	2.95	2.68	2.53	2.44
100	.05	3.94	2.31	1.93	1.68	1.57	1.52	1.48
	.01	6.90	3.21	2.50	2.07	1.89	1.80	1.74
	.001	11.15	4.29	3.12	2.42	2.15	2.00	1.90

TABLE A4. *Critical Values of the Pearson (r) Correlation*

One-tailed	.05	.025	.01	.005
Two-tailed	.10	.05	.02	.01
df				
1	.988	.997	.9995	.9999
2	.900	.950	.980	.990
3	.805	.878	.934	.959
4	.729	.811	.882	.917
5	.669	.754	.833	.874
6	.622	.707	.789	.834
7	.582	.666	.750	.798
8	.549	.632	.716	.765
9	.521	.602	.685	.735
10	.497	.576	.658	.708
20	.360	.423	.492	.537
30	.296	.349	.409	.449
40	.257	.304	.358	.393
50	.231	.273	.322	.354
100	.164	.195	.230	.254

TABLE A5. *Critical Values of the Spearman Rho Correlation*

One-tailed	.05	.025	.01	.005
Two-tailed	.10	.05	.02	.01
n				
4	1.000			
5	.900	1.000	1.000	
6	.829	.886	.943	1.000
7	.714	.786	.893	.929
8	.643	.738	.833	.881
9	.600	.700	.783	.833
10	.564	.648	.745	.794
20	.380	.447	.520	.570
30	.306	.362	.425	.467
40	.264	.313	.368	.405
50	.235	.279	.329	.363
100	.165	.197	.233	.257

TABLE A6. *Critical Values of Chi-Square (χ^2)*

	Significance Level		
df	.05	.01	.001
1	3.84	6.64	10.83
2	5.99	9.21	13.82
3	7.81	11.34	16.27
4	9.49	13.28	18.47
5	11.07	15.09	20.52
6	12.59	16.81	22.46
7	14.07	18.48	24.32
8	15.51	20.09	26.12
9	16.92	21.67	27.88
10	18.31	23.21	29.59
20	31.41	37.47	45.32
30	43.77	50.89	59.70
40	55.76	63.69	73.40
50	67.50	76.15	86.66
100	124.34	135.81	149.45

TABLE A7. *Critical Values for the Mann-Whitney U Test*

N_B/N_A	2	3	4	5	6	7	8	9	10	11	12	13	14	15
2						0	0	0	0	1	1	1	1	
3				0	1	1	2	2	3	3	4	4	5	5
4			0	1	2	3	4	4	5	6	7	8	9	10
5		0	1	2	3	5	6	7	8	9	11	12	13	14
6		1	2	3	5	6	8	10	11	13	14	16	17	19
7		1	3	5	6	8	10	12	14	16	18	20	22	24
8	0	2	4	6	8	10	13	15	17	19	22	24	26	29
9	0	2	4	7	10	12	15	17	20	23	26	28	31	34
10	0	3	5	8	11	14	17	20	23	26	29	33	36	39
11	0	3	6	9	13	16	19	23	26	30	33	37	40	44
12	1	4	7	11	14	18	22	26	29	33	37	41	45	49
13	1	4	8	12	16	20	24	28	33	37	41	45	50	54
14	1	5	9	13	17	22	26	31	36	40	45	50	55	59
15	1	5	10	14	19	24	29	34	39	44	49	54	59	64

Note: *Unlike most other tests learned in this book, reject the null hypothesis if the test value is less than or equal to the value given in the table above.

Sources: Sani and Todman (2006); and Milton (1964).

TABLE A8. ***Critical Values for the Wilcoxon Matched-Pairs Signed-Ranks Test***

	One-tailed, α = .025	α = .005	α = .0005
N	Two-tailed, α = .05	α = .01	α = .001
6	0	—	—
7	2	—	—
8	3	0	—
9	5	1	—
10	8	3	—
11	10	5	0
12	13	7	1
13	17	9	2
14	21	12	4
15	25	15	6
20	52	37	21
25	89	68	45
30	137	109	78

Note: Unlike most other tests learned in this book, reject the null hypothesis if the test value is less than or equal to the value given in the table above.

TABLE A9. ***Summary of Nonparametric Tests***

TEST/METHOD	WHEN TO USE	DEGREES OF FREEDOM	TEST STATISTIC	HOW TO EVALUATE
Chi-square goodness-of-fit	Categorical data Comparing one distribution to a known distribution	$k - 1$	χ^2	If obtained χ^2 is greater than critical value on chi-square table (for specific *df*), reject the null. Report Fisher's exact test if available.
Chi-square test for independence	Categorical data Testing for association between two variables	$(k_1 - 1)(k_2 - 1)$	χ^2	If obtained χ^2 is greater than critical value on chi-square table (for specific *df*), reject the null. Report Fisher's exact test if available.
Independent samples chi-square test	Categorical data Testing for association between two variables when you only have summary statistics (totals)	$(k_1 - 1)(k_2 - 1)$	χ^2	If obtained χ^2 is greater than critical value on chi-square table (for specific *df*), reject the null. Report Fisher's exact test if available.
Wilcoxon signed rank *t* test	Nonparametric alternative to related-samples *t* test Converts measured values to ordinal data	n/a	*y* (smaller absolute value of rank summations)	Compare to critical value on Wilcoxon's signed rank *y* table for given sample size. If *t* (obtained) is less than or equal to critical value, reject the null.
Mann-Whitney *U* test	Nonparametric alternative to two-independent samples *t* test Converts measured values to ordinal data	n/a	U_X or U_Y (whichever is smaller)	Compare to critical value on Mann-Whitney *U* table for two-tailed test (usually) and for given *n*. If *U* (obtained) is less than or equal to critical value, reject the null.
Kruskal-Wallis *H* test	Nonparametric alternative to a one-way between-subjects ANOVA Converts measured values to ordinal data	$k - 1$	*H*	If obtained *H* is greater than critical value on chi-square table (for specific *df*), reject the null.

APPENDIX B

INTEGRATED SUMMARY OF STATISTICAL FORMULAS AND CONCEPTS ACROSS CHAPTERS

In this appendix, we provide a consolidated reference of the key statistical formulas, including sums of squares (*SS*), test statistics, and related computational components introduced throughout the chapters. Rather than presenting each formula in isolation, we organize them to reflect how these statistical tools are conceptually connected and how they contribute to the overall process of data-informed decision-making.

While each chapter focused on specific methods, this appendix brings together the core computational elements to illustrate their shared foundations and practical applications. This summary is intended to serve as a quick reference for readers and to reinforce the idea that statistical procedures are not merely mechanical calculations but are also essential tools for generating meaningful insights from data.

By compiling these formulas in one place, we aim to support readers in understanding the structure of statistical analysis and its role in guiding thoughtful, evidence-based inquiry in social science and education research.

UNDERSTANDING THE CENTER: MEASURES OF CENTRAL TENDENCY (CHAPTER 2)

MEAN

Population mean (μ):

$$\mu = \frac{\sum_{i=1}^{n} x_i}{n}$$

Sample mean ($\bar{X}$):

$$\bar{X} = \frac{\sum_{i=1}^{n} x_i}{n}$$

NOTATION KEY

SYMBOL	DESCRIPTION
x_i	The observed value of variable X for the *i*th case
$\bar{X}$	The sample mean of variable X
μ	The population mean of variable X
N	Population size (i.e., the number of cases in the population)
n	Sample size (i.e., the number of cases in the sample)

EXPLORING DATA SPREAD: MEASURES OF VARIABILITY (CHAPTER 3)

Range = Largest value - smallest value
Interquartile range (IQR) = The third quartile (Q_3) - the first quartile (Q_1)

VARIANCE AND STANDARD DEVIATION IN POPULATION AND SAMPLE DATA

	POPULATION	SAMPLE
Variance	$\sigma_x^2 = \frac{\sum (x_i - \mu)^2}{N}$	$s_x^2 = \frac{\sum (x_i - \bar{X})^2}{n-1}$
Standard Deviation (*SD*)	$\sigma_x = \sqrt{\frac{\sum (x_i - \mu)^2}{N}}$	$s_x = \sqrt{\frac{\sum (x_i - \bar{X})^2}{n-1}}$

Degrees of freedom (df) for sample variance $df = n - 1$
The sum of squares (SS) in a population $= \Sigma (x_i - \mu)_2$
The sum of squares (SS) in a sample $= \Sigma (x_i - \bar{X})_2$

NOTATION KEY

SYMBOL	DESCRIPTION
x_i	The observed value of variable X for the *i*th case
$\bar{X}$	The sample mean of variable X
μ	The population mean of variable X
N	Population size (i.e., the number of cases in the population)
n	Sample size (i.e., the number of cases in the sample)
Σ_x	The population standard deviation of variable X
s_x	The sample standard deviation of variable X
σ_x^2	The population variance of variable X
s_x^2	The sample variance of variable X

THE NORMAL DISTRIBUTION (CHAPTER 5)

The standard normal distribution is useful for determining the likelihood of obtaining observed scores from any normal distribution. Using the z transformation, any normal distribution can be converted to the standard normal distribution.

Population z transformation:

$$Z = \frac{x - \mu}{\sigma}$$

Sample z transformation:

$$Z = \frac{x - \bar{X}}{s}$$

NOTATION KEY

SYMBOL	DESCRIPTION
z	z score
x	The observed value of variable X in its original metric
$\bar{X}$	The sample mean of variable X
μ	The population mean of variable X
Σ	The population standard deviation
s	The sample standard deviation

SAMPLING DISTRIBUTIONS (CHAPTER 6)

The standard error of the mean, or simply the standard error, is the standard deviation of the sampling distribution of sample means.

The standard error of sample mean (S_M or SEM):

$$S_M = \frac{s}{\sqrt{n}}$$

NOTATION KEY

SYMBOL	DESCRIPTION
S_M	The standard error of sample mean
n	Sample size (i.e., the number of cases in the sample)
s	The sample standard deviation

FOUNDATIONS OF STATISTICAL INFERENCE: UNDERSTANDING HYPOTHESIS TESTING (CHAPTER 7)

The one-sample z test is used to determine whether the observed sample mean differs from the population or reference group mean when the population variance is known.

Test statistic for one-sample z test:

$$z = \frac{\bar{X} - \mu}{\frac{\sigma}{\sqrt{n}}}$$

NOTATION KEY

SYMBOL	DESCRIPTION
z	z statistic
$\bar{X}$	The sample mean of variable X
μ	The population mean of variable X
Σ	The population standard deviation
n	Sample size

EVALUATING GROUP AVERAGES AGAINST A BENCHMARK: ONE-SAMPLE *t* TEST AND GOODNESS-OF-FIT ANALYSIS (CHAPTER 8)

ONE-SAMPLE *t* TEST

The one-sample t test is used to determine whether the observed sample mean differs from the population or reference group mean when the population variance is unknown.

TEST STATISTIC

The test statistic:

$$t = \frac{\bar{X} - \mu}{\frac{s}{\sqrt{n}}}$$

Degrees of freedom (df):

$$df = n - 1$$

EFFECT SIZE

Cohen's d effect size measure:

$$d = \frac{\bar{X} - \mu}{s}$$

ESTIMATION

$$\textit{Confidence Interval (CI) for Population Mean}$$
$$= \textit{Sample Mean} \pm \textit{Level of confidence} \times \textit{Standard Error}$$
$$CI(\mu) = \bar{X} \pm t_{critical} \times \frac{s}{\sqrt{n}}$$

NOTATION KEY

SYMBOL	DESCRIPTION
z	z statistic
d	Cohen's d effect size
$\bar{X}$	The sample mean of variable X
$CI\ (i)$	Confidence interval for the population mean of variable X
μ	The population mean of variable X
s	The sample standard deviation
$t_{critical}$	Critical t value for a given confidence level
n	Sample size

CHI-SQUARE TEST FOR THE GOODNESS-OF-FIT TEST

The chi-square goodness-of-fit test is used to determine whether the observed frequencies O_i differ significantly from expected frequencies (E_i) based on a specified distribution.
Test statistic:

$$\chi^2 = \sum \frac{(O_i - E_i)^2}{E_i}$$

Degrees of freedom (df):

$$df = (k - 1)$$

NOTATION KEY

SYMBOL	DESCRIPTION
χ^2	Chi-square statistic
O_i	The observed count for the ith category (i.e., the actual number of cases that fall into the category)
E_i	The expected count for the ith category (i.e., the number of cases that would be expected in the category under the assumption that the null hypothesis is true)
k	The number of categories or levels within a variable

ANALYZING DIFFERENCES BETWEEN TWO GROUPS(CHAPTER 9)

THE INDEPENDENT-SAMPLES *t* TEST

The independent-samples *t* test is used to compare the means of two independent groups to determine whether the difference between them is statistically significant.

TEST STATISTIC

Test statistic:

$$t = \frac{(\bar{X}_1 - \bar{X}_2) - (\mu_1 - \mu_2)}{\sqrt{S_p^2\left(\frac{1}{n_1} + \frac{1}{n_2}\right)}}$$

Degrees of freedom (*df*):

$$df = (n_1 - 1) + (n_2 - 1) = (n_1 + n_2) - 2$$

The pooled variance:

$$S_p^2 = \frac{(n_1 - 1)s_1^2 + (n_2 - 1)s_2^2}{(n_1 - n_2) - 2}$$

If both groups have the same sample size (i.e., $n_1 = n_2$),

$$S_p^2 = \frac{s_1^2 + s_2^2}{2}$$

NOTATION KEY

SYMBOL	DESCRIPTION
t	*t* statistic
$\bar{X}_1$	The sample mean of group 1
$\bar{X}_2$	The sample mean of group 2
μ_1	The population mean of group 1
μ_2	The population mean of group 2
s_p^2	The pooled variance, a weighted average of the two sample variances, assuming equal population variance
s_1^2	The sample variance of group 1
s_2^2	The sample variance of group 2
n_1	The sample size of sample 1 from group 1
n_2	The sample size of sample 2 from group 2

EFFECT SIZE

Cohen's *d* effect size (*d*):

$$d = \frac{\bar{X}_1 - \bar{X}_2}{\sqrt{s_p^2}}$$

Eta-squared (η^2):

$$\eta = \frac{t^2}{t^2 + df}$$

SYMBOL	DESCRIPTION
t	t statistic for independent samples
$\bar{X}_1$	The sample mean of group 1
$\bar{X}_2$	The sample mean of group 2
μ_1	The population mean of group 1
μ_2	The population mean of group 2
s_p^2	The pooled variance, a weighted average of the two sample variances, assuming equal population variance
s_1^2	The sample variance of group 1
s_2^2	The sample variance of group 2
n_1	The sample size of sample 1 from group 1
n_2	The sample size of sample 2 from group 2
df	Degrees of freedom for the independent-samples t test

RELATED-SAMPLES (PAIRED-SAMPLES) *t* TEST

The related-samples *t* test is used to compare the means of two related groups to determine whether there is a statistically significant difference between them. You'll notice several *D*'s in the formula below because we're analyzing difference scores from paired observations, not two separate sample means. The test uses one *distribution of sample differences.*

TEST STATISTIC

Test statistic:

$$t = \frac{\bar{D} - \mu_D}{\frac{S_p}{\sqrt{n}}}$$

Degrees of freedom (*df*):

$$df = n - 1$$

EFFECT SIZE

Cohen's *d* effect size (*d*):

$$d = \frac{\bar{D}}{S_D}$$

NOTATION KEY

SYMBOL	DESCRIPTION
t	t statistic for paired samples
$\bar{D}$	The sample mean of the different scores (i.e., the average of (x_1 - x_2) across n cases)
μ_D	The population mean differences
S_D	The sample *SD* of the difference scores
n	The sample size (i.e., the number of paired observations)

ANALYZING DIFFERENCES AMONG MORE THAN TWO GROUPS (CHAPTER 10)

ONE-WAY ANALYSIS OF VARIANCE (ANOVA) (BETWEEN-SUBJECT DESIGN)

One-way analysis of variance (ANOVA) is used to compare the means of three or more independent groups to determine whether there are statistically significant differences among them. Each participant is in only one group.

TEST STATISTIC

SOURCE	SUM OF SQUARES (SS)	DEGREES OF FREEDOM (*df*) MEAN SQUARES (MS)	TEST STATISTIC	
Between groups	$SS_{BG} = \sum n_j (\bar{X}_j - \bar{X})^2$	k-1	$MS_{BG} = \frac{\sum n_j (\bar{X}_j - \bar{X})^2}{k-1} = \frac{SS_{BG}}{df_{BG}}$	$F = \frac{MS_{BG}}{MS_E}$
Within groups (error)	$SS_E = \sum (\bar{X}_{ij} - \bar{X}_j)^2$	n-k	$MS_E = \frac{\sum (\bar{X}_{ij} - \bar{X}_j)^2}{N-k} = \frac{SS_E}{df_E}$	
Total	$SS_{Total} = \sum (\bar{X}_{ij} - \bar{X}_j)^2$	n-1		

Test statistic:

$$F = \frac{MS_{BG}}{MS_E}$$

Degrees of freedom (*df*):

$$df = (k\text{-}1, n\text{-}k)$$

EFFECT SIZE

Eta squared (η^2):

$$\eta^2 = \frac{SS_{BG}}{SS_{total}}$$

Omega squared (ω^2):

$$\omega^2 = \frac{SS_{BG} - [(k-1)MS_E]}{SS_{total} + MS_E}$$

NOTATION KEY

SYMBOL	DESCRIPTION
x_{ij}	Score of the *i*th case in group *j*
$\bar{X}_j$	The mean of group *j*
$\bar{X}$	The overall mean across *n*
n_j	The number of cases in group *j*
k	The number of groups
n	Total number of cases across all groups

TWO-WAY ANALYSIS OF VARIANCE (BETWEEN-SUBJECT DESIGN)

Two-way ANOVA is used to examine the effect of two independent variables on a dependent variable, including their main effects and possible interaction effect. Each participant is assigned to only one combination of the independent variable levels.

TEST STATISTIC

SOURCE	SS	df	MS	Test Statistic
Factor A (Main effect A)	SS_A	$a-1$	$\frac{SS_A}{df_A}$	$F_A = \frac{MS_A}{MS_E}$
Factor B (Main effect B)	SS_B	$b-1$	$\frac{SS_B}{df_B}$	$F_B = \frac{MS_B}{MS_E}$
A×B (Interaction effect)	SS_{AB}	$(a-1)(b-1)$	$\frac{SS_{AB}}{df_{AB}}$	$F_{AB} = \frac{MS_{AB}}{MS_E}$
Error (within groups)	SS_E	$ab(n-1)$	$\frac{SS_E}{df_E}$	
Total	SS_{total}	$abn-1$		

MAIN EFFECTS

Factor A:

$$F_A = \frac{MS_A}{MS_E}$$
$$df = (a - 1, ab(n - 1))$$

Factor B:

$$F_B = \frac{MS_B}{MS_E}$$
$$df = (b - 1, ab(n - 1))$$

Interaction effect:

$$F_{AB} = \frac{MS_{AB}}{MS_E}$$
$$df = ((a - 1)(b - 1), ab(n - 1))$$

EFFECT SIZE

ETA SQUARED (η^2)

MAIN EFFECTS

Factor A:

$$\hat{\eta}^2_A = \frac{SS_A}{SS_{total}}$$

Factor B:

$$\hat{\eta}^2_B = \frac{SS_A}{SS_{total}}$$

INTERACTION EFFECT

$$\hat{\eta}^2_{AB} = \frac{SS_{AB}}{SS_{total}}$$

OMEGA SQUARED (ω^2)

MAIN EFFECTS

Factor A:

$$\hat{\omega}^2_A = \frac{SS_A - [df_A * MS_E]}{SS_{total} + MS_E}$$

Factor B:

$$\hat{\omega}^2_B = \frac{SS_B - [df_B * MS_E]}{SS_{total} + MS_E}$$

INTERACTION EFFECT

$$\hat{\omega}^2_{AB} = \frac{SS_{AB} - [df_{AB} * MS_E]}{SS_{total} + MS_E}$$

NOTATION KEY

SYMBOL	DESCRIPTION
a	The number of levels in Factor A
B	The number of levels in Factor B
N	The number of cases per cell, assuming all group have the same number of observations
SS_A	Sum of squares for Factor A
SS_B	Sum of squares for Factor B
SS_{AB}	Sum of squares for the interaction between Factor A and B
SS_E	Sum of squares for error (within-group variability)
SS_{total}	Total sum of squares
df_A	Degrees of freedom for Factor A (i.e., a-1)
df_B	Degrees of freedom for Factor B (i.e., b-1)
df_{AB}	Degrees of freedom for the interaction term (i.e., (a-1)(b-1))
df_E	Degrees of freedom for error (i.e., ab(n-1))
df_{total}	Total degrees of freedom (i.e., abn-1)

EXAMINING RELATIONSHIPS BETWEEN TWO VARIABLES (CHAPTER 11)

PEARSON'S CORRELATION COEFFICIENT (*R*)

Pearson's correlation coefficient (r) measures the strength and direction of the linear relationship between two continuous variables, X and Y.

$$r = \frac{\text{Cov}_{XY}}{\sqrt{S_X^2 S_Y^2}}$$

$$\text{where Cov}_{XY} = \frac{\sum_{i=1}^{N}(x_i - \bar{X})(y_i - \bar{Y})}{n-1}$$

EFFECT SIZE

The coefficient of determination = r^2

NOTATION KEY

SYMBOL	DESCRIPTION
r	Pearson's correlation coefficient
Cov_{XY}	Covariance between variable X and Y.
$\bar{X}$	The mean of variable X
$\bar{Y}$	The mean of variable Y
S_X^2	Variance of variable X
S_Y^2	Variance of variable Y
n	Sample size

CHI-SQUARE TEST FOR INDEPENDENCE

The chi-square test for independence is used to determine whether there is a significant association between two categorical variables.
Test statistic:

$$\chi^2 = \sum \frac{(f_0 - f_e)^2}{f_e}$$

Degrees of freedom (*df*):

$$df = (r - 1)(c - 1)$$

NOTATION KEY

SYMBOL	DESCRIPTION
$\div_2$	Chi-square statistic
f_o	The observed count for the *i*th category (i.e., the actual number of cases that fall into the category)
f_e	The expected count for the *i*th category (i.e., the number of cases that would be expected in the category under the assumption that the null hypothesis is true) $f_e = \frac{\text{row total} \times \text{column total}}{N}$
r	The number of categories or levels in the first categorical variable
c	The number of categories or levels in the second categorical variable

PREDICTING AN OUTCOME (CHAPTER 12)

SIMPLE LINEAR REGRESSION

The simple linear regression is used to predict the value of a dependent variable based on the value of one independent variable while also examining the strength and direction of their linear relationship.

MODEL EQUATION

$$Y = bX + a$$

The slope (b):
The intercept (a):

$$a = \bar{Y} - b\bar{X}$$

NOTATION KEY

SYMBOL	DESCRIPTION
Y	Predicted value of the dependent variable Y
X	Value of the independent variable (predictor) X
b	The slope (or regression coefficient) associated with the predictor X that reflects the expected change in Y for each one-unit increase in X
a	The intercept of the regression line (i.e., the value of Y when $x = 0$)
$\bar{Y}$	The mean of the dependent variable Y
$\bar{X}$	The mean of the independent variable X
y_i	The ith value of the dependent variable Y
x_i	The ith value of the dependent variable Y
SS_{XY}	Sum of products of deviation in X and Y, $\Sigma(x_i - \bar{X})(y_i - \bar{Y})$
SS_X	Sum of squares of variable X, $\Sigma(x_i - \bar{X})^2$

ANALYSIS OF SIMPLE REGRESSION

TEST STATISTIC OF THE SLOPE (*B*)

Test statistic:

$$t = \frac{b - B}{SE_b}$$

where

$$SE_b = \frac{S_{y-\hat{y}}}{\sqrt{\Sigma(x_i - \bar{X})^2}}$$

$$S_{y-\hat{y}} = \sqrt{\frac{\Sigma(y_i - \hat{Y})^2}{n - p - 1}}$$

and p = the number of predictor = 1.

Degrees of freedom (*df*):

$$df_{regression} = 1$$
$$df_{residual} = n - 2$$

EFFECT SIZE

$$R^2 = \frac{SS_{regression}}{SS_{total}} = \frac{\sum(\hat{y}_i - \bar{Y})^2}{\sum(y_i - \bar{Y})^2}$$

$$R^2_{adjusted} = 1 - \left(\frac{(1 - R^2)(n - 1)}{n - p - 1}\right)$$

NOTATION KEY

SYMBOL	DESCRIPTION
t	t statistic to test the null hypothesis about the slope (regression coefficient) b
B	The hypothesized value of the slope b, typically 0 under the null hypothesis
SE_b	Standard error of the slope b
$S_{y-\hat{y}}$	Standard deviation of residuals (error in predicting Y)
p	The number of the predictors in the model; in the simple regression, $p = 1$.
$\hat{Y}$	The predicted value of the dependent variable Y
y_i	The ith observed value in the dependent variable Y
x_i	The ith observed value in the dependent variable Y
n	The number of cases (observations) used in the regression analysis
R^2	The proportion of variance in Y explained by X
$R^2_{adjusted}$	Adjusted R-squared, correcting for the number of predictors and sample size
$SS_{regression}$	Sum of squares regression: the portion of the total variability in Y is explained by the regression model.
SS_{total}	Total sum of squares: the total variability in the dependent variable Y around its mean

MULTIPLE REGRESSION

Multiple regression is used to predict the value of a dependent variable based on two or more independent variables.

MODEL EQUATION

$$Y_i = b_0 + b_1 X_{i1} + b_2 X_{i2} + \ldots + b_p X_{ip} + \varepsilon_i$$

Conversion from unstandardized coefficient to standardized coefficient:

$$\beta_p = b_p \frac{SD_Y}{SD_{xp}}$$

NOTATION KEY

SYMBOL	DESCRIPTION
Y_i	The score of the *i*th observation ($i = 1, 2, \ldots, n$) on the dependent variable
b_0	The intercept in multiple regression, which is the predicted value of the dependent variable when all predictor variables are zero
b_1	The slope (or regression coefficient) associated with the predictor X_1 that reflects the expected change in Y for each one-unit increase in X with the other predictors held constant
b_2 through b_p	The regression coefficient for predictors 2 through p. A p refers to the number of predictors in the model.
ε_i	The prediction error for *i*th observation
$\hat{a}_p$	The standardized coefficient for predictor X_p
SD_Y	The standard deviation of the dependent variable Y
SD_{xp}	The standard deviation of the *p*th independent variable X

ANALYSIS OF MULTIPLE REGRESSION

	SUM OF SQUARES (SS)	*df*	MEAN SQUARES (MS)	TEST STATISTIC
Regression (model)	$\sum(\hat{Y}_i - \bar{Y})^2$	p	$\frac{SS_{Regression}}{df_{Model}}$	$\frac{MS_{Model}}{MS_{Error}}$
Error (residual)	$\sum(Y_i - \hat{Y}_i)^2$	$n - p - 1$	$\frac{SS_{Error}}{df_{Error}}$	
Total	$\sum(Y_i - \bar{Y}_i)^2$	$n - 1$	$\frac{SS_{Total}}{df_{Total}}$	

TEST STATISTICS

OMNIBUS TEST (OVERALL MODEL FIT)

Test statistic and degrees of freedom:

$$F = \frac{MS_{Regression}}{MS_{Residual}} \sim F_{(df_{regression},\, df_{Error})}$$

EFFECT SIZE

$$R^2 = \frac{SS_{Regression}}{SS_{Total}}$$

$$R^2_{adjusted} = 1 - \left(\frac{(1 - R^2)(N - 1)}{n - p - 1}\right)$$

SENSITIVITY ANALYSIS

The variance inflation factor (VIF) for the pth predictor:

$$VIP_p = \frac{1}{1 - R_p^2}$$

Tolerance:

$$\text{Tolerance} = (1 - R_p^{\,2}) = \frac{1}{VIP_p}$$

ANALYZING CHANGE AND ADJUSTING FOR COVARIATES(CHAPTER 13)

ONE-WAY REPEATED-MEASURES ANOVA

The one-way repeated-measures ANOVA is used to test for differences in means across three or more related conditions or time points measured on the same participants.

TEST STATISTIC (WHEN SPHERICITY IS ASSUMED)

SOURCE	SS	*df*	MS	Test Statistic
Between groups (conditions)	$SS_{BG} = \Sigma n_k(\bar{X}_k - \bar{X}_{grand})^2$	$k - 1$	$\frac{SS_{BG}}{df_{BG}}$	$\frac{MS_{BG}}{MS_E}$
Between persons	$SS_{Subject} = \Sigma k(\bar{X}_{Subject} - \bar{X}_{grand})^2$	$n - 1$	$\frac{SS_{Subjects}}{df_{Subjects}}$	
Within groups (error)	$SS_E = SS_{Total} - SS_{BG} - SS_{subjects}$	$(k - 1)(n - 1)$	$\frac{SS_E}{df_E}$	
Total	$SS_{Total} = \Sigma(\bar{X}_{ik} - \bar{X}_{grand})^2$	$kn - 1$		

Test statistic:

$$F = \frac{MS_{BG}}{MS_E}$$

Degrees of freedom (*df*):

$$df = ((k - 1),(k - 1)(n - 1))$$

EFFECT SIZE

Partial eta squared (η_p^2):

$$\eta_p^2 = \frac{SS_{BG}}{SS_{total} - SS_{persons}}$$

or

$$\eta_p^2 = \frac{SS_{BG}}{SS_{BG} + SS_E}$$

Partial omega squared (ω_p^2):

$$\omega_p^2 = \frac{SS_{BG} - (df_{BG} * MS_E)}{SS_{BG} + SS_E + MS_E}$$

NOTATION KEY

SYMBOL	DESCRIPTION
x_{ik}	Score of the *i*th case in the *k*th conditions
n_k	The number of the participants in the *k*th condition
$\bar{X}_k$	The mean of the *k*th condition
$\bar{X}_{grand}$	The mean of all scores across all cases and conditions
$\bar{X}_{subject}$	The mean for a particular case *i* across conditions
k	Number of the condition

ANALYSIS OF COVARIANCE

Analysis of covariance (ANCOVA) is used to compare group means while statistically controlling for the effects of one or more continuous covariates.

Model equation (simple ANCOVA model with one dichotomous group variable and one covariate):

$$\hat{Y}_i = a + b_1 X_1 i + b_2 C_i$$

NOTATION KEY

SYMBOL	DESCRIPTION
$\hat{Y}_i$	The predicted score of the *i*th observation ($i = 1, 2, \ldots, n$) on the dependent variable Y
a	The intercept of the regression, which is the predicted value of the dependent variable when all predictor variables are zero
b_1	The slope (or regression coefficient) associated with the group variable X_1 on Y, adjusted for the covariate
b_2	The regression coefficient representing the effect of the covariate C on Y, controlling for group membership
X_1i	The *i*th observation on a binary categorical variable X_1
C_i	The *i*th observation on a covariate (i.e., the continuous variable statistically controlled in the analysis)

ANSWER KEY

Correct answers for each chapter are **in boldface**.

CHAPTER 1

1. Which scale of measurement has an absolute zero point?
   a. Nominal
   b. Ordinal
   c. Interval
   d. Ratio

2. What type of variable can take on an infinite number of possible values within a given range?
   a. Discrete variable
   b. Continuous variable
   c. Nominal variable
   d. Ordinal variable

3. Which research method combines both quantitative and qualitative approaches in a single study?
   a. Quantitative research
   b. Qualitative research
   c. Mixed-methods research
   d. Observational research

4. In research, random assignment primarily enhances which type of validity?
   a. External validity
   b. Internal validity
   c. Construct validity
   d. Content validity

5. Which of the following is an example of a nominal scale?
   a. Temperature in Celsius
   b. Class ranking (e.g., first, second, third)
   c. College major (e.g., psychology, biology, history)
   d. Height in centimeters

CHAPTER 2

1. An educational researcher is investigating if any relationship exists between the hours that students sleeps at night and their math performance across one school district with more than 1,000 students. However, the researcher can only survey 50 students. What notation below would be used to indicate the number of students surveyed?
   a. $N = 50$
   b. $n = 50$. A lowercase n is used to represent a sample size, while a capital N represents the entire population of interest. Since only 50 students (sample) were surveyed out of the 1,000 students (population) attending the school district, a lowercase n should be used.

2. The ______ is the value or values in a dataset that appear most frequently.
   a. Mean
   b. Median
   c. Mode

3. An educational researcher records the following sample of scores: 1, 2, 4, 4, 4, 5, 5, 7, 8, 10. Calculate the sample mean, median, and mode.
   a. Mean = 5
   b. Median = 4.5
   c. Mode = 4

4. An educational researcher obtains the following sample scores: 1, 3, 3, 4, 6, 100. Using what method of central tendency to describe these sample scores might be problematic?
   a. Mean. The outlier of 100 is clearly pulling the mean upward to 19.5—far above the median and mode.
   b. Median
   c. Mode

5. What direction of skewness is created by the data 1, 1, 2, 2, 10?
   a. Positive skewness. The mean is greater than the median, so the tail of the curve would be pulled to the right.
   b. Negative skewness
   c. No skewness

CHAPTER 3

1. What is the range of the dataset 1, 2, 3, 4, 5, 6, 7, 8, 9, 10?
   a. 10
   b. 9, 10 (largest value); 1 (smallest value) = 9
   c. 1
   d. 0

2. The ______ is the square root of the variance?
   a. Interquartile range
   b. Range
   c. Standard Deviation
   d. Mean

3. What is the population variance of the dataset 1, 2, 3, 4, 5?
   a. 1
   b. 2, mean = 3. Subtract each value from the mean, square the results, sum the squares, and divide by 5.
   c. 2.5
   d. 3

4. What is the population standard deviation of the dataset 1, 2, 3, 4, 5?
   a. 1.41, $\sqrt{2} = 1.41$
   b. 1.73
   c. 4.00
   d. 9.00

5. What percentage of values fall within 1 *SD* of the mean if data are normally distributed?
   a. 50%
   b. 68%. The empirical rule states that 68% of values fall within one standard deviation of the mean.
   c. 95%
   d. 99.7%

CHAPTER 4

1. What is the primary purpose of using a histogram in educational research?
   a. To compare categories of data.
   b. To display the relationship between two variables.
   c. To summarize the frequency of continuous data grouped into intervals.
   d. To show proportions of a whole.

2. Which visualization method is best suited for identifying relationships or correlations between two variables?
   a. Pie chart
   b. Box plot
   c. Scatterplot
   d. Stem-and-leaf plot

3. In a box-and-whisker plot, the interquartile range is represented by
   a. The lines extending from the box to the smallest and largest values.
   b. The box itself, showing the range between the 25th and 75th percentiles.
   c. The line in the center of the box representing the median.
   d. The individual dots outside the whiskers.

4. What is a key advantage of tables in presenting data?
   a. They simplify relationships between variables visually.
   b. They display trends and patterns intuitively.
   c. They highlight outliers in large datasets.
   d. They summarize data with precise values for comparison.

5. What is the key difference between a bar chart and a histogram?
   a. Bar charts use touching bars to represent continuous data, while histograms use separated bars for discrete data.
   b. Bar charts use separated bars for discrete data, while histograms use touching bars for continuous data.
   c. Bar charts are only used for percentages, while histograms are used for counts.
   d. There is no difference; the terms are interchangeable.

CHAPTER 5

1. A distribution with high kurtosis has lighter tails than a normal distribution.
   a. True
   b. False. High kurtosis indicates heavier tails than a normal distribution, which means there is a higher likelihood of extreme values (outliers).

2. Which of the following values for skewness and kurtosis indicate that a distribution is approximately normal?
   a. Skewness = 0.5, kurtosis = 2.5
   b. Skewness = 0.25, kurtosis = 0.5. For a distribution to be considered approximately normal, the values for skewness and kurtosis should fall between -1 and 1.
   c. Skewness = 1.0, kurtosis = 3.5
   d. Skewness = -2.0, kurtosis = 4.0

3. What is the probability of a data point falling within 1 standard deviation of the mean in a normal distribution?
   a. 68%. The empirical rule (or 68-95-99.7 rule) describes how data is distributed in a normal distribution.
   b. 95%
   c. 99.7%
   d. 50%

4. A standardized test has a mean score of 500 and a standard deviation of 100. What is the probability that a randomly selected student scores higher than 650? Use the z score formula and the standard normal distribution table to find the answer.
   a. 0.1056 (10.56%)
   b. 0.1587 (15.87%)
   c. **0.0668 (6.68%). Calculate the z score: $(650 - 500) / 100 = 1.5$. The cumulative probability for $z = 1.5$ is .9332. Subtract the cumulative probability from 1: $1 - .9332 = .0668$.**
   d. 0.0228 (2.28%)

5. What is the purpose of converting raw data to z scores?
   a. To reduce the range of the data.
   b. **To standardize data for comparison across distributions. This allows for meaningful comparisons between datasets that may have different scales, units, or ranges.**
   c. To eliminate outliers.
   d. To calculate the mean and median more accurately.

CHAPTER 6

1. What is a sampling distribution?
   a. A distribution of a population's characteristics
   b. **A distribution of a statistic (such as a mean or proportion) calculated from all possible samples of a specific size, a sampling distribution represents the variability of a statistic, such as the mean, across all possible random samples of the same size from a population.**
   c. A list of all individual data points in a population
   d. A random collection of data from a single sample

2. According to the central limit theorem, what happens to the sampling distribution of the sample mean as the sample size increases?
   a. It becomes more skewed.
   b. **It approximates a normal distribution, regardless of the population distribution. The central limit theorem states that as the sample size grows larger (usually $n \geq 30$), the sampling distribution of the sample mean will approach a normal distribution even if the population distribution is not normal.**
   c. It moves closer to the smallest value in the sample.
   d. It diverges from the population mean.

3. Which sampling technique ensures that every member of the population has an equal chance of being selected?
   a. Convenience sampling
   b. Purposive sampling
   c. **Random sampling. Random sampling ensures fairness by giving each member of the population an equal probability of being chosen, reducing bias and improving representativeness.**
   d. Snowball sampling

4. What is the standard error of the mean (SEM), and how is it calculated?
   a. It is the variability of individual data points and is equal to the population standard deviation.
   b. It is the standard deviation of the sampling distribution of sample means, calculated by dividing the sample standard deviation by the square root of the sample size. SEM quantifies how much the sample mean is likely to vary from the true population mean. SEM decreases as the sample size increases, which makes the estimate more precise.
   c. It measures the deviation of a single data point from the sample mean, using the formula *SD/n*.
   d. It is a fixed value representing the population variance.

5. What happens to the sampling distribution of the sample mean if the population distribution is normal?
   a. The sampling distribution will always be skewed, regardless of sample size.
   b. The sampling distribution will only be normal if the sample size is large.
   c. The sampling distribution will match the population distribution for any sample size.
   d. The sampling distribution will be normal, regardless of sample size. If the population distribution is already normal, the sampling distribution of the sample mean will also be normal, even for small sample sizes.

CHAPTER 7

1. What is the purpose of setting an alpha level (α) in hypothesis testing?
   a. To determine the sample size required for the study
   b. To set the threshold probability for rejecting the null hypothesis
   c. To calculate the confidence interval for the population mean
   d. To ensure the sample data are normally distributed

2. A researcher hypothesizes that students from a particular school will outperform the state average on a given test. What type of test should be conducted?
   a. Nondirectional test (two-tailed/two-sided)
   b. Directional test (one-tailed/one-sided)

3. A critical value defines the cutoff point that separates the rejection region from the nonrejection region in a hypothesis test.
   a. True
   b. False

4. What does a p value indicate in hypothesis testing?
   a. The probability that the null hypothesis is true
   b. The likelihood of obtaining the observed result, or something more extreme, if the null hypothesis is true
   c. The size of the effect in the population
   d. The risk of committing a Type II error

5. What is the primary purpose of hypothesis testing in research?
   a. To organize data into meaningful groups
   b. To differentiate real patterns from random chance in data
   c. To calculate the average of a sample population
   d. To ensure data is normally distributed

CHAPTER 8

1. When should a one-sample *t* test be used instead of a *z* test?
   a. When the population mean is known
   b. When the population standard deviation is unknown
   c. When the same size is less than 30
   d. When comparing two sample means

2. What is the formula for the degrees of freedom (*df*) in a one-sample *t* test?
   a. $df = n + 1$
   b. $df = n \times 2$
   c. $\mathbf{df = n - 1}$
   d. $df = \frac{n}{2}$

3. Which of the following is not an assumption of the one-sample *t* test?
   a. Random sampling
   b. Independence of observations
   c. Population variance of 0
   d. Dataset is approximately normally distributed

4. What is the purpose of calculating effect size, such as Cohen's *d*?
   a. To determine statistical significance
   b. To quantify the magnitude of a difference
   c. To check if assumptions are met
   d. To calculate degrees of freedom

5. What is the risk of using a one-tailed test when the actual effect is in the opposite direction?
   a. Type I error
   b. Type II error
   c. Type III error
   d. Increased power

CHAPTER 9

1. What is the main difference between an independent-samples *t* test and a related-samples *t* test?
   a. **Independent-samples *t* tests compare two separate groups, while related-samples *t* tests compare two sets of scores from the same group or matched individuals.**
   b. Independent-samples *t* tests compare means to a population value, while related-samples *t* tests compare variances.
   c. Independent-samples *t* tests require categorical data, while related-samples *t* tests require continuous data.
   d. There is no difference; they are used interchangeably.

2. Which assumption is unique to independent-samples *t* tests but not required for related-samples *t* tests?
   a. Normality
   b. Random sampling
   c. **Equal variances**
   d. Independence of scores within groups

3. When would you use a one-sample *t* test instead of an independent- or related-samples *t* test?
   a. When you have two independent groups to compare
   b. **When you want to compare a single group to a known population mean**
   c. When you compare pre- and posttest scores from the same group
   d. When you have categorical data

4. You are conducting a study to compare the effectiveness of two teaching methods on student performance. You have a small sample size of just 16 students (8 students taught by different teachers in different classrooms—there is no reason to suggest they are related), and the scores in your dataset are not normally distributed. Which test should you use to compare these scores?
   a. **Mann-Whitney *U* test**
   b. Wilcoxon matched-pairs signed-ranks test
   c. Independent-samples *t* test
   d. One-sample *t* test

5. Which effect size measure is commonly reported for independent-samples *t* tests?
   a. r^2
   b. Phi coefficient
   c. **Cohen's *d***
   d. Omega

CHAPTER 10

1. What is the primary purpose of a one-way ANOVA?
   a. To compare the means of two groups
   b. To compare the means of three or more groups
   c. To analyze relationships between continuous variables
   d. To assess correlations between independent and dependent variables

2. Which of the following increases the risk of a Type I error when comparing multiple groups?
   a. Failing to apply a correction for multiple comparisons (e.g., Bonferroni correction)
   b. Running multiple *t* tests on the same dataset
   c. Including covariates in the analysis
   d. Increasing sample size

3. Which test is the most conservative post hoc method for controlling Type I error?
   a. Tukey's HSD
   b. Fisher's LSD
   c. Bonferroni test
   d. Kruskal-Wallis Test

4. The Kruskal-Wallis test is best suited for which type of data?
   a. Ordinal data or nonnormal distributions
   b. Normally distributed interval data
   c. Binary data
   d. Continuous data

5. What does homogeneity of variance mean in ANOVA?
   a. All groups have the same sample size
   b. The dependent variable is normally distributed
   c. The variances of the dependent variable are similar across groups
   d. There is no interaction effect between variables

6. What distinguishes a factorial ANOVA from a one-way ANOVA?
   a. It compares means between more than two groups.
   b. It examines the effects of two or more independent variables and their interactions on a dependent variable.
   c. It includes covariates to control for variability in the dependent variable.
   d. It is used for nonparametric data.

7. Which of the following is an example of a factorial ANOVA design?
   a. Comparing the mean scores of students across three different teaching methods.
   b. Assessing the correlation between test scores and student motivation.

   c. Testing for differences in a single group's test scores across three different time points.
   d. Analyzing the effects of teaching method (lecture vs. inquiry-based) and student age group (young vs. adult) on test scores.

8. What is the effect size measure commonly used in ANOVA to estimate the proportion of variance explained by the independent variable?
   a. Pearson's *r*
   b. Eta squared
   c. Cronbach's alpha
   d. Chi squared

CHAPTER 11

1. What is the main purpose of Pearson's correlation coefficient (*r*) in correlational analysis
   a. To compare means across groups
   b. To measure the extent of causation between two variables
   c. To measure the direction and strength of a relationship between two variables
   d. To test for the significance of regression equations

2. Which of the following is NOT an assumption of correlational analysis?
   a. Linearity
   b. Homoscedasticity
   c. Causality
   d. Normality

3. What does a Pearson's *r* value of 0 indicate?
   a. A strong positive correlation
   b. No correlation
   c. A strong negative correlation
   d. A perfect correlation

4. Which statistical method is used when investigating a relationship between a continuous variable and a dichotomous variable?
   a. Pearson's *r*
   b. Spearman's rho
   c. Point-biserial correlation
   d. ANOVA

5. What does the assumption of homoscedasticity in correlational analysis require?
   a. The data must form a straight line.
   b. The variables must be normally distributed.
   c. The sample size must be large enough.
   d. The spread of data points must be consistent above and below the line across the entire range.

CHAPTER 12

1. What does the slope (*b*) in a simple linear regression equation represent?
   a. The predicted value of Y when $X = 0$
   b. The change in X relative to Y
   c. The change in Y for a one-unit change in X
   d. The coefficient of determination

2. What does R_2 represent in simple linear regression analysis?
   a. The correlation between independent variables
   b. The proportion of variance in Y explained by X
   c. The slope of the regression line
   d. The standard error of the residuals

3. What does tolerance measure in regression analysis?
   a. The extent of multicollinearity among independent variables
   b. The proportion of variance in the dependent variable explained by predictors
   c. The normality of residuals
   d. The degree of model fit

4. What happens if a model includes highly correlated independent variables?
   a. The dependent variable becomes unreliable.
   b. The coefficients of the independent variables become unstable.
   c. The model fit R^2 will always decrease.
   d. The assumption of linearity is violated.

5. What does homoscedasticity mean in regression analysis?
   a. The residuals have a consistent spread across all predicted values.
   b. The independent variables are not correlated.
   c. The predictors have no relationship with the error term.
   d. The regression coefficients are stable across models.

CHAPTER 13

1. What is the key advantage of using a repeated-measures ANOVA?
   a. It eliminates the need for a dependent variable.
   b. It reduces error variance by accounting for individual differences.
   c. It allows for the inclusion of multiple dependent variables.
   d. It does not require any statistical assumptions.

2. What does the term "sphericity" refer to in the context of repeated-measures ANOVA?
   a. Equal variances within each group

b. Equal sample sizes across groups
c. Equal variances of differences between all combinations of related groups
d. Equal means across all groups

3. What is the recommended post hoc procedure for a repeated-measures ANOVA to control for Type I error?
   a. Tukey's HSD
   b. Bonferroni correction
   c. Fisher's LSD
   d. Scheffé test

4. What does a significant result in a repeated-measures ANOVA indicate?
   a. All group means are equal.
   b. The dependent variable is unrelated to the conditions tested.
   c. The assumptions of sphericity have been violated.
   d. There is a difference in at least one pair of group means.

5. What does "partial eta squared" (η_f^2) represent in repeated-measures ANOVA?
   a. The total variation in the dependent variable.
   b. The proportion of total variance explained by between-groups effects after accounting for between-persons variance.
   c. The difference between the group means.
   d. The degree of correlation between independent and dependent variables.

6. What is the role of a covariate in ANCOVA?
   a. To control for variability in the dependent variable that is unrelated to the independent variable(s)
   b. To act as an additional independent variable
   c. To replace the independent variable in the model
   d. To reduce between-group variance

CHAPTER 14

1. Which of the following is NOT a purpose of meta-analysis?
   a. Identifying the central tendency or common effect across studies
   b. Exploring variability in effect sizes
   c. Developing new experimental theories
   d. Explaining why variation in effect sizes occurs

2. A propensity model is typically used to
   a. Measure the direct effect of a treatment variable.
   b. Estimate the likelihood of receiving a treatment based on observed characteristics.
   c. Compare independent variables across different models.
   d. Simplify complex data into manageable clusters.

3. What is a key requirement for using multilevel modeling (MLM) effectively?
   a. A minimum sample size of 50 participants
   b. Randomized controlled trials only
   c. Large datasets with at least 20–30 groups of 20–30 individuals per group
   d. A focus on theoretical over empirical studies

4. Which of the following is a key benefit of structural equation modeling (SEM) over traditional regression analysis?
   a. It requires fewer participants to yield accurate results.
   b. It eliminates the need for latent variables.
   c. It allows the inclusion of multiple dependent variables and mediating variables.
   d. It reduces data complexity by eliminating endogenous predictors.

5. What is the minimum number of studies generally required to conduct a meta-analysis that explores variation?
   a. Two studies
   b. Three studies
   c. Four studies
   d. Twenty studies

REFERENCES

American Psychological Association. (2018, April 19). *Hypothesis*. In APA Dictionary of Psychology. https://dictionary.apa.org/hypothesis

American Psychological Association. (2019). *Publication manual of the American Psychological Association* (7th ed.). American Psychological Association.

Bognar, M. (2021). *Normal distribution applet*. University of Iowa. https://homepage.divms.uiowa.edu/~mbognar/applets/normal.html

Cochran, W. G. (1954). The combination of estimates from different experiments. *Biometrics, 10*(1), 101–129.

Cohen, J. (1988). *Statistical power analysis for the behavioral sciences* (2nd ed.). Routledge.

Foltz, B. (2017, August 16). *Statistics 101: ANCOVA, An Introduction*. YouTube. https://youtu.be/0e8BI2u6DU0?si=vWQT-uxlzgJjTFm6

Huang, F. L. (2018). Multilevel modeling myths. *School Psychology Quarterly, 33*(3), 492–499.

Kutner, M. H., Nachtsheim, C. J., Neter, J., & Li, W. (2013). *Applied linear statistical models* (5th ed.). McGraw Hill.

Laerd Statistics. (n.d.). *Repeated measures ANOVA: Calculating a repeated measures ANOVA.* https://statistics.laerd.com/statistical-guides/repeated-measures-anova-statistical-guide-2.php

Levesque-Bristol, C., Knapp, T. D., & Fisher, B. J. (2011). The effectiveness of service-learning: It's not always what you think. *Journal of Experiential Education, 33*(3), 208–224.

Maas, C. J., & Hox, J. J. (2005). Sufficient sample sizes for multilevel modeling. *Methodology, 1*(3), 86–92.

Milton, R. C. (1964). An extended table of critical values for the Mann-Whitney (Wilcoxon) two-sample statistic. *Journal of the American Statistical Association, 59*(307), 925–934. https://doi.org/10.1080/01621459.1964.10480740

Moore, D. S., McCabe, G. P., & Craig, B. A. (2021). *Introduction to the practice of statistics* (10th ed.). Macmillan Learning.

Privitera, G. J. (2018). *Statistics for the behavioral sciences* (3rd ed.). SAGE.

Sani, F., & Todman, J. B. (2006). *Experimental design and statistics for psychology: A first course* (1st ed.). Blackwell.

Savory, Paul. (2008). *How do you interpret a confidence interval?* University of Nebraska, Industrial and Management Systems Engineering, Instructional Materials. https://digitalcommons.unl.edu/imseteach/11

Smith, J., Hurwitz, M., & Avery, C. (2017). Giving college credit where it is due: Advanced Placement exam scores and college outcomes. *Journal of Labor Economics, 35*(1), 67–147.

INDEX

Page numbers in italics indicate Figures and Tables

ABOUT THE AUTHORS

YUKIKO MAEDA, PHD, is a professor of educational psychology and research methodology at Purdue University. Her areas of expertise include quantitative research methodology in education, with a focus on research synthesis, meta-analysis, and multilevel modeling.

JOHN GIPSON, PHD, serves as chancellor at Ivy Tech Community College in Lake County, Indiana. He previously served as chief of staff to the provost and associate vice president for academic operations at Purdue University, where he also held a courtesy faculty appointment in educational psychology and research methodology.

SHEILA HURT, PHD, is the registrar at Purdue University, where she also holds a courtesy faculty appointment in educational psychology and research methodology. Her research aims to apply advanced methods to decision-making in higher education.

KATIE H. DUFAULT is the director of the Helen Bass Williams Academic Success Center at Purdue University and is a doctoral student in Purdue's College of Education. Her research focuses on learning and support for college students.

www.ingramcontent.com/pod-product-compliance
Lightning Source LLC
Chambersburg PA
CBHW080635220726
48566CB00021B/566